Human Viruses in Sediments, Sludges, and Soils

T0203989

Editors

V. Chalapati Rao, M.Sc., Ph.D.
Associate Professor
Department of Virology and Epidemiology
Baylor College of Medicine
Houston, Texas

Joseph L. Melnick, Ph.D.
Distinguished Service Professor and Chairman
Department of Virology and Epidemiology
Baylor College of Medicine
Houston, Texas

CRC Press
Taylor & Francis Group
Boca Raton London New York

CRC Press is an imprint of the
Taylor & Francis Group, an **informa** business

CRC Press
Taylor & Francis Group
6000 Broken Sound Parkway NW, Suite 300
Boca Raton, FL 33487-2742

Reissued 2019 by CRC Press

© 1987 by Taylor & Francis Group, LLC
CRC Press is an imprint of Taylor & Francis Group, an Informa business

No claim to original U.S. Government works

A Library of Congress record exists under LC control number:

Publisher's Note
The publisher has gone to great lengths to ensure the quality of this reprint but points out that some imperfections in the original copies may be apparent.

Disclaimer
The publisher has made every effort to trace copyright holders and welcomes correspondence from those they have been unable to contact.

ISBN 13: 978-0-367-22036-5 (hbk)
ISBN 13: 978-0-367-22044-0 (pbk)
ISBN 13: 978-0-429-27047-5 (ebk)

Visit the Taylor & Francis Web site at http://www.taylorandfrancis.com and the
CRC Press Web site at http://www.crcpress.com

THE EDITORS

V. Chalapati Rao, Associate Professor of Environmental Virology at Baylor College of Medicine, Houston, died on July 6, 1986. He was born on October 6, 1930 in Andhra, India. He was educated at Andhra University, and received the B.S., M.S., and Ph.D. degrees. He first came to the U.S. as a Fulbright Scholar in 1966, then as a Visiting Scientist of the U. S. Food and Drug Administration in 1976, and finally as a faculty member of Baylor College of Medicine in 1979. While in India, almost single-handedly and under extremely difficult conditions he developed the field of environmental virology. In 1978, he co-chaired the World Health Organization meeting on Human Viruses in Water, Wastewater, and Soil. He was an active participant in many conferences sponsored by the U. S. Environmental Protection Agency, and at the 1986 Annual Meeting of the American Society for Microbiology he convened and chaired a conference on Human Gastroenteritis Viruses in the Environment.

The textbook entitled *Environmental Virology*, for which he was principally responsible, was published in 1986 in the U.K. (Van Nostrand Reinhold) and in the U.S. (American Society for Microbiology). His authoritative review, "Human Viruses in Sediments, Sludges, and Soils", was the lead article in the *Bulletin of the World Health Organization* (64(1), 1986), and was regarded as of sufficient global importance that it was translated and appeared in French in Volume 64, Issue 3. He was the driving force behind the present, expanded monograph on this subject.

Dr. Rao's research interests centered on methods for detecting viral contamination of the water environment. Not content with the limitations imposed by the use of available methods for detection of cytopathogenic enteroviruses, he was successful in developing methods for concentrating and detecting rotaviruses and was working actively on hepatitis A virus — both of which are important human pathogens that may pollute our water sources.

Chalapati Rao was a person of high intelligence and integrity, always eager to carry on with his own investigations and to assist others in their work. He remains an inspiration to all who knew him. Dr. Rao is survived by his wife, Tara Rao, and four daughters, Nagu, Anu, Suja, and Hema.

Joseph L. Melnick is Distinguished Service Professor of Virology and Epidemiology and Dean of Graduate Sciences at Baylor College of Medicine, Houston. Before coming to Baylor in 1958 he earned his doctorate at Yale University and began his teaching career there, remaining on the medical school faculty for 15 years. Wesleyan University, where he earned his A.B. degree, awarded him an honorary D.Sc. degree in 1971 and its highest honor, the Baldwin Medal, in 1986.

Among his many awards and honors are inclusion in the National Foundation for Infantile Paralysis "Polio Hall of Fame" for contributions leading to the development of poliovaccines, the Modern Medicine Distinguished Achievement Award for Contributions to Medical Science, the Freedman Foundation Award of the New York Academy of Sciences, the Eleanor Roosevelt Humanities Award, and the Maimonides Award of the State of Israel. He is an honorary member of the Microbiological Societies of Israel and of the Soviet Union. In 1986 he was elected to be the first honorary member of the new Chinese Society for Medical Virology.

Dr. Melnick is recognized internationally as a leader in virology. He was elected to be the first Chairman of the Virology Section of the International Association of Microbiological Societies. Since 1957 he has been a member of the World Health Organization Expert Advisory Panel on Viral Diseases. In 1974 he served as Chairman of the International Conference on Viruses in Water held in Mexico City under the joint sponsorship of the American Public Health Association and the World Health Organization, and he has continued to be an advisor to the U. S. Environmental Protection Agency and to the World Health Organization in this area.

Dr. Melnick's record of contributions to virus research extends over 40 years; he is the author of more than 1000 research papers in virology and co-author of textbooks in the field. He was among the first to demonstrate that poliovirus is only one of a large group of enteroviruses, and only rarely an invader of the central nervous system. During the development of live polio vaccines, his studies contributed means for evaluating the attenuation and genetic stability of candidate vaccine strains. He developed thermostabilized live polio vaccines, permitting their use for immunization of children living in remote and underdeveloped regions of the world. For four decades, his concerns have included the presence of enteric viruses in sewage and the dangers of waterborne disease. He was among the first to recover human pathogenic viruses from surface waters, and his laboratory has played a leading role in developing methods for detecting and monitoring viruses in our environment.

CONTRIBUTORS

John B. Anderson, Ph.D.
Associate Professor
Geology Department
Rice University
Houston, Texas

James J. Bertucci, Ph.D.
Virology Group Leader
Research and Development Department
Metropolitan Sanitary District of Greater
 Chicago
Chicago, Illinois

Gabriel Bitton, Ph.D.
Professor
Department of Environmental Engineering
 Science
University of Florida
Gainesville, Florida

Yee-Ming Chan, Ph.D.
Staff Chemist
Department of Environmental Engineering
University of Houston
Houston, Texas

Dean O. Cliver, Ph.D.
Professor
Food Research Institute
University of Wisconsin
Madison, Wisconsin

Rob B. Dunbar, Ph.D.
Assistant Professor
Department of Geology
Rice University
Houston, Texas

R. D. Ellender, Ph.D.
Professor
Department of Biological Sciences
University of Southern Mississippi
Hattiesburg, Mississippi

Samuel R. Farrah, Ph.D.
Associate Professor
Department of Microbiology and Cell
 Science
University of Florida
Gainesville, Florida

Charles P. Gerba, Ph.D.
Professor
Department of Microbiology and
 Immunology
University of Arizona
Tucson, Arizona

Steven M. Lipson, Ph.D.
Chief, Virology Laboratory
Division of Microbiology
Department of Pathology
Nassau County Medical Center
East Meadow, New York

Carol Lowry, B.S.
Cullen College of Engineering
University of Houston
Houston, Texas

Cecil Lue-Hing, D.Sc., P.E.
Director
Research and Development Department
Metropolitan Sanitary District of Greater
 Chicago
Chicago, Illinois

Jack V. Matson, Ph.D.
Associate Professor
Department of Civil Engineering
University of Houston
Houston, Texas

Joseph L. Melnick, Ph.D.
Distinguished Service Professor
Department of Virology and
 Epidemiology
Baylor College of Medicine
Houston, Texas

T. G. Metcalf, Ph.D.
Professor
Department of Virology and
  Epidemiology
Baylor College of Medicine
Houston, Texas

V. Chalapati Rao, Ph.D. (Deceased)
Associate Professor
Department of Virology and
  Epidemiology
Baylor College of Medicine
Houston, Texas

Salvador J. Sedita, Ph.D.
Head, Biology Section
Research and Development Department
Metropolitan Sanitary District of Greater
  Chicago
Chicago, Illinois

Mark D. Sobsey, Ph.D.
Professor
Department of Environmental Sciences
  and Engineering
School of Public Health
University of North Carolina
Chapel Hill, North Carolina

Guenther Stotzky, Ph.D.
Professor
Department of Biology
New York University
New York, New York

Richard L. Ward, Ph.D.
Associate Director of Clinical Virology
Department of Clinical Virology
Gamble Institute of Medical Research
Cincinnati, Ohio

Flora Mae Wellings, Sc.D.
Director
Epidemiology Research Center
Health and Rehabilitative Services
Tampa, Florida

Mary Ellen Whitworth, M.S.
Department of Environmental Engineering
University of Houston
Houston, Texas

TABLE OF CONTENTS

Chapter 1

HUMAN ENTERIC VIRUSES IN SEDIMENTS, SLUDGES, AND SOILS: AN OVERVIEW

Joseph L. Melnick

Wastewater and sludge treatment and disposal policies as currently practiced may result in the discharge of any of more than 120 human enteric virus pathogens into aquatic or terrestrial domains. Enteroviruses, including hepatitis virus A, reovirus, adenovirus, and rotavirus, Norwalk virus, and other gastroenteritis viruses have been detected in polluted waters. In addition, non-A, non-B hepatitis viruses have recently been implicated as water-transmitted pathogens. Virus concentrations of 100 plaque-forming units (pfu) per liter of sewage in the U.S. and 100 to 1000 times greater in developing countries have been reported.

Suspended solids-associated virus in raw sewage is partially transferred into sludge during sewage treatment. Virus concentrations of 5000 to 28,000 pfu/ℓ are found in raw sludge, but are greatly reduced during treatment. An average of about 50 pfu/ℓ can be expected in waste treatment plant effluents. Solid-associated virus in wastewater effluents discharged into an aquatic environment settles through water columns onto bottom sediments, where concentrations 10 to 10,000 times greater than those found in water may exist. Sediments represent a reservoir from which viruses can spread when the sediments are disturbed and resuspended in the water column. In the case of estuarine waters, resuspension is common as a result of such factors as storms, dredging, and tides. Resuspended solids-associated virus can be transported from polluted to nonpolluted recreational or shellfish waters with potential health hazards. Ingestion of virus-polluted water by swimmers or bathers or consumption of virus-polluted shellfish may result in viral disease. Oysters, mussels, and clams most commonly are involved as a result of their bioaccumulation of virus through filter feeding and the subsequent consumption of the entire shellfish in a raw or inadequately cooked condition. Outbreaks of infectious hepatitis and viral gastroenteritis transmitted through water or shellfish continue to occur on a worldwide basis.

Determination of the viral quality of recreational and shellfish waters by fecal coliform indexes is unsatisfactory. Virus pathogens have been found repeatedly in such indicator-approved waters, and no consistent relationship between numbers of viruses and fecal coliforms in water, shellfish, or sediments exists.

Direct application of wastewater or sludge to land carries a potential of viral pollution which may lead to contamination of crops and groundwater. Application by sprinkler irrigation introduces a risk of virus aerosols to exposed persons, particularly farm workers and those in nearby residential areas.

Sludge treatment by drying, pasteurization, anaerobic digestion, and composting reduces but does not eliminate virus. Enteroviruses in sludge have survived 30 days of digestion at 50°C. Hepatitis A virus, which is even more thermoresistant, may be expected to survive for longer periods of time.

Virus contamination of groundwater is influenced by the rate of application of wastewater and by soil composition and structure. Other factors include pH level, organic content, and ionic strength of the effluent. While normal rates of water application for agricultural irrigation are about 1 m^3 of water per 1 m^2 of land per year, hydraulic loading rates of effluent disposal on land have been as high as 100 m^3/m^2/year. As expected, virus removal declines with higher loading rates. Viruses are readily adsorbed onto clays so that the higher the content of the soil, the greater the removal. Sandy loams and soils containing organic matter are also favorable for virus removal. Soils containing sand or sand and gravel mixtures do not achieve good removal, and fissured limestone aquifers under shallow soil allow virus transport over greater distances and favor groundwater pollution.

Movement of viruses in soils has been studied at a wastewater reclamation project near St. Petersburg, Fla. Chlorinated secondary effluent was applied by a sprinkler irrigation system to a sandy soil containing little or no silt or clay. Polioviruses and echoviruses were detected in drains well below the surface, demonstrating that viruses survive aeration and sunlight during spraying as well as percolation through 1.5 m of soil. Although at first no viruses were detected in wells 3 m and 6 m below the surface, they were detected after heavy rainfall, as a result of desorption and migration through the soil. Fecal coliform bacteria were not detected in the water containing viruses.

Low pH favors adsorption, while high pH results in elution. High concentrations of soluble organic matter in wastewater compete with viruses for adsorption sites on soil particles, resulting in decreased virus adsorption or even the liberation of previously adsorbed viruses. High concentrations of cations tend to enhance virus retention. Viruses retained near soil surfaces may be eluted and washed down to lower strata by heavy rainfall.

When wells are located in the vicinity of wastewater irrigation or land disposal sites, it would seem the better part of wisdom to prohibit their use as a source of drinking water pending tests of water quality. Since factors influencing virus movement in soil are not yet fully elucidated, and effluent and soil conditions vary considerably, careful study of local conditions is required. Reasonable safety measures should include the siting of such wells at suitable distances from likely sources of contamination, and viral monitoring of water quality may have to be considered.

ACKNOWLEDGMENT

We are grateful to the late Dr. V. C. Rao and the co-authors of this volume for reviewing and summarizing the scattered literature dealing with these aspects of environmental virology.

Chapter 2

PHYSICAL AND CHEMICAL CHARACTERISTICS OF SEDIMENTS, SLUDGES, AND SOILS

Jack V. Matson, C. L. Lowry, Yee Ming Chan, and Mary Ellen Whitworth

TABLE OF CONTENTS

I. INTRODUCTION

This chapter is for those who are unfamiliar with the nature of sediments, sludges, and soils. It is important to understand the principles governing the interactions of these materials with human viruses for the development of methods in virus recovery and in studies relating to virus persistence and inactivation. The relationship of these materials to the water environment has been studied for years by geologists, soil scientists, and environmental engineers. These disciplines have developed a vocabulary for describing the physical and chemical properties of the materials. The vocabulary terms are carefully explained so that a concept of the meaning is established. Some examples are used to illustrate the principles involved. Finally, the relevance of the materials to the environment is explored.

Our ignorance of the complexities and nature of the interactions is manifest. The simplistic way many things are described is a mask for our ignorance. Much work is ahead for a more complete understanding of the science. References are liberally used so that the subject matter can be explored more deeply.

II. PHYSICAL, CHEMICAL, AND BIOLOGICAL CHARACTERISTICS OF SEDIMENTS

A. Sediment Classification

Sediment is solid matter suspended in or deposited from a liquid by physical, chemical, or biological agents. Physical or clastic sediments are eroded solid materials from preexisting rock mechanically transported and deposited by wind, water, or ice in response to gravity. They are generally classified by mineral composition, texture, mass properties, and chemical composition. Nonclastic sediments are formed by chemical and biological processes from material in solution. Chemical sediments are primarily precipitates from water. Chemical sediments are classified by chemical processes, pH, E_H, weathering, and inorganic chemical processes. Biological sedimentation results from interactions involving sediments and organisms. Biological sediment classifications include secretion and degradation of calcium carbonate skeletons, trapping and baffling of sedimentary particles by organisms, pelletization, burrowing, and various effects of microorganisms.[1]

B. Physical Characteristics of Sediments

1. Sediment Mineral Composition

Inorganic clastic sediments contain rock fragment, quartz (SiO_2), feldspars $MAl(Si_3O_8)$, clay minerals, and heavy minerals. Rock fragments are weathered from source rocks whose textural characteristics are identifiable before the fragment is broken down to its basic mineral constituents. Quartz is frequently, but not always, the dominant mineral in sediment. While feldspars are the most common mineral in the earth's crust, they generally constitute less than 15% of clastic sediments. The actual amount of feldspar in sediment varies with weather conditions. Unaltered feldspars imply arid conditions. Clay minerals are produced by the chemical weathering of feldspars, and their abundance suggests a humid climate. Clay minerals are rock-making minerals with tendencies toward flake forming. Clays are composed of silica, alumina, and water with quantities of iron, alkalis, and alkaline earth.[2] A small percentage of denser minerals from some igneous and metamorphic rock sources, called heavy minerals, is included among clastic sediments. Heavy minerals are thus classified because they have relatively high densities.

Density is a ratio of mass to volume and a function of the size, weight, and bonding mechanisms of the particles. The specific gravity of a particle is a ratio of the density of the particle to the density of the same amount of water. The minerals with high specific gravity are heavy compared to water, and this accounts for their settling. Water has a specific gravity between 0.917 and 1.000, and the specific gravity of seawater is 1.026. Quartz and feldspar have a specific gravity of 2.65. The characteristic specific gravity values for clay range from 2.08 to 2.96. It is difficult to measure the density of clays in their natural state because they absorb water. Heavy minerals such as magnetite, zircon, garnet, and tourmaline have specific gravities ranging between 3.07 and 5.18.[1] In general, the higher the specific gravity of a sediment, the faster it settles in water.

2. Sediment Texture

Sediment texture is a measure of uniformity of particle size, sorting, shape, and surface texture. Primarily clastic sediments are distinguished on the basis of particle size. Particles of each grain size are transported based on current strength. The larger particles require stronger currents to carry them to their area of deposition. Table 1 lists the particle diameter range for the three basic size classes of sediment: mud, sand, and gravel.[1] Mud is fine-grained sediment ranging in diameter from 0.002 mm for clay to 0.5 mm for silt. It settles out of suspension slowly in quiet waters such as ebbing floods with minimum wave action. Sedimentary deposits of muds are the grain size most likely to contain organic matter and chemically precipitated rocks. Sand grains range between 0.0625 mm and 2.0 mm in diameter and are large enough to be seen with the naked eye. Sand is carried in suspension by moderate currents or strong winds. Gravel ranges in diameter from 4 mm for pebbles to 2048 mm for boulders. Gravel is transported by strong currents and rapid water flow or intense wave action.[3]

Chemical and physical changes after accumulation cause soft sediment to turn to sedimentary rock in the long term. The aged equivalent of mud is shale, sand turns into sandstone, and gravel becomes conglomerate.

The second measure of sediment texture is sorting. Sorting is a quantitative measure, unique to sedimentary deposits, which determines whether particles in a sample are within range of some average size. Sorting is directly related to the kind of depositing current.[4] Wind and water act on sediment by selecting and removing particles for transportation. In a well- sorted sediment the variation of size from the average is small. Poorly sorted sediment represents many size classes. Heavy minerals are sorted using specific gravity as well as particle shape.

Shapes of sedimentary particles further define their texture on a geometric basis. Basic

Table 1
STANDARD SIZE CLASSES OF SEDIMENT

Limiting (mm)	(Microns μ)	Size Class			
2048		V. Large			
1024		Large	Boulders		1 m
512		Medium			
256		Small		G	
128		Large	Cobbles	R	
64		Small		A V	10^{-1}
32		V. Coarse		E	
16		Coarse		L	
8		Medium	Pebbles		10^{-2}
4		Fine			
2		V. Fine			
1		V. Coarse			10^{-3}
1/2	500	Coarse			
1/4	250	Medium	Sand		
1/8	125	Fine			10^{-4}
1/16	62	V. Fine			
1/32	31	V. Coarse			
1/64	16	Coarse		M	10^{-5}
1/128	8	Medium	Silt	U	
1/256	4	Fine		D	
1/512	2	V. Fine			
		Clay			

particle shape categories are sphericity and roundness. Sphericity defines the degree to which particle shape approaches a sphere. Roundness refers to smoothness or curvature of edges and corners of a particle. Although the sphericity and roundness characterize basic particle shapes, the influence of particle dynamics, water turbulence, and particle content can produce a variation from these basic shapes.

Surface texture of sand- and gravel-size particles is created by abrasion during transport. Particle collision is another major contributor of fractures and pits on a particle. While surface texture can be a clue to particle origins, different surface patterns often reflect a complex history of physical and chemical changes. Therefore, the use of surface texture to define particle origin is only partially effective.

3. Sediment Structure

Sediment structure is based on the mass properties of accumulated sedimentary particles. Sedimentary deposits give indications as to the strength of the transporting agent and the environment of deposition. Some examples of sedimentary environments are lagoons, beaches, deserts, and rivers. The accumulation of sediments is based on the mass properties fabric, porosity, and permeability.[5]

Fabric refers to the arrangement of the accumulated particles in a sedimentary deposit. The important aspects of fabric are particle shape and packing. The shape controls spatial orientation. Figure 1 shows various shapes of particles;[1] each axis may be oriented parallel

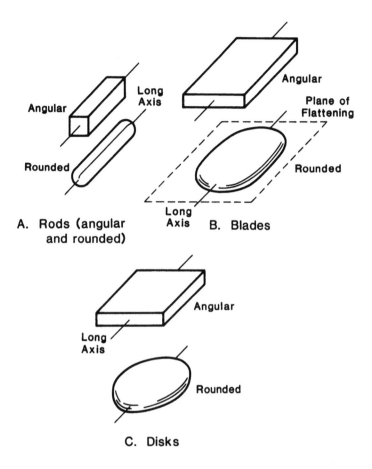

FIGURE 1. Shape as an element affecting the fabric of particles.

to the direction of the current or at right angles to the current. Packing is related to the spatial density of the particles. For instance, if sand particles are infiltrated by mud such that the sand particles no longer support themselves, the fabric will be termed mud-supported.[1]

Porosity is the percentage of the total volume of void space to the total volume of an accumulated sediment. Sediment pores hold water and other fluids. Porosity varies with size, shape, and arrangement of the particles.

Permeability is a measure of the flow of fluids through pore openings in a sediment. The ability of a sedimentary deposit to transmit fluid through its pores depends on the pore space, the fluid type, and the driving force of the fluid. For example, if a fluid flow is interrupted by an impermeable layer of clay, the fluid must either flow parallel to that layer or stay on that side of it.

C. Chemical Characteristics of Sediments
1. Chemical Processes
Various chemical processes occur within sediments in solution. In the absence of organisms, a purely inorganic chemical reaction may take place. All processes and reactions influence pH and E_H. pH and E_H define the reactions that will take place and the reaction products. Some processes result in mineral formation.

2. pH
pH is defined as the negative logarithm to the base 10 of the hydrogen-ion activity of a solution:

$$pH = -\log_{10(H^+)}$$

where (H^+) is the hydrogen ion concentration. pH is measured on a scale from 0.0 to 14.0 where 7 is neutral, a lower pH is acidic, and a higher pH is basic. The pH of lakes and rivers varies between 6 and 9, whereas the pH of surface seawater is about 8.3. Ocean pH varies with depth to 7.5. Carbon dioxide (CO_2) and water (H_2O) act to provide a constant pH in the water environment.

Calcium carbonate ($CaCO_3$) responds to varying levels of pH. At pH levels of 8.3 carbonate particles adsorb an organic coating which prevents dissolution.[1] Reduced pH in the water reflects dissolution of uncoated particles. CO_2 levels, temperature, and pressure are involved to varying degrees in dissolving solid calcium carbonate. The organic coatings are removed when pH in water is high, allowing precipitation of calcium carbonate.

3. E_H

E_H, or redox potential, is a measure of the relative intensity of oxidation or reduction in solution or of the electron concentration in a solution.[1] The potential for electron transfer from one ion to another is measured in volts. Oxidation involves losing an electron, while reduction means gaining an electron. For example, sulfate-reducing bacteria break down sulfate to dissolved sulfur in the form of H_2S or HS.[1] Although pH is accurately categorized as a chemical process, E_H processes can be biological. E_H processes are those such as photosynthesis, respiration, bacterial reactions in the sulfur cycle, and the decomposition of organic matter.[1]

4. Weathering

The process of weathering produces debris in the form of mud, sand or gravel by mechanical and chemical processes. Fragmentation, the physical breakdown of solid rock, is a mechanical process usually occurring in cold weather and promoted by chemical weathering. In return, fragmentation works to aid chemical weathering as physical breakups of rocks open channels for water and air to penetrate. Chemical weathering varies widely with climate and material in its destruction and alteration of existing minerals and formation of new minerals.

5. Inorganic Chemical Processes

Inorganic chemical processes occur in environments free of simple organisms. These processes include formation of evaporite minerals and precipitation of minerals such as zeolite and feldspar.

Examples of evaporite minerals are precipitants. They are composed of minerals that precipitate from brines and then are concentrated by evaporation. The abundant evaporites in nature are gypsum ($CaSO_42H_2O$), anhydrite ($CaSO_4$), and halite ($NaCl$).

During purely inorganic chemical reactions zeolites and feldspars are precipitated. Zeolites contain calcium, sodium, and potassium. The composition of zeolites and feldspars is similar except that zeolites are hydrated.

D. Biological Characteristics of Sediments

1. Calcium Carbonate Skeletons

Organisms partially control chemical sediment by changing the environmental conditions. Calcium carbonate as skeletal debris is the main chemical sediment; it is extracted from seawater by soft-shelled organisms and secreted as hard shells. The degradation of hard shells by predators accounts for particles of calcium carbonate sediment.

2. *Trapping and Baffling*

Trapping and baffling are biological processes which cause sediment accumulation. Organisms causing physical changes to waves and currents, such as various grasses or invertebrates on reefs, are considered sediment baffles. Filamentous blue-green algae trap sediment directly using a sticky surface film.

3. *Pelletization*

Sediment-feeding organisms eat muds, compress them in their systems, and excrete them as sand-sized pellets. These pellets then behave in the physical environment as sand instead of mud. The pellets are produced in various sizes and have a high organic matter content.

4. *Burrowing*

Organisms frequently burrow into sediment searching for food or shelter. Effects of burrowing include breaking of fragile particles and destruction of sedimentary structures. Homogeneous sedimentary rocks such as limestone reflect extensive burrowing by organisms.

5. *Effects of Microorganisms*

Microorganisms such as bacteria participate in the weathering of rocks and create large quantities of fine-grained skeletal debris. Bacteria act as catalysts for inorganic chemical reactions at ordinary temperatures that would require considerable time, heat, pressure, or UV light to complete.[1]

Escherichia coli fecal bacteria readily attach themselves to sediment particles, and streambeds can be good environments for bacterial growth. Any disturbance stirs the sediment and resuspends the bacteria in the water. The concentration of *E. coli* in the bottom sediments of streams is sometimes 2 to 760 times greater than the concentration in the water itself.[6]

III. PHYSICAL AND CHEMICAL CHARACTERISTICS OF SLUDGE

Sludge is the unwanted residual of wastewater treatment processes which must be handled and disposed of properly. This residual is most often a semisolid material. The characteristics of sludges can vary greatly, due to the different types of wastes from which they originate and the types and degree of wastewater treatment.

Prior to discussing the characteristics of sludges, it is necessary to discuss the terms used in describing the different types of sludges. In wastewater treatment the first step is called primary treatment. Its main objective is to remove the settleable solids from wastewater. A primary clarifier is a settling basin that allows heavier solids to settle to the bottom and lighter solids to float to the top. The materials removed from the bottom of the clarifier are known as raw primary sludge. This sludge is often digested and is known as primary digested sludge. Digestion is one of the stabilization processes of sludges in which organic solids are biodegraded (biologically decomposed).

Biological processes are commonly used for secondary treatment. Trickling filters (a bed of media covered with a biological slime over which wastewater is trickled) are widely employed as a biological treatment method. The solids that slough off the filter media are captured. This sludge is called filter humus. Activated sludge is also an aerobic process which utilizes a culture in which bacterial cells are agglomerated together in flocs. Flocs are maintained in suspension either by diffused air added at the bottom of a tank or by mechanical agitation of the tank liquid. The biomass cultured in the aeration tank must be settled out in the final clarifier and recycled to the tank. When the amount of microorganisms produced in the activated sludge process exceeds the amount required by the system, the excess, called waste-activated sludge, must be disposed of. Figures 2 and 3 depict a trickling filter and an activated sludge aeration tank.

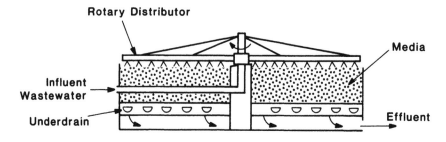

FIGURE 2.   Drawing of a typical trickling filter.

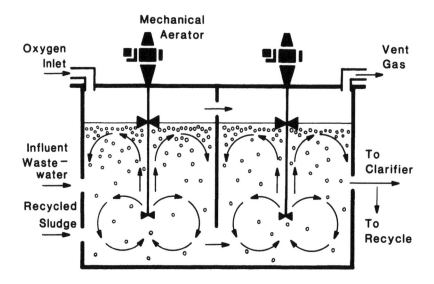

FIGURE 3.   Drawing of an aeration tank in an activated sludge process.

Both filter humus and waste-activated sludge are often mixed with raw primary sludge and digested by aerobic or anaerobic digesters. The resulting material is called mixed digested sludge.

Chemical precipitation is the addition of chemicals to wastewater to facilitate the removal of its dissolved and suspended solids by sedimentation. The most common chemicals used are alum, ferrous sulfate, lime, ferric chloride, and ferric sulfate. The waste from this process is called chemical sludge. The solid concentration of this sludge is usually higher due to the added chemicals.

Table 2 shows the variation in amounts of sludge for various wastewater treatment processes.[9]

A. Physical Characteristics
1. Specific Gravity
Specific gravity is defined as the ratio of the weight of the material to that of an equal volume of water. Reported values of specific gravity for sludge dry solids range from 1.2 to 1.9, while the overall wet sludge specific gravities range from 0.95 to 1.25.[10]

2. Solid Concentration
The solid concentration is used to describe the relative amounts of solids and liquids in a slurry, and is expressed as milligrams per liter or percent solids. Total solids in a sludge are the sum total of dissolved and suspended solids. Suspended solids make up the fraction that remains after filtering through a fiber filter. Each can be further divided into fixed and

Table 2
VARIATION IN AMOUNTS OF SLUDGE

Treatment processes	Amt. dry solids/ million gallons (lb)	Amt./millon gal sewage treated (gal)	Water in sludge (%)
Primary settling	900—1,200	2,500—3,500	93—95
Activated sludge	600—900	15,000—20,000	98—99
Trickling filters (low loading)	400—500	400—700	93—95
Trickling filters (high loading)	600—900	1,200—1,500	96—98
Chemical precip. of raw sewage	3,000—4,500	4,000—6,000	90—93

Table 3
TYPICAL CHEMICAL COMPOSITION OF RAW AND ANAEROBICALLY DIGESTED SLUDGE

	Raw primary sludge		Digested sludge	
	Range	Typical	Range	Typical
Total dry solid (TS, %)	2.0—7.0	4.0	6.0—12.0	10.0
Volatile solid (%TS)	60—80	65	30—60	40
Grease and fats (%TS)	6.0—30.0	—	5.0—20.0	—
Protein (%TS)	20—30	25	12—20	18
Cellulose (%TS)	8.0—15.0	10.0	8.0—15.0	10.0
Silica (SiO_2, %TS)	15.0—20.0	—	10.0—20.0	—
pH	5.0—8.0	6.0	6.5—7.5	7.0
Alkalinity (mg/ℓ as $CaCO_3$)	500—1500	600	2500—3500	3000
Organic acid (mg/ℓ as HAc)	200—2000	500	100—600	200

volatile solids. Volatile solid is the fraction that is lost on ignition of the dry solids at 550 ± 50°C. Some typical solid concentration values are listed in Table 3.[9]

3. Specific Resistance

The specific resistance of a sludge is a quantitative measure of its intrinsic filtration properties. It can be defined as the resistance of a cake having unit dry weight of sludge solid per unit area, to unit rate of flow of liquid having unit viscosity, and is usually expressed as meters per kilogram. Sludges with high specific resistance are difficult to dewater. Typical specific resistance values of primary, activated, and digested sludge are 1.5 to 10 × 10^{14}, 1 to 10 × 10^{13}, and 1 to 6 × 10^{14} m/kg, respectively.[11]

4. Settling

Sludges can be characterized by how well they settle. Sludges with higher solid concentration settle slower than dilute sludges. Sludge settleability is an important parameter in the operation of a treatment plant. The sludge volume index (SVI) is used to measure sludge settleability. SVI is defined as the volume in milliliters occupied by 1 g of sludge solids after quiescent settlement for 30 min of 1ℓ of sludge in a 1-ℓ Imhoff cone, a special measuring cylinder used for this test. The higher the SVI is, the poorer the sludge settleability. For a well-operated activated sludge process, the SVI is normally 100 or less.

Table 4
TYPICAL NUTRIENT VALUES OF DOMESTIC
SEWAGE SLUDGES (% TOTAL SOLID)

Sludge	Nitrogen	Phosphorus (as P_2O_5)	Potassium (as K_2O)	Ref.
Raw primary	2.5	1.6	0.4	9
	2.4	1.1	—	15
	3.0	1.6	0.4	16
Trickling filter	2.9	2.8	—	17
	3.0	3.0	0.5	16
Activated	5.6	7.0	0.6	17
	3.0	3.6	—	15
	5.6	5.7	0.4	16
Mixed digested	5.9	3.5	—	15
	3.7	1.7	0.4	16
	2.5	3.3	0.4	17

5. Particle Size

Sludge is composed of particles of different sizes, consistencies and shapes. The size of the particles changes with time and test conditions. Thus, it is very difficult to characterize sludges by particle size. By assuming the particles were spherical, Mueller et al.[12] reported that the nominal diameters of particles of eight activated sludges ranged from 21 to 115 μm.

6. Distribution of Water

Water in sludge can be divided into four categories[13]:

1.   Free water, which is not attached to solids in any way and can be separated from solids by simple settling
2.   Floc water, which is trapped within the floc
3.   Capillary water, which adheres to the individual particles and can be separated only if the particles are squeezed out of shape
4.   Bound water, which is chemically bound to the individual particles

For a typical activated sludge, about 75% of the water is free water, 20% is floc water, 2% is capillary water, and 2.5% is bound water.[13]

B. Chemical Characteristics

1. Mineral Nutrients

Knowledge about the chemical nature of sludges is important in the consideration of the various alternative disposal methods. Most interest in the chemical properties of sludges has been in relation to their fertilizing value and thus their disposal on agricultural land. The elements of primary concern in fertilizer are nitrogen, phosphorus, and potassium. A common fertilizer is 8-8-8, which contains 8% nitrogen, 8% phosphorus (as P_2O_5) and 8% potassium (as K_2O). Municipal sewage sludges seldom contain such a high level of nutrients. They contain about one fifth of the commercial level of nitrogen and phosphorus. The potassium content is low in relation to nitrogen and phosphorus content. Activated sludges have higher nutrient contents than primary sludges, and digestion appreciably reduces the nutrient value of sludge solids. Table 4 shows some typical nutrient values of sludges.

2. Metals

One concern in using sludges as fertilizer is the metal content because some metals (e.g.,

Table 5
CONCENTRATIONS OF VARIOUS METALS IN SEWAGE SLUDGE

Component	Sample type	Range (%)	Median (%)	Mean (%)
K	Anaerobic	0.02—2.64	0.30	0.52
	Aerobic	0.08—1.10	0.38	0.46
Na	Anaerobic	0.01—2.19	0.73	0.70
	Aerobic	0.03—3.07	0.77	1.11
Ca	Anaerobic	1.9—20.0	4.9	5.8
	Aerobic	0.6—13.5	3.0	3.3
Mg	Anaerobic	0.03—1.92	0.48	0.58
	Aerobic	0.03—1.10	0.41	0.52

Component	Sample type	Range (mg/kg)	Median (mg/kg)	Mean (mg/kg)
Mn	Anaerobic	58—7,100	280	400
	Aerobic	55—1,120	340	420
As	Anaerobic	10—230	116	119
	Aerobic	—	—	—
Mo	Anaerobic	24—30	30	29
	Aerobic	30—30	30	31
Hg	Anaerobic	0.5—10,600	5	1,100
	Aerobic	1.0—22	5	7
Pb	Anaerobic	58—19,730	540	1,640
	Aerobic	13—15,000	300	720

Cd and Ni) either may accumulate in the edible portion of agronomic crops or may cause phytotoxicity. Sommers[17] did an extensive analysis of the metal contents of different sludges. Table 5 is a partial list of his results.

3. Priority Pollutants

Other constituents which may cause damage to the environment and should be considered when characterizing municipal sludges are the toxic organics commonly known as priority pollutants. The concentrations of 24 of the more prevalent or more concentrated of the organic priority pollutants are listed in Table 6.[18]

4. Heat Content

Many wastewater sludges have a fuel value because of their high concentration of organic material. The heat content of untreated primary sludge is the highest. It ranges from 4.4 to 7.2 kcal/g of dry solids.[10] This compares favorably with coal, which has a fuel value of about 7.7 kcal/g. However, only about 4% of the sludge is solid.

5. Particle Aggregation

The net electrical charge on particles is measured as the zeta potential. Large particle charge tends to inhibit aggregation, which is important for efficient settling. The surface charge can be reduced or the effect of it can be overcome by a number of ways: (1) addition of potential-determining ions, (2) addition of electrolytes, (3) addition of long-chained organic molecules, and (4) addition of chemicals that form hydrolyzed metal ions.

Table 6
PRIORITY POLLUTANTS IN MUNICIPAL SLUDGE

	Conc in combined sludges[a]			
	Wet (μg/ℓ)		Dry (mg/kg)	
Chemical	Median	Range	Median	Range
bis-2-Ethylhexyl phthalate	3,806	157—11,257	109	4.1—273
Chloroethane	1,259	517—2,000	19	14.5—24
1,2-*trans*-Dichloroethylene	744	42—54,993	21	0.72—865
Toluene	722	54—26,857	15	1.4—705
Butylbenzylphthalate	577	1—17,725	15	0.52—210
2-Chloronaphthalene	400	400	4.7	4.7
Hexachlorobutadiene	338	10—675	4.3	0.52—8
Phenanthrene	278	34—1,565	7.4	0.89—44
Carbon tetrachloride	270	270	4.2	4.2
Vinyl chloride	250	145—3,292	5.7	3—110
Dibenzo(*a,h*) anthracene	250	25	13	13
1,1,2,-Trichloroethane	222	3—441	3.5	0.04—6.9
Anthracene	272	34—1,565	7.6	0.89—44
Naphthalene	238	23—3,100	7.5	0.9—70
Ethylbenzene	248	45—2,100	5.5	1.0—51
Dibutylphthalate	184	10—1,045	3.5	0.32—18
Phenol	123	27—4,310	4.2	0.9—113
Methylene chloride	89	5—1,055	2.5	0.06—30
Pyrene	125	10—734	2.5	0.33—18
Chrysene	85	15—750	2.0	0.25—13
Fluotanthene	90	10—600	1.8	0.35—7.1
Benzene	16	2—401	0.32	0.05—11
Tetrachloroethylene	14	1—1,601	0.38	0.024—42
Trichloroethylene	57	2—1,927	0.98	0.048—44

[a]   Primary plus secondary sludges.

IV. PHYSICAL AND CHEMICAL CHARACTERISTICS OF SOILS

A. Soil Classification

Soil is composed of solids, liquids, and gases whose composition is determined by geologic, biologic, atmospheric, and anthropogenic influences. Soils are classified by their physical and chemical properties, such as color, texture, structure, consistence, pH, and inorganic and organic matter content.[19] For example, classes of soil texture have been separated by the Soil Survey staff (USDA) according to particle size and percentages of clay, silt, and sand (see Figure 4). A layer of soil with distinct properties is called a soil horizon. In 1960, the Soil Survey staff developed revised names and descriptions for the major and minor horizons.[20] The O (organic) horizons are located above the mineral horizons and contain ≥20% organic matter, depending on the clay content. The mineral horizons contain less organic matter and more clay or other inorganic compounds. Presently the U.S. classifies soils according to six categories: orders, suborders, great groups, subgroups, families, and series. The orders are distinguished by factors such as percentage of organic carbon, clay, base saturation, $CaCO_3$, and the bulk density (g/cm^3)[19] (see Figure 5).

B. Physical Characteristics of Inorganic Solids

1. Soil Texture

The physical properties of soil include texture, structure, consistency, and color. As noted in Figure 4, soil texture is described according to the relative proportions of clay, silt, and

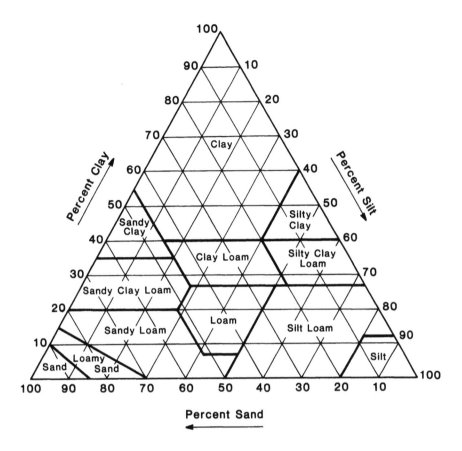

FIGURE 4. Chart showing the percentages of clay (< 0.002 mm), silt (0.002 to 0.05 mm) and sand (0.05 to 2.0 mm) in the basic soil textural classes.[1]

sand below 2.0 mm in diameter. Table 7 gives the diameter range of mineral soil grains. Sand and silt are composed primarily of quartz (SiO_2) and feldspar $MAl(Si_3O_8)$, where M represents Na^+ or K^+, or $MeAl_2Si_2O_8$, where Me represents Ca^{2+} or Mg^{2+}. These compounds are called *primary minerals* and their structure consists of a silicon-oxygen tetrahedron. Sand and silt create large pore spaces because of their size. This allows liquids to rapidly pass through sandy soils.

Clay is primarily composed of the secondary minerals, which include the layer silicates. These compounds are made up of three basic units: the silicon tetrahedral sheet, the aluminum hydroxide sheet, and the magnesium hydroxide sheet. Various combinations of these structures form crystals which can retard or absorb water depending on the properties of that clay. Kaolinite is the most common clay mineral. Clays can also be amorphous, i.e., have a structure that is not defined.

2. Soil Structure

Soil structure is defined as the grouping or aggregating of primary minerals. How close the particles are to each other will determine the mass transfer properties of the soil. There are four primary types of structure: platy, prismlike, blocklike, and spheroidal (see Figure 6). Structure can also be graded on degree of aggregation.

a. Soil Consistence

Soil consistence is a measure of the strength of the bonds that hold the natural aggregate of the soil together. Because these forces are affected by the water content of the soil, consistence is defined when wet, moist, dry, or brittle.

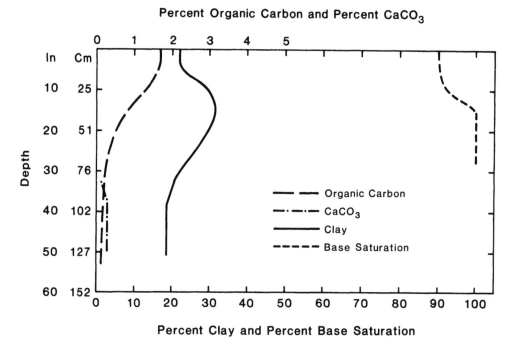

FIGURE 5.   Percentages of clay, organic carbon, base saturation, and calcium carbonate as functions of depth in a representative Mollisol.[1]

Table 7
SIZE LIMITS OF SOIL SEPARATES FROM
TWO SCHEMES OF ANALYSIS[1]

	USDA scheme		International scheme	
Name of separate	Diameter range (mm)	Fraction		Diameter range (mm)
Very coarse sand	2.0—1.0			
Coarse sand	1.0—0.5	I		2.0—0.2
Medium sand	0.5—0.25			
Fine sand	0.25—0.10	II		0.20—0.02
Very fine sand	0.10—0.05			
Silt	0.05—0.002	III		0.02—0.002
Clay	<0.002	IV		<0.002

b. Color

The color of the soil can be an indication of the mineral content. For example, red or yellow soil may indicate the presence of iron oxide. Soil color may also be used to estimate the organic content of the soil.

C. Organic Material in Soils and Its Effect on Physical Properties

Soil organic matter is formed from the biologic degradation of plants and animals. Microorganisms convert complex carbohydrates, fats, and proteins to intermediate compounds and end products. When the material becomes resistant to further biological decay, the end product is called humus. Not all the compounds present in humus have been identified, but

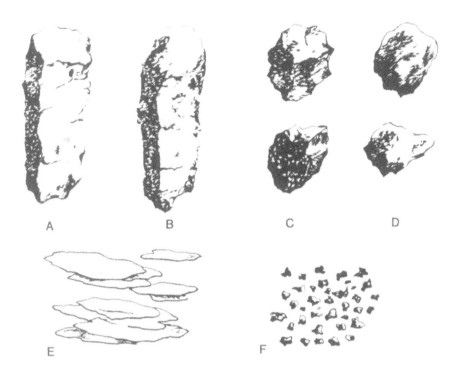

FIGURE 6. Drawings illustrating some of the types of soil structure. (A) Prismatic; (B) columnar; (C) angular blocky; (E) platy; (F) granular.[1]

90% of the amorphous material is believed to be composed of humic acids and polysaccharides.[21] The organic content of soils can vary from 0.5 to 5% by weight, but can range up to much higher values depending on the soil.[21]

Humus has a significant impact on the ability of the soil to retain moisture and to aggregate. This in turn affects the soil structure and consistency. The color of the soil will also become darker as humic compounds increase.

D. Organic and Inorganic Colloids

A colloid is defined by its size and not its chemical structure. Any particle ranging from 10^{-4} to 10^{-7} cm in size qualifies as a colloid. The particle may be crystalline or amorphous in structure and can be derived from organic or inorganic material. Colloids determine most of the physicochemical properties of the soil because of their large surface area. Ghosh and Schnitzer[22] found that the shape of organic colloids is influenced by the sample concentration, the pH, and the ionic strength of the soil. Under normal soil conditions colloids are flexible and linear.

E. Chemical Characteristics of Soils

1. Cation Exchange Capacity

The most important characteristics of soils are their ability to exchange cations and to adsorb compounds onto their surface. Clay particles are normally negatively charged due to the internal substitution of cations. For example, if a magnesium ion (Mg^{2+}) is substituted for an aluminum ion (Al^{3+}), a negative charge results. This charge is permanent and constant, and its magnitude ranges from 1 to 2×10^{-7} meq/cm^2.[23] A variable, pH-dependent charge is also present in organic and inorganic colloids due to the presence of functional groups at the surface of the particles. An exposed oxygen with an unsatisfied valence will attract a hydrogen ion (H^+) to its surface. The sign of the variable charge can vary from negative

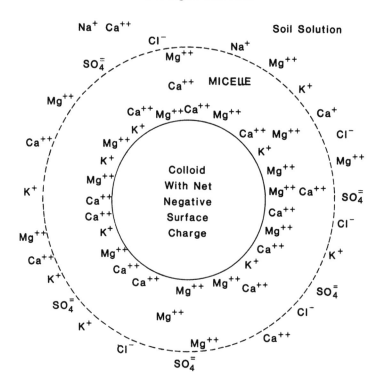

FIGURE 7.   A micelle containing a colloidal particle, a diffuse layer of cations, and water molecules.

to neutral to positive, and the magnitude of the charge can range from zero to several times that of the permanent charge.[23] A measure of the negative charges of a soil under specified conditions is called the cation exchange capacity (CEC) and is expressed as milliequivalents per 100 g of soil. If a soil has variable charges as well as permanent charges, its cation exchange capacity will increase with increasing pH. Organic matter has a higher CEC than clay, although there is considerable variation in the capacity of clays. Drake and Motto[24] analyzed the cation exchange capacity of New Jersey soils and found that 59% of the variation in the CEC of soils was due to differences in clay and organic matter.

Not all ions have the same affinity for the negative colloidal surface charge. For example, an aluminum ion (Al^{3+}) is more strongly bound to a negative charge than is a sodium ion (Na^+). In general the higher the valence, the stronger the bonding. Thus, the outer layer of cations surrounding the colloid may vary as a more preferred ion is exchanged for one less preferred. In wet or moist soils, colloids will also bind water molecules. The resulting particle, having an inner anionic surface surrounded by an outer layer of cations and water molecules, is called a micelle (see Figure 7).

2. Percentage Base Saturation

Another property related to cation exchange capacity is the percentage base saturation. This is a measure of the percentage of basic cations, such as calcium (Ca^{2+}), potassium (K^+), magnesium (Mg^{2+}), and sodium (Na^+), that are bound to the negative surface of the colloid. If all of the cation exchange capacity is occupied by hydrogen ions (H^+) or aluminum ions (Al^{3+}), then the soil has a zero base saturation. This property is pH related and tends to increase with decreasing rainfall.

3. Adsorption

The second important chemical property of soil is its ability to adsorb liquids, solids, and gases onto its surface. The adsorptive capacity of soil is determined by its surface area and its surface functional groups. Cation exchange is one example of the adsorption mechanism. Both organic and inorganic particles can be adsorbed by clays and humus. The ability of a soil to retain water is increased by the presence of organic matter. However, inorganic clays are also capable of attracting and binding water molecules to their surface with ionic forces. These forces impart a distinct and rigid structure to the adsorbed water.[25]

The adsorption of organic matter onto soil colloids involves several mechanisms, including cation exchange. Van der Waals interactions, attractions between two uncharged molecules, are thought to be significant in organic adsorption.[26]

The adsorption of inorganic cations has been discussed previously. When several species of organic and inorganic compounds are present in the soil solution at the same time, competition for adsorption sites onto the soil can occur. This competition results in a lower adsorption of a less preferred species in the presence of other preferred species. Solution concentration, pH, and the nature of the competing species influence competitive adsorption.[27]

4. Soil pH

Soil pH is determined by placing 2.5 cm^3 of soil in 4 mℓ of 0.015 M CaCl$_2$, mixing, and measuring the pH after 1 hr with a pH meter or pH paper.[19] A solution of CaCl$_2$ is added because soil pH is influenced by two factors: the ratio of soil to water and the equilibrium salt concentration.[21] As discussed earlier, a colloid has a layer of exchangeable cations surrounding its surface, including aluminum (Al^{3+}) and hydrogen (H$^+$) ions. If the pH measurement is taken close to this concentrated region, a lower reading will be obtained than one taken away from the surface of the colloid. In addition, calcium ions (Ca^{2+}) will exchange with bound hydrogen and aluminum ions, releasing them into the solution. In solution aluminum ions will hydrolyze and produce additional hydrogen ions, resulting in a lower pH.

Many soil properties are influenced by soil pH, including percentage base saturation, cation exchange capacity, adsorption potential, biological activity, fertility, and toxicity of the soil. The pH of a soil may decrease with increasing rainfall due to the leaching of basic cations or the addition of hydrogen ions (H$^+$) from acid rain. Acid soils will dissolve aluminum and other mineral precipitates. The increased concentration of these ions may cause the soil to become infertile.

The ability of a soil to resist a change in pH is called its buffering capacity. For example, it is common practice to add lime (CaCO$_3$) to raise the pH of a soil. Lime is hydrolyzed to produce an hydroxyl ion (OH$^-$), which in turn neutralizes a hydrogen ion (H$^+$) in solution. However, colloids will release hydrogen ions from their surfaces to replace the hydrogen ions in solution, and the pH will remain unchanged. This process is repeated with the addition of more lime until the buffering capacity is exhausted and the pH of the solution increases.

5. Weathering Reactions

Soil composition may change with time due to the effects of weathering. A clay soil composed primarily of kaolinite may be hydrolyzed to a solution of aluminum hydroxide ions. Weathering reactions are divided into two types. Congruent dissolution is the dissolving of a solid into solute ions, and incongruent dissolution is the full or partial precipitation of solute ions. Excessive rainfall will leach ions from the soil, but the decay of plant and animal matter will replenish the soil.

6. Redox Reactions

Another important chemical reaction in soil involves the transfer of electrons from one ion to another ion, or a redox reaction. Oxidation is the loss of an electron, and reduction is the acceptance of an electron. Redox reactions in soils can have an effect on soil pH. The reduction of metal oxides often results in an increase in the soil pH, while the oxidation of organic matter reduces the pH of the soil.

REFERENCES

1. **Friedman, G. M. and Sanders, J. E.,** *Principles of Sedimentology,* John Wiley & Sons, New York, 1978, 35, 64, 66, 67, 120, 130, 133, 136-138.
2. **Grim, R. E.,** *Clay Minerology,* McGraw-Hill, New York, 1953, 1.
3. **Press, F. and Siever, R.,** *Earth,* W. H. Freeman, San Francisco, 1982, 271.
4. **Krumbein, W. C. and Sloss, L. L.,** *Stratigraphy and Sedimentation.* W. H. Freeman, San Francisco, 1963, 75.
5. **Davis, R. A., Jr.,** *Depositional Systems,* Prentice-Hall, Englewood Cliffs, N.J., 1983.
6. Agricultural Research Service, *Agricultural Research,* Vol. 31, No. 12, U.S. Department of Agriculture, USDA, Washington, D.C., 1983, 9.
7. **Golterman, H. L., Sly, P. G., and Thomas, R. L.,** *Study of the Relationship Between Water Quality and Sediment Transport,* United Nations Educational, Scientific and Cultural Organization, Paris, 1983.
8. **Wheaton, F. W.,** *Aquacultural Engineering,* John Wiley & Sons, New York, 1977.
9. Municipal Sludge Management: Environmental Factors, EPA 430/9-77-004, Office of Water Program Operations, U.S. Environmental Protection Agency, Washington, D.C., 1977.
10. Sludge Processing Transportation and Disposal/Resource Recovery, a Planning Perspective, Water Planning Division Rep. WPD 12-75-01, U.S. Environmental Protection Agency, Washington, D.C., 1979.
11. **Metcalf and Eddy, Inc.,** *Wastewater Engineering,* McGraw-Hill, New York, 1979, 642.
12. **Mueller, J. A., Voelkel, K. G., and Boyle, W. C.,** Nominal diameter of floc related to oxygen transfer, *J. San Eng. Div. Am. Soc. Civ. Eng.,* 92(SA2), 9, 1966.
13. **Vesilind, P. A.,** *Treatment and Disposal of Wastewater Sludges,* Ann Arbor Science, Ann Arbor, Mich., 1979, chap. 2.
14. **Burd, R. S.,** A Study of Sludge Handling and Disposal, FWPCA (EPA) 41Publ. WP-20-4 Environmental Protection Agency, Washington, D.C., 1968.
15. **Metcalf and Eddy, Inc.,** *Wastewater Engineering,* McGraw-Hill, New York, 1972, chap. 11.
16. **Anderson, M. S.,** Fertilizing characteristics of sewage sludge, *Sewage Ind. Wastes,* 31, 6, 1959.
17. **Sommers, L. E.,** Chemical composition of sewage sludge and analysis of their potential use as fertilizers, *J. Environ. Qual.,* 6, 225, 1977.
18. **Naylor, L. M. and Loehr, R. C.,** Priority pollutants in municipal sewage sludge, *BioCycle,* 23, 18, 1982.
19. Soil Survey staff, Soil Taxonomy, Agric. Handb. No. 436, Soil Conservation Service, U.S. Department of Agriculture, Washington, D.C., 1975, 463.
20. Soil Survey staff, *Soil Classification,* Soil Conservation Service, U.S. Department of Agriculture, Washington, D.C., 1960, 25.
21. **Bohn, H. L., McNeal, B., and O'Connor, G. A.,** *Soil Chemistry,* John Wiley & Sons, New York, 1979, 90 and 205.
22. **Ghosh, K. and Schnitzer, M.,** Macromolecular structures of humic substances, *Soil Sci.,* 129(5), 266, 1980.
23. **Uehara, G. and Gillman, G. P.,** Charge characteristics of soils with variable and permanent charge minerals. I. Theory, *Soil Sci. Soc. Am. J.,* 44(2), 250, 1980.
24. **Drake, E. H. and Motto, H. L.,** An analysis of the effects of clay and organic matter content on the cation exchange capacity of New Jersey soils, *Soil Sci.,* 133(5), 281, 1982.
25. **Hillel, D.,** *Soil and Water: Physical Principles and Processes,* Academic Press, New York, 1971, 44.
26. **Sposito, G.,** *The Surface Chemistry of Soil,* Oxford University Press, New York, 1984, 143.
27. **Murali, V. and Aylmore, L. A. G.,** Competitive adsorption during solute transport in soils. III. A review of experimental evidence of competitive adsorption and an evaluation of simple competition models, *Soil Sci.,* 136(5) 279, 1983.

Chapter 3

METHODS OF RECOVERING VIRUSES FROM AQUATIC SEDIMENTS

V. Chalapati Rao and Gabriel Bitton

TABLE OF CONTENTS

I. INTRODUCTION

Discharge of both treated and untreated domestic wastes and sewage sludges into coastal estuarine waters results in the entry of more than 120 human enteric virus pathogens including those causing poliomyelitis, infectious hepatitis, and acute gastroenteritis.

A certain percentage (3 to 49%) of viruses associated with solids in sewage effluents.[1] Viruses associated with large-sized organic and inorganic particulates soon leave the water column and settle down in the bottom sediments, while viruses adsorbed on colloidal particles tend to stay afloat in the water for a longer period of time. Suspended solids-associated virus that recently settled out of the water column accumulates in a loose, fluffy layer over the compact bottom sediments. Sediments in coastal waters serve as a reservoir of human enteric viruses from which the virus can be released into the water column by storm action, dredging, and boating. Viruses from fluffy sediments can be more easily resuspended by mild turbulence or water movements. These resuspended viruses from polluted waters can be transported to remote nonpolluted areas used for shellfish production and bathing, depending on the current velocity, water circulation pattern, and prevailing winds. Shellfish (oysters, mussels, and clams, which are filter-feeding animals, i.e., they sieve out suspended food particles from a current of water passing through the shell cavity) could concentrate enteric viruses from the water in their tissues. Since the entire shellfish is often consumed raw or inadequately cooked, it can serve as a passive carrier of human pathogenic enteric viruses.

Viruses transported to bathing beaches can cause a health hazard during recreational activities, primarily swimming (especially if the head is immersed), but infection is also possible as a result of wading.

The magnitude of a health hazard associated with human viruses in estuarine waters depends upon virus survival and quantity in sediments and overlying water and their transport to relatively nonpolluted recreational and shellfish-growing waters.

II. SURVIVAL OF HUMAN VIRUSES IN ESTUARINE SEDIMENTS

Data from several laboratory experiments indicate prolonged survival of enteric viruses when they are associated with solids. Potential absorbents found in marine waters include sand, clays (montmorillonite, kaolinite, and illite), aquatic life forms (algae and bacteria), silts and sediments.

In a comparison of the survival times of four virus types in seawater and in seawater containing sediments collected from Galveston Bay, Smith et al.[2] demonstrated that in the presence of sediments, enteroviruses (echovirus 1, coxsackievirus B3, and coxsackievirus A9) survive longer in sediment than in the overlying water.

Studies on the survival of enteroviruses under field conditions were also carried out at two sites near a sewer outfall in Galveston Bay (Texas).[2] Cellulose dialysis tubes filled with seawater or a mixture of sediment and seawater seeded with known numbers of polio- and echoviruses were held in a plastic bait bucket with perforated sides at a depth of 1.5 m. The quantity of virus remaining in the samples was determined at 1-day intervals for 7 days. Both polio- and echovirus were inactivated much faster in seawater alone than in the presence of sediment and confirmed the protective effect of sediment toward viruses.

In a recent study, Rao et al.[3] examined the survival of poliovirus 1 and rotavirus SA11 seeded in seawater supplemented with fluffy sediments and suspended solids collected from Galveston Bay. Since these two fractions of solids have the greatest chance of being transported to distant areas, data on the survival of viruses associated with them are especially significant. Test results indicated that both viruses survived longer when associated with both solid fractions than in seawater alone. Both virus types could be detected in the fluffy sediment and suspended solids even on day 19, but could not be detected beyond day 9 in seawater. Prolonged survival of enteric viruses associated with sedimentary particulates and the significance attached to their ecology in the marine environment prompted the development of quantitative methods of enumerating viruses in estuarine sediments.

III. SEPARATION OF VIRUSES FROM SEDIMENTARY PARTICULATES

Earlier work related to the recovery of viruses from estuarine sediments was confined to extraction of naturally occurring viruses from sediments using commonly known eluents without controlled virus-seeded experiments to determine the efficiency of the procedure.

Glycine buffer (0.05 M: pH, 8.5),[4] nutrient broth (pH, 8.5),[5] and sterilized seawater[6] have been used for recovering naturally occurring enteroviruses from estuarine sediments. These results, in general, indicated the presence of higher numbers of viruses in sediments than in the overlying water.

Following recent investigations in the development of quantitative methods for concentration and detection of viruses seeded into sediments, it was recognized that the initial elution (separation) of virus from sediments is crucial for the overall efficiency of a method. The effectiveness of this step is clearly related to both the eluent and the type of sediment tested. Eluents examined include (1) glycine-EDTA, (2) urea-lysine, (3) beef extract-sodium nitrate, (4) beef extract-glycine, (5) 6% beef extract, (6) casein-lecithin, and (7) tryptose phosphate broth.

A. Glycine-EDTA

Gerba et al.[7] developed a quantitative method for elution of enteroviruses from estuarine sediments and their subsequent concentration. In laboratory experiments, poliovirus 1 was adsorbed to 500-g amounts of dried sediment (21% sand, 25% clay, 54% silt, and 3.8% organic matter) suspended in artificial seawater. After centrifugation, supernatant seawater was assayed to determine the quantity of virus adsorbed to the sediment. Then the sediment was resuspended in 1500 mℓ of 0.25 M glycine (pH, 11.5) containing 0.05 M EDTA. After addition to the sediment, the eluent had a final pH of 11.0. The sediment-eluent mixture was mixed for 10 min on a shaker table. The sample was then centrifuged at 2510 × g for 4 min and the supernatant assayed. Over $^3/_4$ (79%) of seeded poliovirus has been recovered from the sediment.

The glycine-EDTA method of Gerba and colleagues has been evaluated subsequently. Results of these studies indicated very low recovery of seeded poliovirus: <1%,[8] 0 to 0.1%,[9] 3.3 to 17%,[10] and 0.1 to 4.2%.[10a]

B. Urea-Lysine

Bitton et al.[8] investigated the efficacy of ten different eluents for desorbing viruses from marine and freshwater sediments: the marine sediments used in this study were a sandy sediment (99.7% sand and 0.3% clay) and an organic muck sediment. Two freshwater sediments (99.6% sand and 0.3 to 0.4% clay) were also included in the study. Sediment samples of 10 g were mixed with 20 mℓ of seawater or lake water which had been seeded with poliovirus type 1. The mixtures were mechanically shaken for 30 min and then centrifuged at 4000 rpm for 4 min. The supernatants were poured off and assayed for virus. Adsorption to marine sediments was 99% and 100%, compared to 37% and 45% for the two freshwater sediment types examined.

After an initial evaluation of ten eluents for virus recovery from sediments, the investigators selected urea-lysine, beef extract, and casein. Results of their detection efficiency are presented in Table 1. Virus recovery was generally higher for freshwater sediments than for marine sediments. Among the three eluents tested on marine sediments, virus recovery with

Table 1
UREA-LYSINE METHOD FOR POLIOVIRUS RECOVERY
FROM AQUATIC SEDIMENTS[8]

Type of sediment	Eluent[a]	Virus recovery (%) after		
		Primary elution	Conc of eluate	Overall procedure
Marine sediment	4 *M* Urea + 0.05 *M* lysine	44	50	22
	3% Beef extract	9	93	8
	1% Purified casein	34	41	14
Freshwater sediment	4 *M* Urea + 0.05 *M* lysine	43	92	39
	3% Beef extract	56	92	51
	1% Purified casein + 0.1% Tween® 80	47	125	59

[a] All eluents were adjusted to pH 9.0.

beef extract was 9%, while urea-lysine recovered 44% of input virus. With respect to freshwater sediments, improved virus recoveries of 43, 56, and 47% have been obtained for the three eluents, respectively. Mixing time required for elution of virus adsorbed to sediments was 1 min for urea-lysine (prolonged contact inactivates virus), compared to 30 min for the other eluents. Best results were obtained with urea-lysine that was prepared a few hours before use. The authors speculated that the mechanism of virus elution from sediments may be due to interference by urea with hydrophobic interactions between viruses and aquatic sediments.

The urea-lysine method was also tested for elution of poliovirus 1 inoculated into marine and freshwater sediments of different compositions.[11] Virus recovery ranged from 1.9% to 31.1% for marine sediments and from 0.1% to 33.7% for freshwater sediments.

C. Beef Extract-Sodium Nitrate

The potential of chaotropic compounds for the recovery of enteric viruses from estuarine sediments has been investigated by Wait and Sobsey.[9] A chaotrope is a low-molecular-weight ionic compound which alters the thermodynamics of a solution in such a way as to favor the solubilization of hydrophobic substances and improve the recovery of enteric viruses from highly organic estuarine sediments. Chaotropic agents such as $NaNO_3$, $NaCl$, and KCl alone were poor eluents of poliovirus from sediment but were effective when combined with 3% beef extract. Estuarine sediment, a black organic muck (10 mℓ), suspended in 10 mℓ of seawater seeded with 4×10^4 to 4×10^5 plaque-forming units (pfu) of poliovirus 1 per milliliter was mixed on a rotary shaker for 15 min and then centrifuged at $1850 \times g$ for 10 min. Under these conditions 99.9% of seeded poliovirus was adsorbed to sediment. For eluting the virus, 30 mℓ of 3% beef extract supplemented with 2 *M* $NaNO_3$ (BE-NO_3) adjusted to pH 5.5 was added to the sediment deposit and was mixed vigorously on a rotary shaker for 15 min. After centrifugation at $5140 \times g$ for 5 min, the eluate was recovered and tested for virus. Efficiency of the elution step for poliovirus recovery was 71%. BE-NO_3 was compared with two more eluents for recovering poliovirus 1, echovirus 1, and rotavirus SA11 from an estuarine sediment, and these results are presented in Table 2. Average overall recoveries of 42, 30, and 23% for the three virus types, respectively, were obtained. Beef extract acts by competing with virus for adsorption sites, thus reducing the electrostatic interaction between virus and sediments. Furthermore, the chaotropic ions may increase desorption by decreasing the hydrophobic interaction between viruses and sediments. Thus, both electrostatic and hydrophobic interactions may play a role in the desorption of enteric viruses from estuarine sediments.

Table 2
COMPARISON OF METHODS FOR RECOVERY OF
ENTERIC VIRUSES FROM ESTUARINE SEDIMENT

Virus	Virus input (pfu/50 g of sediment)	No. of trials	Mean overall virus recovery (%)[a]		
			BE-NO$_3$	BE-GLY	GLY-EDTA
Poliovirus 1	10^4	2	39	9.7	0.1
	218	2	44	18.0	0
Echovirus 1	10^4	2	43	4.4	0.1
	615	2	16	2.6	1.5
Rotavirus SA11	370	2	23	0	0

[a] BE-NO$_3$, beef extract-sodium nitrate method; BE-GLY, beef extract-glycine method; GLY-EDTA, glycine-EDTA method.

From Wait, D. A. and Sobsey, M. D., *Appl. Environ. Microbiol.*, 46, 369, 1983. With permission.

Usefulness of beef extract-sodium nitrate has also been examined for recovering poliovirus 1 experimentally adsorbed to five different sediment samples collected from Galveston Bay and the New York Bight sludge dump site.[10] Virus recoveries ranged from 58 to 88%.

D. Beef Extract-Glycine

Cooper et al.[12] used 3% beef extract prepared in 0.25 *M* glycine (BE-GLY; pH, 10.5) for elution of enteric viruses from sediments collected from San Francisco Bay. Efficacy of BE-GLY was subsequently evaluated by Wait and Sobsey[9] in recovering poliovirus 1 seeded into sediment samples containing organic muck and obtained from the Newport River estuary in North Carolina. Virus recoveries of 9.7 and 18% have been made in two trials. None of the rotavirus SA11 was recovered. Echovirus 1 recovery was 2.6 and 4.4% in two trials (Table 2). Lesser recoveries (0.2 to 8.2%) of poliovirus seeded into five different sediment samples collected from Galveston Bay and the New York Bight have been obtained using BE-GLY (pH, 10.5; Table 3).[10] Rao et al.[10] demonstrated 21 to 53% recovery of seeded virus in the sediment-eluate mixture prior to centrifugation, and this indicates that the virus was still solids-associated. Only 0.2 to 8.2% of the original virus could be recovered in the supernate after centrifugation of the sediment.

Beef extract (3%) was also used for eluting viruses from estuarine sediments by Tsai et al.[10a] Two sediment types with clay content of 42 and 9%, respectively, collected from the Mississippi estuary were seeded with known amounts of poliovirus and extracted with 3 and 10% beef extract solutions (Difco). Mean recovery of virus was 13% (high clay) and 3.7% (low clay). Strength of the beef extract did not make any difference in the amount of virus recovered. The use of 3% beef extract powder (Inolex) led to the recovery of 40% of added virus from both the sediment types.

E. 6% Beef Extract

This particular beef extract was examined for its efficiency in separating poliovirus 1 from freshwater and marine sediments.[11] Six aquatic sediments collected on a single occasion and stored at 4° were used in assessing viral adsorption and recovery. The mean adsorption to marine sediments was 98.3%, but adsorption to freshwater sediments was slightly lower (67.5%). For virus elution, 10 g of sediments to which virus had been adsorbed was mixed with three times the amount (w/v) of 6% beef extract (Difco) at pH 9.0. This mixture was vortexed for 30 sec and then incubated at 4°C for 1 hr for virus elution. The samples were

Table 3
BEEF EXTRACT-GLYCINE METHOD FOR POLIOVIRUS ELUTION FROM ESTUARINE SEDIMENTS[10]

Sediment type[a]	Virus in sediment + seawater[b]	Virus adsorbed to sediment (%)[c]	Eluent	Virus in sediment-eluate mixture[d]	Virus in eluate after centrifuging sediment (%)[e]
1	1.1×10^7	99.95	3% BE-GLY	40.9	1.4
2	5.1×10^7	99.95		36.0	0.3
3	1.3×10^7	99.97		53.0	0.2
4	1.1×10^7	—		—	8.2
5	6.0×10^6			21.2	7.2
1	2.6×10^7	99.93	GLY-EDTA	19.2	7.6
2	2.1×10^7	99.91		36.0	3.3
3	2.3×10^7	99.99		16.0	17.0
4	1.4×10^7	99.95		7.4	6.2
5	1.4×10^6	98.20		—	4.5

[a] 1, 2, and 3 were collected from Chocolate Bayou; 4 and 5 were collected from the New York Bight during the Delaware Cruise.
[b] After a 30-min shaking at 200 rpm, 1 mℓ of the sample was diluted and assayed.
[c] The sediment-seawater mixture was centrifuged at 1000 rpm for 30 min, the supernatant was decanted, and 1 mℓ was diluted and assayed.
[d] The sediment is suspended in three to four times its volume with either beef extract or EDTA and shaken at 300 rpm for 10 min in EDTA and 30 min in beef extract. A 1-mℓ sample is diluted and assayed. In the case of EDTA, the sample is neutralized with $MgCl_2$.
[e] For EDTA, the sample is centrifuged at 4000 rpm for 4 min, and for beef extract at 1500 rpm for 15 min. A 1-mℓ supernatant is diluted and assayed.

Table 4
RECOVERY OF ENTEROVIRUSES WITH SEMIPURIFIED SOYBEAN LECITHIN-SUPPLEMENTED CASEIN

Virus type	Recovery (%) from sediment no.[a]				Av. of all sediments
	1	2	3	4	
Poliovirus	75.1	9.7	61.4	2.6	37.2
Coxsackievirus	54.4	91.5	78.4	55.7	70.0
Echovirus	58.9	78.5	69.0	56.8	65.8
Average for all viruses	62.8	59.9	69.6	38.3	

[a] Clay content of sediment: type 1, 3%; type 2, 26%; type 3, 7%; type 4, 34%.

From Johnson, R. A., Ellender, R. D., and Tsai, S. C., *Appl. Environ. Microbiol.*, 48, 581, 1984. With permission.

then centrifuged at $10,000 \times g$ for 20 min, and virus in the supernatant was tested. The investigators also examined 2% skim milk (pH, 9.0) and 4 *M* urea plus 2 *M* lysine (pH, 9.0) as eluents in comparative tests along with 6% beef extract. Six percent beef extract was the best of the three eluents in this study. Recovery efficiency ranged from 15 to 53% for marine sediments and from 36 to 153% for freshwater sediments (Table 5). Efficiency

Table 5
EFFICIENCY OF VARIOUS ELUENTS IN THE SEPARATION OF ENTERIC VIRUSES ADSORBED TO SEDIMENTS

Eluent	Source of sediment	Composition of sediment	Virus type	Virus recovery (%)	Ref.
Glycine-EDTA	Galveston Bay, Tex.	21% sand, 25% clay, 54% silt, 3.8% organic matter	Polio 1	79	7
	Florida coast	99.7% sand, 0.3% clay	Polio 1	<1	8
	Newport River estuary, N.C.	Black organic muck	Polio 1	0.1, 0	9
			Echo 1	4.4, 2.6	
			Rota SA11	0	
	Chocolate Bayou, near Houston, Tex.	40% sand, 38% clay, 20% silt	Polio 1	3.3—7.0	10
	New York Bight sludge dump site	—	Polio 1	4.5—6.2	
	Mississippi estuary	10% sand, 42% clay, 12% silt	Polio 1	0.1—4.2	10a
			Cox B3	4.2—11.1	
		79% sand, 9% clay, 12% silt	Cox B3	3.8—4.3	
Urea-lysine	Florida coast				
	Marine sediment	99.7% sand, 0.3% clay	Polio 1	44	8
	Freshwater sediment	99.6% sand, 0.4% clay	Polio 1	43	
	New Zealand				
	Marine sediment	Fine	Polio 1	1.9	11
		Sandy		31.1	
		Gravel		11.2	
	Freshwater sediment	Fine	Polio 1	33.7	
		Sandy		27.2	
		Gravel		0.1	
Beef extract-nitrate	Newport River estuary	Black organic muck	Polio 1	39, 44	9
			Echo 1	43, 16	
			Rota SA11	23	
	Galveston Bay	30% sand, 40% clay, 30% silt	Polio 1	88	10
	Chocolate Bayou	40% sand, 38% clay, 20% silt	Polio 1	62	
	N.Y. Bight sludge dump site	—	Polio 1	58	
	Lake Houston, Tex.	30% sand, 60% clay, 10% silt	Polio 1	100, 40, 30	
			Rota SA11	22	
			Hep A	20	
3% Beef extract	Newport River estuary	Black organic muck	Polio 1	9.7, 16.0	9
	Chocolate Bayou	40% sand, 38% clay, 20% silt	Polio 1	0.2—1.4	10
	N.Y. Bight sludge dump site	—	Polio 1	7.2, 8.2	
	Mississippi estuary	10% sand, 42% clay, 48% silt	Polio 1	40	10a
			Cox B3	29	
		79% sand, 9% clay, 12% silt	Polio 1	41, 46	
			Cox B3	14	
6% Beef extract	New Zealand				
	Marine sediment	Fine	Polio 1	15.5	11
		Sandy		53.3	
		Gravel		25.5	

Table 5 (continued)
EFFICIENCY OF VARIOUS ELUENTS IN THE SEPARATION OF ENTERIC VIRUSES ADSORBED TO SEDIMENTS

Eluent	Source of sediment	Composition of sediment	Virus type	Virus recovery (%)	Ref.
	Freshwater sediment	Fine		43.1	
		Sandy		35.8	
		Gravel		153.0	
	Galveston Bay	30% sand, 40% clay, 30% silt	Polio 1	0.1—1.8	10
	Lake Houston	30% sand, 60% clay, 10% silt	Polio	4.8—9.0	
Casein-lecithin	Mississippi estuary	99% sand, 0% clay, 1% silt	Polio 1	75	13
			Cox B1	54	
			Echo 11	59	
		52% sand, 17% clay, 30% silt	Polio 1	10	
			Cox B1	91	
			Echo 11	78	
		89% sand, 5% clay, 6% silt	Polio 1	61	
			Cox B1	78	
			Echo 11	69	
		37% sand, 23% clay, 39% silt	Polio 1	3	
			Cox B1	56	
			Echo 11	57	
10% Tryptose phosphate broth	Lake Houston	30% sand, 60% clay, 10% silt	Rota SA11	56	10
			Hep A	70	

of 6% beef extract for eluting poliovirus 1 from five sediment types was also examined by Rao and colleagues.[10] Virus recovery ranged from 0.1 to 1.8% for estuarine sediments and from 4.8 to 9% for freshwater sediments (Table 5).

F. Casein-Lecithin

Recovery of enteric viruses from estuarine sediments with lecithin-supplemented eluents has been examined by Johnson et al.[13] Lecithin (phosphatidylcholine), a component of beef extract, is found in mammalian cell membranes as part of the lipid bilayer and is available commercially as soybean, egg, and synthetic lecithins. Four sediment types collected from the Mississippi estuary and differing in clay content from 3 to 34% were tested for adsorption and elution of polio-, coxsackie-, and echoviruses. Four lecithin preparations mixed with 4% nutrient broth, 5% beef extract, and 0.5% isoelectric casein have been tested. A 3% semipurified soybean lecithin mixed with 0.5% isoelectric casein recovered an average of 37, 70, and 66% of the three viruses, respectively, from all sediment types (Table 4). An increase in the clay content of a sediment generally resulted in lower virus recovery. Recoveries of virus were considerably lower (10 to 24%) with the three eluents when tested without lecithin supplement. The investigators attributed the release of virus by lecithin-containing eluents to a decrease in hydrophobic bonding and a lowering of surface tension energy. Hydrophobic bonding of virus to clay surface may have been disrupted by the surfactant effect of lecithin, allowing greater contact with water molecules and virus release into suspending eluent.

G. Tryptose Phosphate Broth

Several eluents have been tested for recovering poliovirus 1, rotavirus SA11, and hepatitis

A virus experimentally added to freshwater sediments collected from Lake Houston.[10] Three percent beef extract-2 M NaNO$_3$ (pH, 5.5) and 2% isoelectric casein (pH, 8.5) recovered an average of 66% of poliovirus; 10% tryptose phosphate broth could recover both rotavirus and hepatitis A virus with efficiencies of 56 and 70%, respectively. A single eluent for recovering all three virus types efficiently from freshwater sediments has not been identified yet.

IV. CONCENTRATION OF VIRUSES FROM SEDIMENT ELUATES

Isolation of viruses from field samples requires processing of 100 to 500 g of sediment using at least 300 to 1500 mℓ of eluate. Concentration of the eluate to a smaller volume is necessary for detection of low numbers of viruses. Methods commonly used in water virology have been adapted for recovering viruses from sediment eluates. The methods evaluated include (1) membrane filtration, (2) organic flocculation, (3) iron oxide adsorption elution, and (4) polyethylene glycol precipitation.

A. Membrane Filtration

Gerba et al.[7] used Filterite filters for concentrating virus from glycine-EDTA eluates (pH, 11.0) obtained from estuarine sediments. The eluate was adjusted to pH 3.5 by addition of 1 M glycine buffer (pH 2.0). Aluminum chloride (1 M) was then added to yield a 0.06 M final concentration, and the solution was passed through 3.0- and 0.45-μm pore-size, 142-mm-diameter epoxy-fiberglass (Duofine series, Filterite Corp.) filters. AlCl$_3$ concentration was kept high enough to overcome any interference from EDTA present in the eluate. An organic floc sometimes forms when the pH of the solution is adjusted to 3.5, but the authors noted that the flocs did not interfere with the passage of the sample through filters or with the virus recovery. Virus was recovered from the filters by passage of two 25-mℓ volumes of 0.25 M glycine buffer (pH, 11.5) through the filter series. The eluate was then quickly neutralized by addition of 1 M glycine buffer (pH, 2.0). A final eluate of 30 to 50 mℓ was obtained with a virus recovery efficiency of 66.6%.

Rao et al.[10] tested the usefulness of Filterite filters in concentrating poliovirus from primary eluates of 3% beef extract-2 M NaNO$_3$ obtained from an estuarine sediment. The procedure is essentially the same as used by Gerba et al.[7] except that the sediment eluate was different. A range of 11 to 34% of the seeded virus was recovered in three trials.

One difficulty with this method was that the filters used in the concentration step often clogged because of organic matter derived from the sediment. The organic flocs also appeared to affect virus elution from filters.

B. Organic Flocculation

Wait and Sobsey[9] examined a simple procedure in which the 3% beef extract-2 M NaNO$_3$ eluate from an estuarine sediment was adjusted to pH 3.5 to flocculate beef extract proteins along with added virus. This resulted in the recovery of only 2% of virus, with the rest remaining unflocculated in the supernatant. To improve flocculation, the investigators added an antichaotrope, 2 M ammonium sulfate, which resulted in a visible increase in protein flocculation and also improved virus recovery to 24%. In a further attempt to enhance the flocculation process, two organic polymers, polyethylene glycol (carbowax PEG 20,000) and cat-Floc T (0.01%) in the presence of 2 M (NH$_4$)$_2$SO$_4$ were tested. Virus recovery in excess of 100% was obtained upon addition of cat-Floc T, while PEG (0.6%) plus 0.5 M MgCl$_2$ resulted in 72% virus recovery. The ability of saturated ammonium sulfate to improve the flocculation of beef extract and hence to increase virus recovery was recently reported by Shields and Farrah.[22] The percent recovery for various phages and enteroviruses varied between 72 and 107%. The application of this method to sediment eluates should be very helpful.

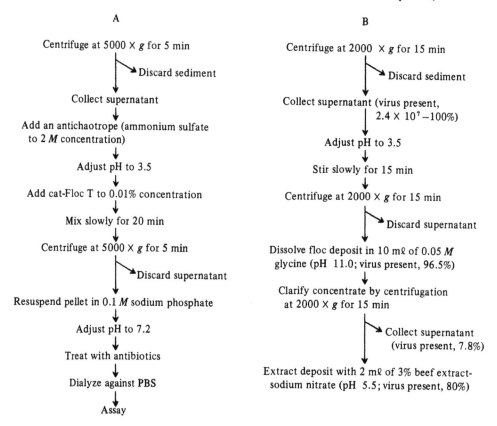

FIGURE 1.   Method of concentrating enteric viruses from estuarine sediments. (A from Wait, D. A. and Sobsey, M. D., *Appl. Environ. Microbiol.*, 46, 379, 1983. With permission. B from Rao, V. C., Rao, T. V., Metcalf, T. G., Dahling, D., and Melnick, J. L.[10])

Rao and colleagues[10] evaluated the organic flocculation method to concentrate poliovirus seeded into beef extract-sodium nitrate eluates from an estuarine sediment. Neither an antichaotrope nor cat-Floc was added to enhance flocculation of proteins. Beef extract paste instead of powdered beef extract was used. The sediment eluate, after centrifugation at 2000 × *g* still contained a large amount of fine suspended solids derived from the sediment, which upon acidification to pH 3.5 produced a good visible floc. The floc was centrifuged at 2200 × *g* for 15 min and the deposit suspended in 10 mℓ of 0.05 *M* glycine (pH, 11.0). Glycine pH drops to 9.6 because of the acidity of the floc. Virus recovery was 96.5%. It is important to note that the floc deposit was highly turbid and had to be diluted 1000-fold before assay on monolayer cultures; however, field samples containing low numbers of viruses do not permit dilution of the sample concentrates. The only alternative is to clarify the sample by centrifugation; but, centrifugation of our glycine concentrates resulted in the recovery of only 7.8% of the original virus in the supernatant. Nearly 88% of the virus was sedimented along with solids. Extraction of this deposit with 2 mℓ of beef extract-sodium nitrate (pH, 5.5) recovered about 80% of the virus. Various steps of our procedure along with the procedure developed by Wait and Sobsey are outlined in Figure 1. Results reported in Table 6 (footnotes c, d, and e) represent the combined recovery of virus in the supernatant and sediment extract of our final sample concentrate.

Concentration of poliovirus 1 added to freshwater sediment eluates of 2% casein and 10%

Table 6
METHODS OF CONCENTRATING ENTERIC VIRUSES FROM SEDIMENT ELUATES

Method	Vol of sediment eluate (mℓ)	Vol of final conc. (mℓ)	Virus type	Virus recovery (%)	Ref.
Membrane filtration	1500	30—50	Polio 1	66	7
	30	10	Polio 1	11—34	10
Organic flocculation	30	3	Polio 1	72[a]	9
				138[b]	
	900	12	Polio 1	62[c]	10
				74[d]	
				96[e]	
	25	4	Polio 1	53[f]	
	25	4	Polio 1	36[g]	
Iron oxide adsorption-elution	25	5	Polio 1	30[h]	10
			Polio 1	49[i]	
Polyethylene glycol precipitation	150	—	Polio 1	76	11
				92	
Aluminum hydroxide flocculation	30	—	Polio 1	50[j]	8
				92[k]	

[a] 3% beef extract-2 M NaNO$_3$ eluent (from an organic muck sediment) was supplemented with antichaotrope 2 M (NH$_4$)$_2$SO$_4$ and PEG (0.6%)-0.5 M MgCl$_2$.

[b] Same eluent as described in note a supplemented with 0.01% cat-Floc T.

[c] Sediment from Chocolate Bayou near Galveston where effluents from an oil refinery are discharged.

[d] Sediment from Galveston Bay where activated sludge effluents are discharged.

[e] Sediment collected from sludge dump sites in the New York Bight.

[f] Freshwater sediment from Lake Houston eluted with 2% casein seeded with poliovirus 1 was concentrated at a pH of 4.5. Floc deposit was suspended in 0.15 M Na$_2$HPO$_4$.

[g] Tryptose phosphate broth, the primary eluent from Lake Houston sediment, seeded with poliovirus was concentrated at pH 3.5. Floc deposit was suspended in 0.15 M Na$_2$PO$_4$.

[h] Casein eluate from Lake Houston sediment.

[i] Tryptose phosphate broth eluate from Lake Houston sediment.

[j] Marine sediment.

[k] Freshwater sediment.

tryptose phosphate broth by organic flocculation resulted in a recovery of 53 and 36% of virus (Table 6, footnotes f and g).

C. Iron Oxide Adsorption-Elution

Magnetic iron oxide (Fe$_2$O$_3$-Fe$_3$O$_4$, Fisher Co., Fair Lawn, N.J.) has been extensively used in our laboratory for reconcentrating polioviruses, rotaviruses, and hepatitis A viruses from primary eluates of microporous filters. In the case of estuarine sediments, the method developed for filter eluates has been adapted for recovering rotaviruses from field samples. No virus input studies were conducted to demonstrate the efficiency of virus recovery; however, with regard to freshwater sediments, virus was seeded into sediment eluates and recovered by iron oxide. In these experiments, 10-g samples of Lake Houston sediment were extracted separately with 50 mℓ volumes of 2% casein (pH, 8.5) or 10% tryptose phosphate broth (pH, 9.5) and centrifuged at 1200 × g for 30 min to obtain clear supernatants. Each extract was divided into two aliquots of 25 mℓ each, seeded with poliovirus 1, and concentrated either by iron oxide or organic flocculation. An outline of the iron oxide procedure is given in Figure 2. Virus adsorption to iron oxide from casein and tryptose phosphate broth was 86 and 97%, respectively, while virus recovery was 30 and 49% (Table 6, footnotes h and i).

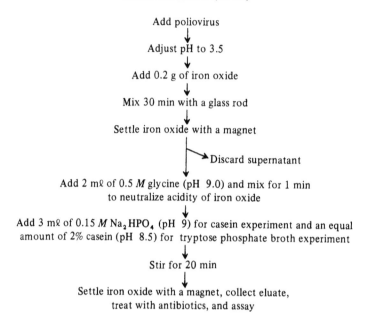

CASEIN OR TRYPTOSE PHOSPHATE BROTH ELUATE
FROM SEDIMENT (25 mℓ)

Add poliovirus

Adjust pH to 3.5

Add 0.2 g of iron oxide

Mix 30 min with a glass rod

Settle iron oxide with a magnet

→ Discard supernatant

Add 2 mℓ of 0.5 *M* glycine (pH 9.0) and mix for 1 min
to neutralize acidity of iron oxide

Add 3 mℓ of 0.15 *M* Na_2HPO_4 (pH 9) for casein experiment and an equal
amount of 2% casein (pH 8.5) for tryptose phosphate broth experiment

Stir for 20 min

Settle iron oxide with a magnet, collect eluate,
treat with antibiotics, and assay

FIGURE 2.   Concentration of poliovirus 1 from freshwater sediment eluates using magnetic iron oxide.

D. Polyethylene Glycol Precipitation

Lewis et al.[11] tested the efficiency of PEG to concentrate poliovirus 1 from 6% beef extract eluates obtained from sediments. PEG 6000 was added to the beef extract (pH, 9.0) to give a final concentration of 8% and mixed at 4°C for 1 hr followed by centrifugation at 10,000 × *g* for 20 min. The pellet was resuspended in tissue culture growth medium containing 5% calf serum. Virus recoveries of 76 and 92% were reported for two sediment types (Table 6).[11] It was found that recovery of viruses from freshwater sediments could be increased by the addition of 2% NaCl, but addition of NaCl did not enhance the recovery of viruses from marine sediments.

E. Aluminum Hydroxide Flocculation

Urea-lysine eluates containing poliovirus derived from both marine and freshwater sediments were concentrated by adsorption to aluminum hydroxide flocs by Bitton and colleagues.[8] In this procedure, $AlCl_3$ was added to urea-lysine eluates to give a final concentration of 0.005 *M*. The eluate was then adjusted to pH 7 by the addition of 1 *M* sodium carbonate and mixed for 5 min. The floc formed was collected by centrifugation at 5000 rpm for 5 min. The supernatant was discarded, and the floc was mixed with 5 vol of 0.1 *M* EDTA-3% beef extract (pH, 9.0). The sample was then centrifuged at 5000 rpm for 5 min, and the supernatant was adjusted to pH 7 and dialyzed against PBS at pH 7.0 overnight at 4°C. Results (Table 6) indicate 50% recovery of virus from a marine sediment and 92% from a freshwater sediment.[8]

V. PROMISING APPROACHES FOR VIRUS RECOVERY FROM AQUATIC SEDIMENTS

Currently it is difficult to suggest any single method for recovery of viruses from sediments. Most of the methods published are adaptations of those used in recovering viruses from water and wastewater. Studies comparing the effectiveness of different eluents on sediments collected from different areas (Table 5) have indicated widely different virus recoveries. The best elution of virus from both marine and freshwater sediments was obtained with beef extract supplemented with sodium nitrate.[9,10] Although comparison of eluent efficacy is difficult, especially when differences exist in the sediment composition, beef extract-sodium nitrate appears effective as judged from its ability to elute poliovirus from different sediment types (Table 5).[9,10]

All the methods tested for concentrating viruses from sediment eluates appear to be more than optimal in their efficiency (Table 6), although a single method has not been evaluated on different eluates.

Because of the complexity and variable composition of sediments, some of the methods should be evaluated with model enteroviruses on well-characterized sediments in a round-robin testing. From such a concerted effort, a suitable and efficient procedure for recovering viruses from field samples of sediments may be available.

VI. FIELD DATA ON VIRUS ISOLATION FROM ESTUARINE AND FRESHWATER SEDIMENTS

Data on the isolation of viruses from field samples of estuarine and freshwater sediments are summarized in Table 7. An examination of the information would reveal that the number of studies on virus detection in sediments are limited, that most of the data pertain to marine sediments, and that little is known about virus occurrence in freshwater sediments. In most of the studies, virus elution from sediments was achieved by extraction with glycine-EDTA at pH 11.0, beef extract at pH 9 to 10.5, or urea-lysine at pH 9.0. In recent studies, beef extract was mostly used because certain viruses (e.g., adenoviruses and rotaviruses) may be inactivated by the high pH of the glycine-EDTA eluent.

Among the viruses isolated, most groups of enteric viruses were found in the sediments. Polioviruses have been detected in sediments in most of the studies.

VII. SUMMARY AND CONCLUSIONS

Many enteric viruses adsorb to marine sediments at ambient pH of seawater with greater than 99% efficiency, perhaps due to high ionic concentration, while virus adsorption to freshwater sediments is slightly lower. Several methods have recently been developed for recovering viruses from sediments. Most of these methods pertain to virus recovery from estuarine sediments. The efficiency of any viral recovery technique involving sediments depends primarily on the initial elution of virus from sediments. The efficacy of many of the eluents tested varied with the type of sediment used. Sediment composition (clay, sand, silt, and organic matter) and ionic composition of the water may influence virus elution from sediments. From the limited data available, beef extract supplemented with a chaotrope such as sodium nitrate at pH 5.5 appears to recover virus effectively from both freshwater and estuarine sediments of varying compositions.

Methods used for concentrating poliovirus from sediment eluates appear to be reasonably efficient, but their efficiency must be evaluated with regard to the recovery of rotavirus and hepatitis A virus from a variety of sediments.

Further evaluations of some of the currently available methods should be conducted using well-characterized sediments in a round-robin study so as to provide a standard method for virus detection in sediments.

Table 7
ISOLATION OF VIRUSES FROM SEDIMENTS

Sediment source	Virus type isolated	Dominant virus	Conc method (sample vol)	Virus conc	Ref.
Houston ship channel	P1, P2, P3 CB5 E7	Pl	Elution with glycine (pH, 8.5); recon. via membrane filtration (100 g)	0.9—4.1 pfu/100 g	4
Near a sewage outfall in Italy; P2 was a vaccine strain	P2 Reoviruses Nonpolio enteroviruses		Elution with seawater; conc via PE60 polyelectrolyte	0.4—40 $TCID_{50}$/ 100 ml of sediment eluate	6
Estuarine sediments, St. Augustine, Fla.	Pl, P3	P3	Elution with urea-lysine (50—100 g estuarine sediment)	0—708 $TCID_{50}$/50 g	8
Freshwater sediments, New Zealand	Pl, P2, P3 CB5	—	Elution with 6% beef extract (pH, 9.0)	4—62 pfu/100 g	11
Marine sediments in New Zealand	Pl, P2, P3 CB5	—	Conc. via PEG 6000 (100 g)	3—2400/100 g	11
Oyster beds, Galveston Bay area	P2 CB1, CB5 CA16	CB5	Elution with glycine-EDTA (pH, 11.0); (4000 ml sediment)	0—480 pfu/20 ℓ	14
Contiguous with deep marine sewage outfall, Miami, Fla.	Pl CB3, CB4 E1, E6, E7 E13, E21, E26	Pl	Elution with glycine-EDTA (pH, 11.0); conc via Filterite filters (400 g)	0—112 pfu/ℓ	15
Marine sediments in New Zealand near sewage outfall off Dunedin	P2 CB4, CB5	—	Elution with 6% beef extract; conc via PEG 6000 (100 g)	0—2400 pfu/100 g	22
Sediments of canals located along upper Texas gulf coast	P1 E1, E7, E19	Pl	Elution with glycine-EDTA (pH, 11.0); conc via ads to filters (400 ml wet sediment)	—	16
Sediments from bed of River Wear, U.K. (in vicinity of sewage treatment plant); 9-mo. sampling period	P1, P2, P3 CB2, CB3, CB4 CB5, CB6	—	Method 1: elution with glycine-EDTA; recon. via membrane filtration Method 2: elution with beef extract; recon. via organic flocculation (45 g)	0—54 pfu/45 g	17
Philadelphia sludge dump (40—60 m deep, 70 km E of Ocean City, Md.)	P2 CB3 E1, E9	—	Elution with 6% beef extract (pH, 10.5); conc via organic flocculation (300—400 g)	7—46 pfu/kg	18
New York Bight sludge dump site (30 m deep)	CB3, CB5 E1, E7	E1	Elution with 6% beef extract (pH, 10.5); conc via organic flocculation (300—400 g)	2—182 pfu/kg	18
Estuarine sediments from polluted area, Galveston Bay	CB4	—	Elution with beef extract (pH, 10.5); conc via organic flocculation (300 g)	7—10 pfu/1000 g	3

Table 7 (continued)
ISOLATION OF VIRUSES FROM SEDIMENTS

Sediment source	Virus type isolated	Dominant virus	Conc method (sample vol)	Virus conc	Ref.
Fluffy estuarine sediments from polluted area, Galveston Bay	P1, P2 CB3,CB4	—	Elution with beef extract (pH, 10.5); conc via organic flocculation	39—398 pfu/1000 g	3
Estuarine sediments from Galveston Bay	Rotaviruses	—	Elution with beef extract	1200 fluorescent foci/1000 g, compact sediment; 800—3800/FF/378 ℓ, fluffy sediment	19
Sediments from Romanian section, Danube River (1972—1977 survey)	P1, P3 CA4 CB5 E7, E11, E12	P3	Elution; conc via PE60 method (100 g)	15.8% + samples (mean)	
Sediments from Romanian section, Danube River (1972—1977 survey)	P1, P3 CA CB3 Adenoviruses	—	Elution; conc via yeast cell method (100 g)	15.8% + samples (mean)	20

P, poliovirus; CA, coxsackievirus A; CB, coxsackievirus B; E, echovirus.

REFERENCES

1. **Gerba, C. P., Stagg, C. H., and Abadie, M. G.,** Characterization of sewage solid-associated viruses and behavior in natural waters, *Water Res.*, 12, 805, 1978.
2. **Smith, E. M., Gerba, C. P., and Melnick, J. L.,** Role of sediment in the persistence of enteroviruses in the estuarine environment, *Appl. Environ. Microbiol.*, 35, 685, 1978.
3. **Rao, V. C., Seidel, K. M., Goyal, S. M., Metcalf, T. G., and Melnick, J. L.,** Isolation of enteroviruses from water, suspended solids and sediments from Galveston Bay; survival of poliovirus and rotavirus adsorbed to sediments, *Appl. Environ. Microbiol.*, 48, 404, 1984.
4. **Metcalf, T. G., Wallis, C., and Melnick, J. L.,** Virus enumeration and public health assessments in polluted surface water contributing to transmission of virus in nature, in *Virus Survival in Water and Wastewater Systems*, Malina, J. F., Jr. and Sagik, B. P., Eds., University of Texas, Austin, 1974, 57.
5. **Vaughn, J. M. and Metcalf, T. G.,** Coliphages as indicators of enteric viruses in shellfish and shellfish raising estuarine waters, *Water Res.*, 9, 613, 1975.
6. **De Flora, S., De Renzi, G. P., and Badolati, G.,** Detection of animal viruses in coastal seawater and sediments, *Appl. Microbiol.*, 30, 472, 1975.
7. **Gerba, C. P., Smith, E. M., Schaiberger, G. E., and Edmond, T. D.,** Field evaluation of methods for the detection of enteric viruses in marine sediments, in *Methodology for Biomass Determinations and Microbial Activities in Sediments*, Litchfield, C. D. and Seyfried, P. L., Eds., American Society for Testing and Materials, Philadelphia, 1979, 64.
8. **Bitton, G., Chou, Y. J., and Farrah, S. R.,** Techniques for virus detection in aquatic sediments, *J. Virol. Methods*, 4, 1, 1982.
9. **Wait, D. A. and Sobsey, M. D.,** Method for recovery of enteric viruses from estuarine sediments with chaotropic agents, *Appl. Environ. Microbiol.*, 46, 379, 1983.
10. **Rao, V. C., Rao, T. V., Metcalf, T. G., Dahling, D., and Melnick, J. L.,** unpublished data.
10a. **Tsai, S. C., Ellender, R. D., Johnson, R. A., and Howell, F. G.,** Elution of viruses from coastal sediments, *Appl. Environ. Microbiol.*, 46, 797, 1983.

11. **Lewis, G. D., Loutit, M. W., and Austin, F. J.**, A method for detecting human enteroviruses in aquatic sediments, *J. Virol. Methods,* 10, 153, 1985.

12. **Cooper, R. C., Johnson, K. M., Straube, D. C., Brown, L. A., and Lysmer, D.**, Development and Evaluation of Methods for the Detection of Enteric Viruses in San Francisco Bay Shellfish, Water and Sediment, Sanitary Engineering Research Laboratory Report UCB/SERL 79-3, School of Public Health and College of Engineering, University of California, Berkeley, 1980.

13. **Johnson, R. A., Ellender, R. D., and Tsai, S. C.**, Elution of enteric viruses from Mississippi estuarine sediments with lecithin-supplemented eluents, *Appl. Environ. Microbiol.*, 48, 581, 1984.

14. **LaBelle, R. L., Gerba, C. P., Goyal, S. M., Melnick, J. L., Cech, I., and Bogdan, G. F.**, Relationships between environmental factors, bacterial indicators, and the occurrence of enteric viruses in estuarine sediments, *Appl. Environ. Microbiol.*, 39, 588, 1980.

15. **Schaiberger, G. E., Edmond, T. D., and Gerba, C. P.**, Distribution of enteroviruses in sediments contiguous with a deep marine sewage outfall, *Water Res.*, 16, 1425, 1982.

16. **Gerba, C. P., Goyal, S. M., Smith, E. M., and Melnick, J. L.**, Distribution of viral and bacterial pathogens in a coastal canal community, *Mar. Pollut. Bull.*, 8, 279, 1977.

17. **Wyn-Jones, A. P. and Edwards, E. R.**, The adsorption of enteroviruses by river sediments, in *Viruses and Disinfection of Water and Wastewater*, Butler, M., Ed., University of Surrey Press, London, 1982.

18. **Goyal, S. M., Adams, W. N., O'Malley, M. L., and Lear, D. W.**, Human pathogenic viruses at sewage disposal sites in the middle Atlantic region, *Appl. Environ. Microbiol.*, 48, 758, 1984.

19. **Rao, V. C., Metcalf, T. G., and Melnick, J. L.**, Development of a method for concentration of rotavirus and its application to recovery of rotaviruses from estaurine waters, *Appl. Environ. Microbiol.*, 52, 484, 1986.

20. **Nestor, I., Lazar, L., Sovrea, D., and Ionescu, N.**, Investigations on viral pollution in the Romanian section of the Danube River during 1972—1977 period. *Zentralbl. Bakteriol. Parasitenkd. Infektionskr. Hyg. Abt. I: Orig. Reihe B*, 173, 517, 1981.

21. **Lewis, G. D., Loutit, M. W., and Austin, F. J.**, Human enteroviruses in marine sediments near a sewage outfall on the Otago coast, *N. Z. J. Freshwater Mar. Sci.*, 19, 187, 1985.

22. **Shields, P. A. and Farrah, S. R.**, Concentration of viruses in beef extract by flocculation with ammonium sulfate, *Appl. Environ. Microbiol.*, 51, 211, 1986.

Chapter 4

ROLE OF SEDIMENT IN THE PERSISTENCE AND TRANSPORT OF ENTERIC VIRUSES IN THE ESTUARINE ENVIRONMENT

R. D. Ellender, John B. Anderson, and Robert B. Dunbar

TABLE OF CONTENTS

I. INTRODUCTION

The contamination of coastal waters by animal wastes is associated with a variety of sources of pollution including farms,[1-3] septic tanks,[4-6] sewage effluents of cities and small towns,[7] sewage sludges,[8,9] and storm runoff.[10,11] Enteric viruses, as part of the microbial flora of sewage, must contend with myriad soluble and particulate, organic and inorganic fractions which serve either to destroy or protect the infectious nature of the virus. More than 120 different virus types are found in sewage,[12] but other animal viruses, originating from domestic and wild animal excreta,[13,14] and viruses of fish, plants, and microorganisms[15-17] are part of the flora which enter polluted waters. Not surprisingly, the greatest research effort has been toward developing an understanding of the ecology of human enteric viruses.[18-22]

The substantial volume of literature on environmental virology has centered on studies which employ cultivable viruses (including hepatitis A virus)[23-28] and bacteriophage, but additional investigations are needed to evaluate the environmental risk of noncultivable agents of epidemic viral gastroenteritis (Table 1).[29] As primary agents of epidemic viral gastroenteritis, they represent a major problem in developing countries with minimal water and wastewater treatment practices. From this group, the Norwalk-like agents appear to be readily transmissible by shellfish.[30,31] Human rotaviruses have not been cultured directly from environmental samples, but they have been grown from stools[32] and detected in sewage by indirect immunofluorescence.[33,34] The prevalence of rotaviruses in the stool of infected persons and the ubiquitous formation of rotaviral antibody in the population[35] strongly suggest that rotaviruses are commonly transferred from person to person. Therefore, the water route can be indirectly associated with human rotaviral disease.[36] The association of the other gastrointestinal viruses with human disease and their detection in the human stool suggest possible transmission by the water route.

All human viruses transmitted by the water route pose a threat to public health. The movement and persistence of enteric viruses, including the gastroenteritis viruses in the environment, have not been sufficiently examined to allow conceptual model development of the overall problem. The purpose of this chapter is to present an overview of the physical, chemical, and biological factors considered responsible for virus persistence and transport in the estuarine environment.

II. ESTUARINE TURBIDITY

A. Constituents of Natural Estuarine Turbidity

Sediments found in coastal waters are a result of rock weathering on land (terrigenous sediment) and organic production. They are transported to continental margins by rivers, wind, or ice, and it is estimated that some 250×10^{14} g of material are carried annually to the ocean.[37] Approximately 80% of this loam is particulate, and the remainder is in the dissolved form. Of the 8 billion tons of sediment annually transported to the sea, most is trapped in estuaries and near-shore areas associated with human activity.[38,39]

The settling of sediments is closely tied to the character of estuarine circulation. Sand particles settle from suspension as flow velocities decrease. Clays and other fine materials tend to flocculate as freshwater mixes with saltwater (particles aggregate as electrolytic forces bring them together).

Estuarine sediments are largely terrigenous in nature, and composed mostly of clays and silts. Descriptive classifications are widely used and often based on grain size. Geometric grade scales are preferred, the most common of which is that of Krumbein.[40] Particle sizes are expressed as phi units, where phi is equal to the negative log to the base 2 of the particle diameter in millimeters (Table 2).

Table 1
DIFFICULT-TO-CULTIVATE AGENTS OF
VIRAL GASTROENTERITIS THAT MAY
BE TRANSMITTED BY POLLUTED
WATER

Virus group	Serotypes	Ref.
Adenovirus	2+	70,71
Astrovirus	2+	72—74
Calicivirus	3?	75,76
Coronavirus	1+?	77
Norwalk virus[a]	3?	78
Parvovirus	?	79
Rotavirus (excluding SA11)	4+	33,36,80
Small round viruses[a]	?	81,82
Small round structured viruses[a]	?	81

[a] Related antigenic and epidemiologic factors.

Table 2
CLASSIFICATION OF SEDIMENT
PARTICLES

Particle size (μm)	Phi	Size class
1000—2000	0.0—1.0	Very coarse sand
500—1000	1.0—0.0	Coarse sand
250—500	2.0—1.0	Medium sand
125—250	3.0—2.0	Fine sand
62.5—125	4.0—3.0	Very fine sand
31—62.5	5.0—4.0	Coarse silt
3.9—31	8.0—5.0	Medium to fine silt
<3.9	>8.0	Clay

By far the largest volume of sediment entering modern estuaries consists of fine-grained material which is maintained in suspension by weak currents and/or fluid turbulence (i.e., wave motion). Because most estuaries are fairly energetic environments, these sediments will not be deposited unless they are incorporated into larger aggregates or agglomerates. Aggregrates include composite particles which are firmly bound by either intermolecular or atomic cohesive forces and agglomerates are weakly bound particles held together by electrostatic fields, surface tension, or organic matter.[41] Flocculation of clays results in composite particles with settling velocities that are still quite small (2 to 18 $\times$ 10^{-3} cm/sec)[42] and is therefore not an important sedimentation mechanism in estuaries.[43-45] In contrast, fecal pellets have settling velocities in the range of 0.04 to 2.0 cm/sec,[46,47] and the volume of sediment produced in this manner may be substantial. For example, Schubel[48] and Schubel and Dana[49] found that copepods and other zooplankton filter a volume of sediment equivalent to the suspended sediment load of Chesapeake Bay in only a few days. Oysters are at least as efficient at processing suspended sediment,[50] and studies in Buzzards Bay[51] have shown that the clam *Yoldia limatula* is capable of reworking sediments as fast as they are deposited. It is these types of observations that have led to a general consensus that biological processes are mainly responsible for the production of composite particles in estuaries[41,45] and provide the means by which fine-grained material is deposited in relatively energetic environments.

Even larger aggregates and agglomerates have settling velocities that are sufficiently small that they are maintained in suspension by relatively weak currents.

The finest sediment particles are those of clay, and the size of the clay particles is established by the properties of clay minerals. Clay minerals are ionic spheres of oxygen, silicon, aluminum, iron, magnesium, and potassium arranged in a regular three-dimensional pattern. The type of mineral (or group) is determined by the arrangement of the elements, and an element can be replaced by another element producing a clay mineral species. There are seven clay mineral groups and about 50 clay mineral species. Clay mineral suites in estuaries are consistent with local river systems. For example, coastal waters east of Mobile, Ala. are dominated by kaolinite; montmorillonite is found in estuaries in Texas, Louisiana, and Mississippi. Illite and chlorite are deposited by riverine systems on the northeast coast of the U.S., and west coast sediments are predominantly illite.

The composition of suspended solids in estuarine waters is variable, and consists of diatom frustules, dinoflagellates, organic aggregates, mineral grains, opaque particles thought to be of anthropomorphic origin, and particles of iron hydrous oxide.[48,52,53] Other metal oxides, especially those of Mn, Cu, and Zn, are commonly observed, and their concentrations suggest a close association between the metals.

Estuarine sediments exhibit considerable variation in organic carbon content. Bottom sediments composed mostly of sand usually contain a maximum of 1% organic carbon, in comparison to silts and clays, which may contain upwards of 5% organic carbon.[54] Anaerobic bottom sediments containing abundant vegetal matter or raw sewage may have organic carbon values as high as 10 to 20%. Thus, organic matter levels increase in estuarine waters polluted with sewage (as a consequence of the fertilizing and nutritive properties) and as a result of photosynthesis (approximately 2% of world primary production).[55] Surface waters carry only a small portion of organic matter; the majority is biodegraded and bioconverted in sediments. The nonliving organic fraction found in water is assumed to be fecal pellets, minute plankton remains, and platelike aggregates possibly formed when air bubbles act as nuclei for adsorption of dissolved organic and inorganic compounds.[56] This process, initiated by air and the polymeric excretions of bacteria and plankton, could account for the adsorption of virus to suspended estuarine particles and their eventual disappearance from the water column.

In addition to natural flocculation and settling, particulates can be removed from estuarine water during shellfish and zooplankton feeding. Oyster reefs process suspended particulates and are important in materials cycling in estuarine ecosystems.[57] The public health significance of shellfish ingestion of particulate-associated virus is well documented[18] and a known threat to human health.

Storms, seasonal effects, and tidal cycles can dramatically alter the composition of the particles present in the suspended solids. Since changes in the composition of suspended particulates may occur rapidly, sampling of the water column for viruses should be coordinated with a rigorous analysis of the events which occur in the water column.

B. Determining Suspended Particle Size

The measurement of size and/or settling velocity of fine-grained sediments has been greatly facilitated in recent years by the development of a new generation of electronic instrumentation such as the Hydrophotometer, Sedigraph, Laser Particle Counter, and Electrozone Particle Counter. The first three instruments measure settling velocity by passing a light beam, X-ray beam or laser beam, respectively, through a water column containing dispersed sediments. The settling velocity distribution for the sample is determined by measuring energy transmission vs. time. The Electrozone Particle Counter is a modification of the Coulter Counter and determines particle size. More detailed reviews of these methods and their reliability are found in the literature.[58-61]

Ongoing research addresses the pros and cons of each of the instruments that are now on the market. These studies have shown these devices to be reasonably accurate in terms of instrument reproducibility, and the results from standards analyzed by each instrument show reasonably good agreement. Thus, the instrumentation needed for fine-grained sediment analysis does exist; however, major problems persist with regard to sampling procedure and sample treatment. Our own results have shown that very different settling velocity and size distributions are obtained, even with the same instrument, for the same sample treated in different ways. This problem stems mainly from the different ways in which samples are disaggregated (i.e., sonic vibration vs. washing in mild detergents).

A longstanding argument among geologists concerns whether grain size or settling velocity is the best property for describing sediments. In studies involving the transport of suspended sediments, settling velocity is a more meaningful parameter, simply because the relationships between suspension and settling velocity are more straightforward. This is because the same grain properties which influence particle suspension — namely, particle size, shape, and density — also control critical suspension velocity, and there is not always a direct correlation between particle size and settling velocity.

Determination of *in situ* settling velocity and size distribution presents a special problem because certain classes of particles begin to alter their size and shape almost immediately upon being sampled. This problem appears to be most severe for loosely bound aggregates of terrigenous detritus and large mucilaginous organic particles. Also, in most cases suspended sediments are not concentrated in natural waters in sufficient quantity to be analyzed directly. Suspended sediments must therefore be concentrated in some manner, and this can result in modification of the original size and settling velocity distribution of the sample. Sediment concentration should be done by allowing the sample to settle without agitation and by avoiding significant changes in temperature. Analyses should be done as soon as the required concentration is obtained. In this regard, instruments which measure settling velocity using light transmission have the combined advantages of requiring only a small sample and being reasonably portable.

C. Nature and Measurement of Estuarine Sedimentation

Estuaries represent complex sedimentary environments because of the dynamic interplay between fluvial and marine processes. Still, in recent years much has been learned about estuarine sedimentation. Excellent reviews on the subject may be found in Ippen,[62] Postma,[42] Meade,[43] Goldberg,[63] and Schubel.[45]

One of the principal features of estuaries is the turbidity maximum, a zone of high suspended sediment concentration situated within the upper reaches of the estuary.[42,45] It is located within that portion of the estuary where seawater encounters and first mixes with fluvial waters. Finer, river-borne sediments are transported down the estuary, and their dispersal is influenced by wind and tidal circulation and wave motion. Those particles whose settling velocities are sufficiently large to enable them to settle out of the surface layer will likely be transported back toward the river mouth as they descend into the deeper, landward flow. In this manner sediments may be recycled within the upper reaches of the estuary, thus contributing to the turbidity maximum.[45,64,65] Lateral migration of this feature within the estuary is mainly controlled by fluvial discharge rates. As fluvial discharge increases, the turbidity maximum migrates seaward, and during periods of peak discharge suspended sediments may even bypass the estuary. During periods of low fluvial discharge marine waters may migrate well up into rivers as a salt wedge, and the locus of deposition is thus shifted toward the head of the estuary.

Superimposed on this fluvial influence are the effects of wind, waves, tides, and Coriolis force. The size, shape, orientation, and depth of an estuary also influence its circulation regime, and thus sedimentation.

Tidal circulation within estuaries can be quite vigorous and may result in complete flushing of fairly large estuaries within a matter of a few days (e.g., Mobile Bay, Ala.).[66] Tidal circulation in large, wide estuaries is influenced by the Coriolis force. For example, a northern hemisphere, South-facing estuary experiences stronger flood currents along its eastern side, whereas ebb currents are stronger along the western side of the estuary. The result is a general counterclockwise flow. Again, Mobile Bay is a good example of this effect.[66] Our own measurements of tidal currents in Galveston Bay (Texas) show peak velocities of several tens of centimeters per second, sufficient to resuspend water-laden, muddy beds and maintain a full size range of fine-grained sediments in suspension. Also, Schubel et al.[67] noted dramatic changes in the concentration and size of suspended sediments in the water column between tidal cycles, especially near the bottom, which they attributed to resuspension by tidal currents.

In larger estuaries, persistent winds can initiate surface currents with velocities of tens of centimeters per second. When these wind-driven currents are coupled with tidal current, the results can be significant transport of suspended sediment within and out of the estuary.

The size, shape, and orientation of an estuary relative to prevailing winds controls wind fetch, and therefore wave geometry. Due to differences in the fetch and prevailing wind patterns, the depth of the storm wave base will undoubtedly vary within a given estuary, and the frequency of resuspension events in different portions of the estuary will likewise vary. These effects may be largely seasonal as wind direction and speed change from season to season. In estuaries with large, shallow platforms situated above storm wave base, storm-related resuspension may lead to sudden and dramatic episodes in which large volumes of fine-grained material are resuspended. On the other hand, a deep estuary or a long, narrow estuary whose long axis is perpendicular to prevailing winds will be less affected by storm-related resuspension events.

Man has had an important effect on many estuaries by construction features which alter natural circulation. Most significant in terms of sedimentation has been the construction of ship channels which typically extend along the entire axis of the estuary from river to sea. These deep channels may behave as salt-wedge estuaries within otherwise partially mixed estuaries,[66] and therefore sedimentation within these features may be quite different from that of the estuary. The main difference is that river-borne sediments tend to be trapped in the upper reaches of the channel and not subjected to resuspension. Ship channels are thus efficient sediment traps and indeed capture most of the sediment entering many modern estuaries. More often than not, this is probably a beneficial effect in terms of water quality.

Estuarine sedimentation may be characterized by measurements of sediment accumulation and time series studies of suspended and sinking particulate concentrations in the water column. Water column studies provide information about sedimentary processes which occur over time scales of days or weeks, while the sediment record is normally used to study estuarine variability over time scales greater than 1 year. Estuarine sediments collected with coring devices may be age-dated by a variety of radiometric techniques. The isotope ^{210}Pb (half-life, 22 years) is useful for developing sediment chronologies for the last century, while ^{226}Ra and ^{14}C (half-lives, 1620 and 5700 years, respectively) are best suited for longer-term studies. Estuarine sedimentation rates determined by radiometric age-dating are high, normally in the range of 1 to 20 mm/year, consistent with previous observations that in many cases estuaries act as sediment traps for a large percentage of the incoming sediment load.[43]

Suspended particulate concentrations may be measured directly by filtration or indirectly using optical transmission or scattering devices. The nephelometer,[68] the turbidity meter, and the transmissiometer, all of which measure light scattering *in situ*, are commonly used in coastal waters. Light scattering is a complex process, dependent on grain size, shape, and composition as well as particulate concentrations, and careful calibration with filtered samples is required before the technique can be used with precision.

Table 3
AVERAGE LEVELS OF
ENTERIC VIRUSES IN
SEDIMENTS AND OVERLYING
WATERS

	Water	Sediment
Estuarine sediments[a]	125	2040
Marine sediments[a]	3.9	5995
Estuarine sediments[b]	65	106

[a] pfu/100 ℓ.[93]
[b] pfu/20 ℓ.[94]

Suspended particulate concentration in most estuaries ranges from 20 to approximately 200 mg/ℓ, although particulate concentrations in excess of 10,000 mg/ℓ have been recorded in the estuaries of large rivers draining mountainous regions.[45] Particulate concentrations vary with seasonal runoff and storm frequency. Thus, most estuaries experience large short-term excursions in total particle surface area accessible for adsorptive interaction with the water column.

Sinking fluxes of particulate matter are measured by field experiments with upward-facing sediment collection chambers known as sediment traps. A variety of different trap designs and operative principles have been field tested during intercalibration experiments.[69] Most traps in use today are conical or cylindrical in shape and employ baffling materials on the collecting surface which are thought to act as false bottoms. Particle interceptor traps collect large, rapidly settling materials which are unlikely to be sampled or observed using techniques for characterization of the suspended load. Most importantly for estuarine studies, sediment trap experiments provide a means for direct examination of particle settling modes (i.e., fecal pellets, flocs, individual grains, organic aggregates, etc.) following sediment resuspension or injection events.

Information about particle aggregation and settling processes is a prerequisite for calculating estuarine clearing times following turbid events, particle residence times in the water column, and mean particle trajectories under the influence of currents. Although sediment trap technology has advanced to the point where oceanic deployments are commonplace, few trapping experiments have focused on estuarine regions.

III. VIRAL PERSISTENCE IN THE ESTUARINE ENVIRONMENT

A. Viral Persistence in Estuaries

The fact that enteric viruses naturally bind to cells, tissues, adsorptive flocs, and other matrices leads us to logically assume that this characteristic will be exhibited when virus enters the marine environment. Long-term persistence of virus in this ecosystem appears to be synonymous with adsorption in regard to previous findings demonstrating a viral protective effect by sediment.[83,84] Smith et al.[85] found that enteric viruses survive from 1.5 to 5 times longer in sediment than in estuarine water. This result is also strengthened by the findings of higher levels of virus in sediments than in overlying waters (Table 3).

Virus entering the marine environment via sewage may exist as single or aggregate particles, or as virus adsorbed to sewage solids. Solids-associated enteroviruses in sewage have been observed by several investigators.[86,87] Schaub and Sorber[88] and Smith et al.[85] have suggested that viruses from sewage adsorb to estuarine particulates and are protected from inactivation. Of the types of particulates found in estuaries, suspended solids were more often associated with virus isolation[89,90] than were compact or fluffy sediments. It is probable that variations in the levels of free vs. adsorbed virus constantly occur as a result of chemical equilibria, changes in salinity, conductivity and pH, and physical mixing by tidal and wind activity as well as natural aggregation properties of suspended matter.

Viral inactivation in seawater can be chemical, physical, or biological phenomena. The presence of sodium chloride is considered destructive to free virus in seawater,[91,92] but salinity alone cannot account for the decrease in virus number. Autoclaving seawater lowers the rate of viral inactivation by possibly reducing the levels of toxic trace metals in solution. Other cations are known to inactivate viruses (e.g., rotavirus SA11 heated to 50°C in 2 M $MgCl_2$), but have an opposite effect on related species (e.g., reovirus type 1 under identical conditions).[95]

The protective effect of clay may be a result of adsorption of antiviral enzymes,[96-98] increased viral protein stability,[99] inactivation of ribonucleases,[100] adsorption of trace metals by clays,[101] or the insertion of virus into clay particle aggregates. Protection also appears to be a result of virus morphology and the placement of the virus on the surface of the suspended particles. For example, tailed bacteriophage which attach to clay via tail proteins are more susceptible to the effects of chlorine.[102] For animal viruses, adsorption to particulates may stabilize the surface proteins, allowing less interaction with the surrounding environment.

The role of organic matter in the protection of viruses in the estuarine environment is not clear, but estuarine waters are known to contain terrestrial, soluble animal and plant proteins and polysaccharides, mucilaginous slimes, microbial proteins and sugars, detritus, the feces and exudates of marine animals, and seafood processing wastes.[103] Virus persistence would appear to be a series of declining steps of varying complexity dependent upon the rate of organic degradation by microbial, chemical, or physical mechanisms.

The most prominent physical factors which contribute to virus inactivation in the estuarine environment are temperature and solar radiation.[104,105] Temperature may be the more important of the two and has received the greatest attention.[106,107] Colwell and Kaper[108] showed that human enteroviruses were stable and detectable after 46 weeks at 4°C, and over a salinity range of 10 to 34 ‰ (g/kg). At 25°C, virus was rarely detected after 8 weeks. Direct exposure to sunlight, particularly the UV and blue wavelengths, is detrimental to viruses.[109] The presence of clay and algae in marine waters retards sunlight penetration and protects virus from inactivation.[110]

The presence of microorganisms adversely influences the persistence of viruses in the estuarine environment.[111] The role of marine vibrios (i.e., *Vibrio fischeri*) has been examined with respect to virus loss in seawater,[112] and a possible role of algae and protozoa in the viricidal activity of seawater has been proposed, but the extent of their influence remains unclear.[105]

The impact of estuarine invertebrates on virus removal from the water column has not been explored. Crustacea such as copepods may be important because they remove suspended solids from the water and produce discrete pellets of fecal material. This material has been of considerable interest to sedimentologists because of the amount of clay in fecal pellets and their apparent contribution to bottom sediments.[113] In shear volume of material produced (200 pellets per individual per day and up to 500,000 organisms per 1 m^3 of water),[114] this material represents a major component in the water column. Since copepod feeding is indiscriminate, large quantities of clay may be found in the pellets. The fate of ingested virus has not been evaluated. Work by Cowey and Corner[115] indicates that fecal pellets tend to reflect the chemical composition of their original food. Fecal pellets are covered by a protective surface membrane[113] whose rate of microbial digestion in the water is governed by temperature.

B. Nature of the Adsorptive Process

Viruses appear to readily adsorb to sand, clay minerals, natural clay mixtures, and other

Table 4
MAJOR AND MINOR IONS OF
RIVER AND SEA WATERS

	Ion	River water (mg/ℓ)	Seawater (mg/ℓ)
Major	Na	6.3	10,770
	K	2.3	399
	Mg	4.1	1,294
	Ca	15	412
	Fe	0.7	<0.011
	Cl	7.8	19,340
	SO_4	11.2	2,712
	HCO_3	58.4	140
	Si	13.1	<0.01 to > 10
Minor	Li	3	170
	F	100	1,300
	Ba	10	10
	U	0.3	3.3
	Mo	1	11
	Cr	1	0.2
	Mn	7	0.4
	Cu	7	1
	Zn	20	2.5
	Sb	1	0.3
	Pb	3	0.03

particulate matter found in the estuarine environment.[54,99,116-121] Viruses, as hydrophilic colloids, may depend for adsorption on the presence of cations, the pH of the suspending medium,[122] the virus isoelectric point, the size and shape of the virus particle, the presence or absence of an envelope, the charge density on the virion, and the nature and chemistry of the adsorbent. The electrical charge of colloidal particles in natural waters, with reference to clay, is influenced by pH, salinity, concentration of cations, and the concentration and type of clay mineral. Carlson et al.[118] speculated that clay, organic matter, and other suspended particulates in estuarine water retain a net electronegative charge at the pH of the natural water; therefore, particulates in suspension repel one another. A lowering of the pH and the addition of metal cations to the mixture lowers the electronegativity of the particles wherein the cation is able to bridge the gap between virus and clay particle.

Adsorption of virus to solids may also proceed in a fashion similar to the process of solids coagulation where the presence of cations reduces the energy of the particle interaction.[123] Interaction energy is the net value of the columbic electrostatic repulsive energy and van der Waals or other energies considered together.[124] The metal cation is thought to reduce the repulsive forces on solids and virus, allowing shorter-range forces such as hydrogen bonding or van der Waals interactions to support binding.[125]

Average global values of major and minor ions in river and seawater (Table 4) allow certain conclusions which affect virus-particle interactions. Differences in ion concentration occur between river and sea waters and have a direct effect on virus adsorption. The net electronegative charge of estuarine particulates results from the dissociation of ionizable groups on the surface of viruses and the unequal charge distribution on clay minerals. A double cation layer develops on the electronegative particle and includes the fixed, or Stern, layer and the outside, diffuse Gouy layer. In the Gouy layer, the ions move freely, producing a decrease in electrical potential as the distance from the surface increases. A compression

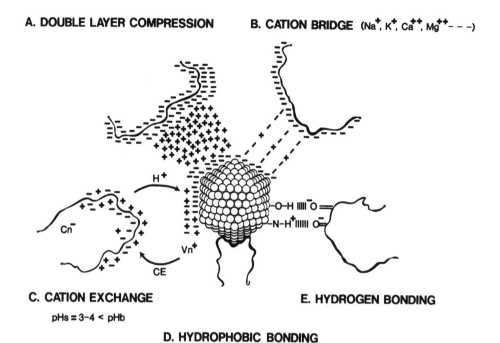

FIGURE 1. Possible mechanisms of virus-particulate bonding.

of the double layer may occur by an increase in the molar cation concentration or an increase in ion valency. This compression allows suspended particulates to move closer and permits other bonding interactions.[126] By extension, compression of the double layer should occur as river water particulates are mixed with saltwater, allowing greater virus-suspended solids interactions.

Lipson and Stotzky[123,127,128] have stressed the interrelationship of reovirus adsorption and the cation exchange capacity of a clay mineral. They demonstrated adsorption occurs mainly to negative sites on montmorillonite and kaolinite, even though the virus is also negatively charged. Adsorption was observed in distilled and saline water, and was not related to the surface area of the clay. They speculate[123] that the pH of the colloidal clay surface is 3 to 4 units below the pH of the suspension and that the acidity results in protonation of the virus particle followed by cation exchange at the clay-virus interface.

Recently, Wait and Sobsey[129] have concluded that in addition to electrostatic and cation exchange interactions hydrophobic residues may be involved in the adsorption of viruses to particulate matter. To minimize the disruptive effects on its hydrogen-bonded network, water forces hydrophobic groups together. This "attraction" allows a weak bonding interaction between any two atoms due to fluctuating electrical charges. This attraction, the van der Waals force, can be important if two macromolecular surfaces fit very closely together. This theory is supported by the demonstration of nonpolar residues on the picornavirus surface and by a dramatic increase in the elution of virus from sediments by chaotrophic agents. Studies by Johnson et al.[117] have also noted a similar increase in virus elution from sediment using an isoelectric casein mixture supplemented with phosphatidyl choline. The advantage of both chaotropes and phosphatidyl choline eluents is the disruption of the H^+ bonding of water and, at the same time, a decrease in the influence of the hydrophobic interaction.

Virus-particulate bonding is apparently complex and subject to continual change. Figure 1 attempts to summarize the noncovalent mechanisms of attractions, but it is obvious that most or perhaps all of the mechanisms presented are interrelated. Two general concepts are presented. First, the requirement for adsorption of metal cations, in conjunction with pH, overcomes the net negative charges on the virus surface and the surface of the suspended particulate. The concepts of double-layer compression, cation bridges, and cation exchange are all generally associated with the same type of interaction. Second, the potential for hydrophobic residue influence exists and is supported by the finding of nonpolar groups on the surface of virus particles. The nature and quantity of hydrophobic groups on suspended estuarine particulates is unknown, but is consistent with the idea that soluble proteins and other natural substances also bind to suspended matter and contain hydrophobic residues.

The adsorption of virus to estuarine particles, therefore, favors the concept of multiple interactions and a variety of bonding mechanisms. The variation which exists in virus surface components probably indicates that adsorption of different virus groups will favor particulate mechanisms, but that the basic mechanisms will remain essentially unchanged.

The stability of the virus-particulate interaction in the natural estuarine environment is poorly understood. Several investigators have demonstrated the ability of suspended solids to protect viruses from inactivation, but this is not a suitable measure of the stability of the bonding. In the natural system, the degree of adsorption probably varies with the continual changing of conditions. Competition for sites on suspended particulates produces virus release; readsorption may be immediate if suitable sites exist.

In the near-shore environment, concentrations of plant and animal protein, polysaccharides, and humic and fulvic acids produce interference with virus adsorption. If a bound virus survives these environmental modifications, its potential for additional transport is increased.

Competition for sites on clay particles has been evaluated,[130] and it appears that different viruses utilize different sites of attachment. Reovirus type 3 binds primarily to negatively charged sites on montmorillonite and kaolinite,[127] whereas coliphages T1 and T7 bind primarily to positively charged sites.[131] Herpesvirus type 1 has not been shown to bind to either of the charged sites;[125] this may indicate the greater importance of hydrophobic bonding for enveloped viruses.

IV. VIRAL TRANSPORT IN THE ESTUARINE ENVIRONMENT

A. Conditions Which Influence Particle Movement

Sediment transport occurs within the water column via suspension and at the sediment-water interface via bedload transport.[132] The latter mechanism involves either very slow movement of grains by rolling and sliding along the bottom (traction) or movement by short trajectory hops (saltation). Fine-grained sediments are more easily transported within the water column via suspension. Suspended grains move at the velocity of the transporting current; therefore, transport rates for suspended material are orders of magnitude faster than for grains transported by bedload processes.

It is the turbulent motion of the fluid that holds grains in suspension. The criterion for suspension is that the shear velocity of the current exceeds the settling velocity of the particle to be suspended. Shear velocity is a convenient way of expressing the shear stress at the bed using the dimensions of velocity (centimeters per second)[133] and in a crude sense can be thought of as the upward component of flow. It is much smaller than the average velocity of the current flowing above the bed (mean velocity). For steady uniform flow, the relationship of shear velocity and the mean velocity of a current flowing immediately above the bed is in the range of $U = 6U*$ (where U is mean velocity and $U*$ is shear velocity) for rough beds[134] and $U = 13U*$ for smooth beds.[135] These relationships should be taken only as approximations, as flow within estuaries is seldom steady and uniform. They represent upper-limit estimates, as greater turbulence, which characterizes most estuaries, will result in higher shear velocities. Where flow is highly turbulent, it may be impossible to relate vertical suspending current energy to the mean current velocity measured some distance above the bed.[41] The critical shear velocity and mean velocity (velocities required for suspension)

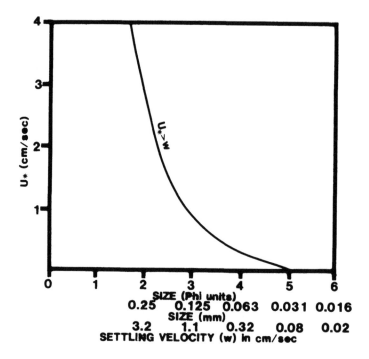

FIGURE 2.   Shear velocities required for suspension of fine-grained sediments.

for fine-grained sediments (very fine sand to clay size particles) are presented in Figure 2. From Figure 2 it is clear that fine-grained particles are transported in suspension at relatively slow current speeds. Again, the speed at which suspended particles are transported is equal to the speed of the current. For example, a 30-cm/sec current is capable of transporting a full range of fine-grained sediment a distance of just over a kilometer in 1 hr. Thus, sediment transport by this mechanism is quite capable of distributing fine-grained sediments and their living passengers throughout fairly large estuaries within the period of a single storm event or even within a single tidal cycle.

Once deposited, it requires high current speeds (several tens of centimeters per second) to resuspend the fine-grained sediment, due mainly to the cohesive nature of fine-grained beds. In contrast to sandy sediments, clay-rich sediments begin to compact within the upper 1 to 2 cm of the sediment-water interface, and individual platelet-shaped clay particles are stacked and interlocked to form a consolidated sediment bed. Very strong currents (> 20 cm/sec) are required to erode muddy beds, and when erosion does occur, the product may be large mud flakes rather than dispersed clay-sized particles.

Factors which influence bed cohesion and, therefore erodability, of fine-grained sediment beds include water content, bed roughness, organic matter content, mineralogy, and texture. Erodability decreases as the water content decreases and as the organic matter content of the bed increases. The latter effect is largely due to the binding influence or organic substances, although this effect is offset to some degree by the increased standing stock of deposit-feeding organisms, and hence greater bioturbation, in organic-rich sediments. Benthic organisms may resuspend sediment directly into the water column and also increase sediment erodability by converting smooth beds into rough beds. As the roughness of the sediment bed is increased, turbulence within the benthic boundary layer (that portion of the water column very near the bottom and whose flow is influenced by the bed) increases.

Clay mineralogy regulates bed cohesion via differences in molecular attractions between individual clay particles, and sediment texture influences grain-to-grain interlocking. More poorly sorted beds are more difficult to erode than well-sorted beds.

Resuspension is important because it provides a means, other than increases in fluvial drainage into the estuary, by which large volumes of sediment can be introduced into the water column. Resuspension is greatly facilitated by orbital wave motion. Thus, sediments that settle onto shallow portions of the estuary are more likely to be resuspended than sediments which settle into deeper basins that lie below the reach even of storm-generated waves. The level of wave resuspension is referred to as wavebase and is roughly situated at a water depth equal to one half the wavelength. There is a fair-weather wavebase situated at relatively shallow depths, where normal waves influence the bottom, and a storm wavebase situated at a greater depth, where only large storm waves influence the bottom. Fine-grained sediments transported through the estuary at depths shallower than storm wavebase may be deposited and resuspended numerous times before being permanently deposited in an area below storm wavebase.

Storm wavebase should be represented in bottom sediments by a transition from sandy bottom sediment to mud, so an examination of bottom sediment distribution coupled with a good bathymetric base map can allow one to measure that portion of the estuary in which resuspension occurs.

B. The Significance of Solids-Associated Virus

The majority of studies which have dealt with viruses as contaminants in estuarine waters have usually stressed the presence of virus in the water column. Sewage solids-associated virus should elute when distributed in saltwater[87,121] resulting in an increase in the number of virus particles free in the water column. This dissociation is rapidly amended, and virus appears to readily reassociate with estuarine particulates (possibly as a consequence of the increased concentration of metal cations) and gain some degree of protection from this relationship.

Reassociated virus has direct public health significance, especially in regard to persistence and transport. Investigations concerning the status of natural enteroviruses and rotaviruses in estuarine water demonstrate the greater likelihood of virus isolation from suspended particulates. Metcalf et al.[90] recovered enteroviruses from 72% of suspended solids and 47% of fluffy sediments, but only 14% from water and 6% from compact sediment samples. Rotavirus recovery from suspended solids (78%) was also greater than the percentage recovered from water samples (50%). Poliovirus and rotavirus (SA11) survived longer when associated with estuarine particulates;[89] both viruses survived for 19 days when sediment-associated, but could not be found in the water past day 9.

The sequence of virus isolations from estuarine samples (suspended solids > fluffy sediment > water > compact sediment)[90] represents both a consequence of natural processes and the need for additional evaluation of estuarine pollution. Suspended particulates, especially those which are < 1 μm, can remain in suspension indefinitely unless removed from the water column by flocculation, ingestion, or other natural mechanisms. Viruses which remain attached to particulates should decrease in number as the particles become part of the fluffy layer and finally the anaerobic compact layer. Virus release to the water column will vary due to fluctuations in pH and salinity, and may occur at any time. However, release should be a predictable consequence. Nonparticulate-associated virus may have a greater impact upon the public health from the viewpoint of infectivity, but solids-associated virus does remain infective[54,128] and is probably of greater importance to both viral persistence and transport.

C. Field Observation of Viral Transport

From a public health standpoint, viral transport in the marine environment is a critical problem. Adsorption to solids is a prerequisite for transport.

Several studies have noted the significant public health problem posed by viral transport to bacteriologically approved estuarine waters. Metcalf et al.[136] examined the numbers and types of viruses in wastewater treatment plant effluents flowing into the Houston ship channel, at downstream sampling sites on the channel, and in Galveston Bay. Virus numbers were shown to decrease in direct proportion to the distance of the sampling location downstream from the plants.

Although an approximately tenfold reduction in virus numbers occurred during transport (3.5 to 23.1 plaque-forming units [pfu] per gallon in discharge; 0.3 to 2.5 pfu/gal in channel waters), viable virus was found 8 mi downstream in channel water. Virus was not detected in water from an upper Galveston Bay sampling site 21 mi from the nearest discharge point. Although no virus was found in 105 gal of Galveston Bay water and no indication of a public health concern was noted in the fecal coliform content of this water, virus was detected in shellfish collected at the same site. Virus was also isolated from bottom muds from the ship channel and from the upper Galveston Bay sampling site, indicating that virus adsorption to natural particulates was responsible, at least in part, for removal of virus from surface waters.

Viral transport is not unique to coastal waters and has been examined in other ecosystems. Recent reports by Schaiberger et al.[137] and Dahling and Safferman[138] illustrate this point. In the former study, performed at Miami Beach, raw sewage was discharged from a deep marine outfall 3.6 km (2.3 mi) offshore and at a depth of 44 m. Enteric viruses were not isolated from the water beyond 200 m from the outfall, but were found in beach sediments. Virus numbers in sediments declined over this distance (78 to 112 pfu/ℓ at the outfall, and zero to 30 pfu/ℓ at the beach). Indicator bacteria were not found in beach sediments, demonstrating the importance of sediment analysis for viruses to determine the health hazard in coastal waters. The latter study demonstrated the presence of virus in the subarctic Tanana River 317 km from the source of domestic pollution. The degree to which virus was solids-associated was not determined in this investigation, but the data indicate that viral transport over long distances is possible when conditions are present which favor virus survival.

Transport of virus has been difficult to analyze under field conditions. Problems arise from the low numbers of viruses in estuarine water, the dilution of virus numbers at sampling sites far removed from the source of pollution, the rate and direction of suspended solids transport, and the fate of virus at a site of public health significance following transport. Investigators at the Gulf Coast Research Laboratory, Ocean Springs, Miss.[139] have examined the movement of hydrocarbons in estuary models and have shown that movement is restricted to a 3- to 4-mi distance from the source of contamination. One mechanism of hydrocarbon transport is adsorption to suspended solids. These studies indicate that virus movement may also be limited by the sedimentation processes in estuaries.

V. SUMMARY AND CONCLUSIONS.

The public health significance of viruses in shellfish and coastal recreational waters is recognized, but the mechanisms of virus movement to recreational and shellfish harvesting waters and the persistence of virus infectivity in the estuarine environment remains poorly understood.

Particulate matter in the water column plays a significant role in viral persistence and transport. There are differences and variations of the problem in coastal systems, and these differences can be traced to the virus type as well as the physical and chemical features of the individual estuarine ecosystems.

A better understanding of this problem must include a comprehensive examination of the nature of the suspended particles, the circulation patterns of water, the influence of local fluvial discharges, the turbidity maximum, the chemistry and physics of virus adsorption to and elution from particulates, and the mechanisms of particulate settling from the water column. In addition, the nature and type of the virus particle, and the potential for human disease are important. The large number of agents of gastroenteritis which are difficult to grow or are presently uncultivatable must be examined from an environmental perspective.

Future investigations of estuarine virology should bring together the expertise of invertebrate and vertebrate biologists, epidemiologists, geologists, sedimentologists, and oceanographers as well as virologists to develop a model of virus transport and persistence which will allow the prediction of potential public health problems.

REFERENCES

1. **Ruane, R. J., Chu, T., and Vandergriff, V. E.,** Characterization and treatment of waste discharged from high density catfish cultures, *Water Res.*, 11, 789, 1977.
2. **Wright, R. C.,** A comparison of the levels of fecal indicator bacteria in water and human feces in a rural area of a tropical developing country (Sierra Leone), *J. Hyg.*, 89, 69, 1982.
3. **Hollow, B. F., Owen, J. R., and Sewell, J. I.,** Water quality in a storm receiving feedlot effluents, *J. Environ. Qual.*, 11, 5, 1982.
4. **Hendry, G. S. and Toth, A.,** Some effects of land use on bacteriological water quality in a recreational lake, *Water Res.*, 16, 105, 1982.
5. **Stewart, L. W. and Reneau, R. B.,** Spacial and temporal variation of fecal coliform movement surrounding septic tank soil adsorption in two atlantic coastal plain soils, *J. Environ. Qual.*, 10, 528, 1981.
6. **Brown, K. W., Wolf, H. W., Donnelly, K. C., and Slowey, J. F.,** The movement of fecal coliforms and coliphages below septic lines, *J. Environ. Qual.*, 8, 121, 1979.
7. **Roberts, D. M. and Willis, R. R.,** Environmental aspects of marine and inland sewage treatment options for coastal towns, *Water Res.*, 13, 15, 1981.
8. **Subrahmanyan, T. P.,** Persistence of enteroviruses in sewage sludge, *Bull. WHO*, 55, 431, 1977.
9. **Norton, M. G.,** The control and monitoring of sewage sludge dumping at sea, *Water Pollut. Control*, 77, 402, 1978.
10. **Simpson, D. E. and Kemp, P. H.,** Quality and quantity of stormwater runoff from a commercial land-use catchment in Natal, South Africa, *Water Sci. Technol.*, 14, 323, 1982.
11. **Culley, J. L. and Phillips, P. A.,** Bacteriological quality of surface and subsurface runoff from manured sandy clay loam soils, *J. Environ. Qual.*, 11, 155, 1982.
12. **WHO Scientific Group,** Human Viruses in Water, Wastewater, and Soil, Tech. Rep. Ser. No. 639, World Health Organization, Geneva, 1979.
13. **Woode, G. N.,** Rotaviruses in animals, in *Virus Infections of the Gastrointestinal Tract*, Tyrell, D. A. and Kapikian, A. Z., Eds., Marcel Dekker, New York, 1982, 295.
14. **Garwes, D. J.,** Coronaviruses in animals, in *Virus Infections of the Gastrointestinal Tract*, Tyrell, D. A. and Kapikian, A. Z., Eds., Marcel Dekker, New York, 1982, 315.
15. **Wolf, K.,** Advances in fish virology: a review 1966—1971, *Symp. Zool. Soc. London* 30, 305, 1972.
16. **McAllister, P. E.,** Fish viruses and viral infections, in *Comprehensive Virology*, Fraenkel-Conrat, H. and Wagner, R. R., Eds., Vol. 14, Plenum Press, New York, 1979, 401.
17. **Matthews, R. E.,** *Plant Virology*, 2nd ed., Academic Press, New York, 1981.
18. **Gerba, C. P. and Goyal, S. M.,** Detection and occurrence of enteric viruses in shellfish: a review, *J. Food Protect.*, 41, 743, 1978.
19. **Goddard, M. and Butler, M.,** *Viruses and Wastewater Treatment*, Pergamon Press, New York, 1981.
20. **Gerba, C. P. and Goyal, S. M.,** *Methods in Environmental Virology*, Marcel Dekker, New York, 1982.
21. **Berg, G., Bodily, H. L., Lennette, E. H., Melnick, J. L., and Metcalf, T. G.,** *Viruses in Water*, American Public Health Association, Washington, D.C., 1976.
22. **Berg, G.,** *Viral Pollution of the Environment*, CRC Press, Boca Raton, Fla., 1983.
23. **Provost, P. J. and Hilleman, M. R.,** Propagation of human hepatitis A virus in cell culture *in intro*, *Proc. Soc. Exp. Biol. Med.*, 160, 213, 1979.
24. **Kojima, H., Shibayoma, T., Sato, A., Suzuki, S., Ichida, F., and Hamada, C.,** Propagation of human hepatitis A virus in conventional cell lines, *J. Med. Virol.*, 7, 273, 1981.
25. **Bitton, G., Chou, Y., and Farrah, S. R.,** Techniques for virus detection in aquatic sediments, *J. Virol. Methods*, 4, 1, 1982.
26. **Gerba, C. P., Smith, E. M., and Melnick, J. L.,** Development of a quantitative method for detecting enteroviruses in estuarine sediments, *Appl. Environ. Microbiol.*, 34, 158, 1977.

27. **Tsai, S. C., Ellender, R. D., Johnson, R. A., and Howell, F. G.,** Elution of viruses from coastal sediments, *Appl. Environ. Microbiol.*, 46, 153, 1983.

28. **Lewis, G. D., Loutit, M. W., and Austin, F. J.,** A method for detecting human enteroviruses in aquatic sediments, *J. Virol. Methods*, 10, 153, 1985.

29. **Cukor, G. and Blacklow, N. R.,** Human viral gastroenteritis, *Microbiol. Rev.*, 48, 157, 1984.

30. **Murphy, A. M., Grohmann, G. S., Christopher, P. J., Lopez, W. A., Davey, G. R., and Millson, R. H.,** An Australia-wide outbreak of gastroenteritis from oysters caused by Norwalk virus, *Med. J. Aust.*, 2, 329, 1979.

31. **Appleton, H.,** Outbreaks of viral gastroenteritis associated with the consumption of shellfish, in *Viruses and Wastewater Treatment*, Goddard, M. and Butler, M., Eds., Pergamon Press, New York, 1981, 287.

32. **Wyatt, R. G., Kapikian, A. Z., Thornhill, T. S., Sereno, M. M., Kim, H. W., and Channock, R. M.,** In vitro cultivation in human fetal intestinal organ culture of a reovirus-like agent associated with non-bacterial gastroenteritis in infants and children, *J. Infect. Dis.*, 130, 523, 1974.

33. **Smith, E. M. and Gerba, C. P.,** Development of a method for detection of human rotavirus in water and sewage, *Appl. Environ. Microbiol.*, 43, 1440, 1982.

34. **Bates, J., Goddard, M. R., and Butler, M.,** The detection of rotaviruses in products of wastewater treatment, *J. Hyg. Cambr.*, 93, 639, 1984.

35. **Follett, E. A. C., Sanders, R. C., Beards, G. M., Hundley, F., and Desselberger, U.,** Molecular epidemiology of human rotaviruses, *J. Hyg.*, 92, 209, 1984.

36. **Goddard, M. R. and Sellwood, J.,** Enteric virus transmission via the water route with particular reference to rotavirus detection, *Monogr. Virol.*, 15, 229, 1984.

37. **Garrels, R. M. and MacKensie, M.,** *Evolution of Sedimentary Rocks*, W. W. Norton, New York, 1971.

38. **Currary, J. R.,** Modes of emplacement of prospective hydrocarbon reservoir rocks of outer continental marine environments, in *Geology of Continental Margins*, Ser. 5, AAPG Continuing Education, 1977, E1.

39. **Curray, J. R.,** Transgressions and regression, in *Papers in Marine Geology*, Miller, R. C., Ed., Macmillan, New York, 1964, 175.

40. **Krumbein, W. C.,** Application of logarithmic moments to size-frequency distributions of sediments, *J. Sediment Petrol.*, 6, 35, 1936.

41. **Drake, D. E.,** Suspended sediment transport and mud deposition on continental shelves, in *Marine Sediment Transport and Environmental Management*, Stanley, D. J. and Swift, D. P., Eds., John Wiley & Sons, New York, 1976, 127.

42. **Postma, H.,** Sediment transport and sedimentation in the estuarine environment, in *Estuaries*, Lauff, G. H., Ed., Publ. No. 83, American Association for Advancement of Science, Washington, D.C., 1967, 158.

43. **Meade, R. H.,** Transport and deposition of sediments in estuaries, *Geol. Soc. Am. Mem.*, 133, 91, 1972.

44. **Schubel, J. R.,** Turbidity maximum of the northern Chesapeake Bay, *Science*, 161, 1013, 1968.

45. **Schubel, J. R.,** An Eclectic Look at Fine Particles in the Coastal Ocean, Proc. Pollutant Transfer by Particles Workshop, Old Dominion University, Norfolk, Va., 1982, 51.

46. **Haven, D. S. and Morales-Alamo, R.,** Biodeposition as a factor in sedimentation of fine suspended solids in estuaries, *Geol. Soc. Am. Mem.*, 133, 121, 1972.

47. **Smayda, T. J.,** Some measurements of the sinking rates of fecal pellets, *Limnol. Oceanogr.*, 14, 621, 1969.

48. **Schubel, J. R.,** Tidal variation of the size distribution of suspended sediment at a station in the Chesapeake Bay turbidity maximum, *Neth. J. Sea Res.*, 5, 252, 1971.

49. **Schubel, J. R. and Dana, T. W.,** Agglomeration of fine-grained suspended sediment in northern Chesapeake Bay, *Powder Technol.*, 6, 9, 1972.

50. **Bernard, F. R.,** Annual biodeposition and gross energy budget of mature Pacific oysters *Crassostrea gigas*, *J. Fish. Res. Bd. Can.* 31, 185, 1974.

51. **Rhoads, D. C.,** Rates of sediment reworking by *Yoldia limatula* in Buzzards Bay, Massachusetts, and Long Island Sound, *J. Sediment. Petrol.*, 33, 723, 1963.

52. **Hirbrunner, W. R. and Wangersky, P. J.,** Composition of the inorganic fraction of the particulate organic matter in seawater, *Mar. Chem.*, 4, 43, 1976.

53. **Manheim, F. T., Meade, R. H., and Bond, G. C.,** Suspended matter in surface waters of the Atlantic continental margin from Cape Cod to the Florida Keys, *Science*, 166, 371, 1970.

54. **Schaub, S. A. and Sagik, B. P.,** Association of enteroviruses with natural and artificially introduced colloidal solids in water and infectivity of solids-associated virus, *Appl. Microbiol.*, 35, 212, 1975.

55. **Stumm, W. and Morgan, J. J.,** *Aquatic Chemistry*, 2nd ed., Wiley Interscience, New York, 1981, 506.

56. **Lal, D.,** The oceanic microcosm of particles, *Science*, 198, 997, 1977.

57. **Dame, R. F., Zingmark, R. G., and Haskin, E.,** Oyster reefs as processors of estuarine materials, *J. Exp. Mar. Biol.*, 83, 239, 1984.

58. **Anderson, J. B., Singer, J. K., Taylor, R. S., and Ledbetter, M. T.,** A comparison of electronic grain size analysis methods, in *Results from a Workshop on Fine-Grained Sediments*, SEPM Mid-Year Meeting, San Jose, Calif., Gorsline, D., Ed., abstr., 1985.

59. **Jordan, C. F., Jr., Fryer, G. E., and Hemmen, E. H.,** Size analysis of silt and clay by hydrophotometer, *J. Sediment. Petrol.*, 41, 489, 1971.

60. **Stein, R.,** Rapid grain-size analyses of clay and silt fraction by Sedigraph 5000D: comparison with Coulter counter and Atterberg methods, *J. Sediment Petrol.*, 1985, in press.

61. **Swift, D. J. P., Schubel, J. R., and Sheldon, R. E.,** Size analysis of fine grained suspended sediments: a review, *J. Sediment. Petrol.*, 42, 122, 1972.

62. **Ippen, A. T.,** Sedimentation in estuaries, in *Estuary and Coastline Hydrodynamics*, Ippen, A. T., Ed., McGraw-Hill, New York, 1966, 648.

63. **Goldberg, E. D.,** *The Biogeochemistry of Estuarine Sediments*, United Nations Educational, Scientific and Cultural Organization, Paris, 1978.

64. **Festa, J. F. and Hansen, D. V.,** Turbidity maxima in partially mixed estuaries: a two-dimensional numerical model, *Estuarine\Coastal Mar. Sci.*, 7, 347, 1978.

65. **Schubel, J. R.,** Size distributions of the suspended particles of the Chesapeake Bay turbidity maximum, *Neth. J. Sea Res.*, 4, 283, 1969.

66. **McPherson, R. M., Jr.,** The hydrography of Mobile Bay and Mississippi Sound, *Alabama J. Mar. Sci.*, 1, 1, 1970.

67. **Schubel, J. R., Carter, H. H., Wilson, W. G., Heaton, M. G., and Gross, M. G.,** Field Investigations of the Nature, Degree, and Extent of Turbidity Generated by Open-Water Pipeline Disposal Operations, Dredged Materials Research Program Tech. Rep. No. D-78-30, U.S. Army Engineering Waterways Experiment Station, Vicksburg, Miss., 1978,245.

68. **Ewing, M. and Thorndike, E. M.,** Suspended matter in deep ocean water, *Science*, 147, 1291, 1965.

69. **Dymond, J., Fisher, K., Clauson, M., Cobbler, R., Gardner, W., Sullivan, I., Soutar, A., Berger, W., and Dunbar, R.,** A sediment trap intercomparison study in Santa Barbara Basin, *Earth Planet. Sci. Lett.* 53, 409, 1981.

70. **Uhnoo, I., Wadell, G., Svensson, L., and Johansson, M. E.,** Importance of enteric adenoviruses 40 and 41 in acute gastroenteritis in infants and young children, *J. Clin. Microbiol.*, 20, 365, 1984.

71. **Brown, M., Petric, M., and Middleton, P. J.,** Diagnosis of fastidious enteric adenoviruses 40 and 41 in stool specimens, *J. Clin. Microbiol.*, 20, 334, 1984.

72. **Kurtz, J. B., Lee, T. W., Craig, J. W., and Reed, S. E.,** Astrovirus infection in volunteers, *J. Med. Virol.*, 3, 221, 1979.

73. **Kurtz, J. B., Lee, T. W., and Pickering, D.,** Astrovirus associated gastroenteritis in a children's ward, *J. Clin. Pathol.*, 30, 948, 1977.

74. **Snodgrass, D. R. and Gray, E. W.,** Detection and transmission of 30 nm virus particles (astroviruses) in feces of lambs with diarrhoea, *Arch. Virol.*, 55, 287, 1977.

75. **Cubitt, W. D. and Barrett, D. T.,** Propagation of human candidate calicivirus in cell culture, *J. Gen. Virol.*, 65, 1123, 1984.

76. **Studdert, M. J.,** Caliciviruses — brief review, *Arch. Virol.*, 58, 157, 1978.

77. **Caul, E. O. and Egglestone, S. I.,** Coronaviruses in humans, in *Virus Infections of the Gastrointestinal Tract*, Tyrrell, D. A. J. and Kapikian, A. Z., Eds., Marcel Dekker, New York, 1982, 179.

78. **Kapikian, A. Z., Greenberg, H. B., Wyatt, R. G., Kalica, A. R., and Chanock, R. M.,** The Norwalk group of viruses — agents associated with epidemic viral gastroenteritis, in *Virus Infections of the Gastrointestinal Tract*, Tyrrell, D. A. J. and Kapikian, A. Z., Eds., Marcel Dekker, New York, 1982, 147.

79. **Murphy, A. M., Grahmann, G. S., and Millson, R. H.,** Parvovirus gastroenteritis — a new entity for Australia, *Med. J. Aust.*, 1, 121, 1978.

80. **Vonderfecht, S. L., Huber, A. C., Eiden, J., Mader, L. C., and Yolken, R. H.,** Infectious diarrhea of infant rats produced by a rotavirus-like agent, *J. Virol.*, 52, 94, 1984.

81. **Caul, E. O. and Appleton, H.,** The electron microscopical and physical characteristics of small round human fecal viruses: an interim scheme for classification, *J. Med. Virol.* 9, 257, 1982.

82. **Paver, W. K., Caul, E. O., Ashley, C. R., and Clark, S. R.,** A small virus feces, *Lancet*, 1, 237, 1973.

83. **Boardman, G. D. and Sproul, O. J.,** Protection of viruses during disinfection by adsorption to particulate matter, *J. Water Pollut. Control Fed.*, 49, 1857, 1977.

84. **Liew, P. F. and Gerba, C. P.,** Thermostabilization of enteroviruses by estuarine sediment, *Appl. Environ. Microbiol.*, 40, 305, 1980.

85. **Smith, E. M., Gerba, C. P., and Melnick, J. L.,** Role of sediment in the persistence of enteroviruses in the estuarine environment, *Appl. Environ. Microbiol.*, 35, 685, 1978.

86. **Wellings, F. M., Lewis, A. L., and Mountain, C. W.,** Demonstration of solids-associated virus in wastewater and sludge, *Appl. Environ. Microbiol.*, 31, 354, 1976.

87. **Gerba, C. P., Stagg, C. H., and Abadie, M. G.,** Characterization of sewage solid-associated viruses and behavior in natural waters, *Water Res.*, 12, 805, 1978.

88. **Schaub, S. A. and Sorber, C. A.,** Viruses on solids in water, in *Viruses in Water*, Berg, G., Bodily, H. L., Lennette, E. H., Melnick, J. L., and Metcalf, T. G., Eds., American Public Health Association, Washington, D.C., 1976, 128.

89. **Rao, V. C., Seidel, K. M., Goyal, S. M., Metcalf, T. G., and Melnick, J. L.,** Isolation of enteroviruses from water, suspended solids, and sediments from Galveston Bay: survival of poliovirus and rotavirus adsorbed to sediments, *Appl. Environ. Microbiol.,* 48, 404, 1984.

90. **Metcalf, T. G., Rao, V. C., and Melnick, J. L.,** Solid-associated viruses in a polluted estuary, *Monogr. Virol.,* 15, 97, 1984.

91. **Matossian, A. M. and Garabedian, G. A.,** Virucidal action of seawater, *Am. J. Epidemiol.,* 85, 1, 1967.

92. **Mitchell, R. and Jannach, H. W.,** Processes controlling virus inactivation in seawater, *Environ. Sci. Technol.,* 3, 941, 1969.

93. **Bitton, G.,** *Introduction to Environmental Virology,* Wiley Interscience, New York, 1980, 87.

94. **LaBelle, R. L., Gerba, C. P., Goyal, S. M., Melnick, J. L., Cech, I., and Bogdan, G. F.,** Relationships between environmental factors, bacterial indicators, and the occurrence of enteric viruses in estuarine sediments, *Appl. Environ. Microbiol.,* 39, 588, 1980.

95. **Estes, M. K., Graham, D. Y., Smith, E. M., and Gerba, C. P.,** Rotavirus stability and inactivation, *J. Gen. Virol.,* 43, 403, 1979.

96. **Cliver, D. O. and Hermann, J. E.,** Proteolytic and microbial inactivations of enteroviruses, *Water Res.,* 6, 797, 1972.

97. **Hermann, J. E. and Cliver, D. O.,** Degradation of coxsackievirus A9 by proteolytic enzymes, *Infect. Immun.,* 7, 513, 1973.

98. **Grabow, W. O., Prozesky, O. W., Applebaum, P. C., and Lecatus, G.,** Absence of hepatitis B antigen from feces and sewage as a result of enzymic destruction, *J. Infect. Dis.,* 131, 658, 1975.

99. **Gerba, C. P. and Schaiberger, G. E.,** Effect of particulates on virus survival in seawater, *J. Water Pollut. Control Fed.,* 47, 93, 1975.

100. **Singer, B. and Fraenkel-Conrat, H.,** Effects of bentonite on infectivity and stability of TMV-RNA, *Virology,* 14, 59, 1961.

101. **Bitton, G. and Mitchell, R.,** Effects of colloids on the survival of bacteriophage in seawater, *Water Res.,* 8, 227, 1974.

102. **Bystricky, V., Stotzky, G., and Schiffenbauer, M.,** Electron microscopy of T1 bacteriophage adsorbed to clay minerals: application of the critical point drying method, *Can. J. Microbiol.,* 28, 1278, 1975.

103. **Brinsfielf, R. B., Winn, P. N., and Phillips, D. G.,** Characterization, treatment and disposal of wastewater from Maryland seafood plants, *J. Water Pollut. Control Fed.,* 50, 1943, 1978.

104. **Block, J.,** Viruses in environmental waters, in *Viral Pollution of the Environment,* Berg, G., Ed., CRC Press, Boca Raton, Fla., 1983, 118.

105. **Kapuscinski, R. B. and Mitchell, R.,** Processes controlling virus inactivation in coastal waters, *Water Res.,* 14, 363, 1980.

106. **Dimmock, N. J.,** Differences between the thermal inactivation of picornaviruses at "high" and "low" temperatures, *Virology,* 31, 338, 1967.

107. **Metcalf, T. G. and Stiles, W. C.,** Survival of enteric viruses in estuary waters and shellfish, in *Transmission of Viruses by the Water Route,* Berg, G., Ed., Wiley Interscience, New York, 1967, 437.

108. **Colwell, R. R. and Kaper, J.,** Distribution, survival and significance of pathogenic bacteria and viruses in estuaries, in *Estuarine Interactions,* Wiley, M. L., Ed., Academic Press, New York, 1978.

109. **Hill, W. F., Hamblet, F. E., Benton, W. H., and Akin, E. W.,** Ultraviolet devitalization of eight selected enteric viruses in estuarine water, *Appl. Microbiol.,* 19, 805, 1970.

110. **Bitton, G., Fraxedas, R., and Gilford, G. E.,** Effect of solar radiation on poliovirus: preliminary experiments, *Water Res.,* 13, 225, 1979.

111. **Shuval, H., Thompson, A., Fattal, B., Cymbalista, S., and Wiener, Y.,** Natural virus interaction processes in seawater, *J. San. Eng. Div. Am. Soc. Civ. Eng.,* 97, 587, 1971.

112. **Magnusson, S., Gundersen, K., Brandberg, A., and Lycky, E.,** Marine bacteria and their possible relation to the virus inactivation capacity of seawater, *Acta Pathol. Microbiol. Scand.,* 71, 274, 1967.

113. **Honjo, S. and Roman, M. R.,** Marine copepod fecal pellets production, preservation and sedimentation, *J. Mar. Sci.,* 36, 45, 1978.

114. **Perry, H. M. and Christmas, J. Y.,** Estuarine zooplankton: Mississippi, in *Gulf of Mexico Estuarine Inventory and Study, Mississippi, Phase IV: Biology,* Christmas, J. Y., Ed., Gulf Coast Research Laboratory, Ocean Springs, Miss., 1973.

115. **Cowey, C. B. and Corner, E. S.,** The amino acid composition of certain unicellular algae and of the fecal pellets produced by *Cananus finimarchicus* when feeding on them, in *Some Contemporary Studies in Marine Science,* Barnes, H., Ed., G. Allen & Unwin, London, 1966, 716.

116. **LaBelle, R. L. and Gerba, C. P.,** Influence of pH, salinity, and organic matter on the adsorption of enteric viruses to estuarine sediment, *Appl. Environ. Microbiol.,* 38, 93, 1979.

117. **Johnson, R. A., Ellender, R. D., and Tsai, S. C.,** Elution of enteric viruses from Mississippi estuarine sediments with lecithin-supplemented eluents, *Appl. Environ. Microbiol.*, 48, 581, 1984.

118. **Carlson, G. F., Woodward, F. E., Wentworth, D. F., and Sproul, O. J.,** Virus inactivation on clay particles in natural waters, *J. Water Pollut. Control Fed.*, 40, R89, 1968.

119. **Hamblet, F. E., Hill, E. F., Akin, E. W., and Benton, W. H.,** Poliovirus uptake and elimination, oysters and human viruses: the effect of seawater turbidity, *Am. J. Epidemiol.*, 89, 562, 1969.

120. **Moore, B. E., Sagik, B. P., and Malina, J. F.,** Virus associated with suspended solids, *Water Res.*, 9, 197, 1975.

121. **Schaub, S. A., Sorber, C. A., and Taylor, G. W.,** The association of enteric viruses with natural turbidity in the aquatic environment, in *Virus Survival in Water and Wastewater Systems*, Malina, J. F. and Sagik, B. P., Eds., Center for Research in Water Resources, University of Texas, Austin, 1974.

122. **Zerda, K. S., Gerba, C. P., Hou, K. C., and Goyal, S. M.,** Adsorption of viruses to charge-modified silica, *Appl. Environ. Microbiol.*, 49, 91, 1985.

123. **Lipson, S. M. and Stotzky, G.,** Adsorption of viruses to particulates: possible effects on virus persistence, in *Virus Ecology*, Misra, A. and Polasa, H., Eds., South Asian Publ., New Delhi, 1984, 165.

124. **Melnick, J. L. and Gerba, C. P.,** The ecology of enteroviruses in natural waters, *Crit. Rev. Environ. Control*, 10, 65, 1980.

125. **Stotsky, G.,** Surface interactions between clay minerals and microbes, viruses, and soluble organics, and the probable importance of these interactions to the ecology of microbes in soils, in *Microbial Adhesion to Surfaces*, Berkeley, R., Lynch, J. M., Melling, J., Rutter, P. R., and Vincent, B., Eds., Ellis Harwood, Chichester, England, 1980, 231.

126. **LaBelle, R. L., and Gerba, C. P.,** Investigation into the protective effects of estuarine sediment on virus survival, *Water Res.*, 16, 469, 1982.

127. **Lipson, S. M. and Stotzky, G.,** Adsorption of reovirus by clay minerals: effect of cation exchange capacity, cation saturation, and surface area, *Appl. Environ. Microbiol.*, 46, 673, 1983.

128. **Lipson, S. M. and Stotzky, G.,** Infectivity of reoviruses adsorbed to homoionic and mixed cation clays, *Water Res.*, 19, 227, 1985.

129. **Wait, D. A. and Sobsey, M. D.,** Method for recovery of enteric viruses from estuarine sediments with chaotropic agents, *Appl. Environ. Microbiol.*, 46, 379, 1983.

130. **Lipson, S. M. and Stotzky, G.,** Specificity of virus adsorption to clay minerals, *Can. J. Microbiol.*, 31, 50, 1985.

131. **Schiffenbauer, M. and Stotzky, G.,** Adsorption of coliphage T1 and T7 to clay minerals, *Appl. Environ. Microbiol.*, 43, 590, 1982.

132. **Krone, R. B.,** Flume Studies of the Transport of Sediment in Estuarial Shoaling Processes, Final Report, Hydraulic Engineering Laboratory and Sanitary Engineering Research Laboratory, University of California, Berkeley, 1962.

133. **Blatt, H., Middleton, G., and Murray, R.,** *Origin of Sedimentary Rocks*, Prentice-Hall, Englewood Cliffs, N.J., 1980.

134. **Middleton, G. V. and Southard, J. B.,** Mechanics of Sediment Movement, short course No. 3, Society of Ecologists, Paleontologists, and Mineralogists, 1977, 10.2.

135. **Southard, J. B., Young, R. A., and Hollister, C. D.,** Experimental erosion of calcareous ooze, *J. Geophys. Res.*, 76, 5903, 1971.

136. **Metcalf, T. G., Wallis, C., and Melnick, J. L.,** Virus enumeration and public health assessment in polluted surface water contributing to transmission of virus in nature, in *Virus Survival in Water and Wastewater Systems*, Malina, J. F. and Sagik, B. P., Eds., Center for Research in Water Resources, University of Texas, Austin, 1974, 57.

137. **Schaiberger, G. E., Edmond, T. D., and Gerba, C. P.,** Distribution of enteroviruses in sediments contiguous with a deep marine sewage outfall, *Water Res.*, 16, 1425, 1982.

138. **Dahling, D. R. and Safferman, R. S.,** Survival of enteric viruses under natural conditions in a subarctic river, *Appl. Environ. Microbiol.*, 38, 1103, 1979.

139. **Lytle, T. and Lytle, J.,** personal communication.

Chapter 5

VIRUS ASSOCIATION WITH SUSPENDED SOLIDS

V. Chalapati Rao

TABLE OF CONTENTS

I. INTRODUCTION

Solids found in natural waters are clays, fecal material of human and animal origin, and biological life forms such as bacteria, algae, and protozoa. Clays and organic colloids in water can adsorb a variety of bacteria and viruses and can serve to concentrate, protect, and transport them in the aquatic environment.

A clay is usually defined on a particle size basis, the upper limit being 2-μm diameter. Soils and sediments in nature will have varying proportions of material that contains clay-mineral components (usually the phyllosilicates), as well as nonclay-mineral material that may include a variety of substances such as iron and aluminum oxides and hydroxides, quartz, amorphous silica, carbonates, and feldspar. The interactions of metal cations such as Al^{3+}, Mg^{2+}, and Fe^{3+} with clays include adsorption by ion exchange and precipitation as hydroxides or hydrous oxides on clay surfaces; these interactions are influenced by the pH and E_H values of the water.

Natural organic materials in soils form complexes with clays. Since the exact chemical and physical nature of these organic materials is not known, the kinds of interaction they have with clays are less well known. It is obvious that clays which are eroded from soil surfaces into streams, lakes, and estuaries will probably be complexed with organic matter to some degree. Humic acid, a constituent of soil organic matter, may be strongly adsorbed by clays, presumably by interaction with positive sites on the edges of clay particles or because polyvalent cations on the cation-exchange complex act as bridges. Fulvic acid (another constituent of soil organic matter) is very strongly bound to Cu^{2+} on the exchange sites of montmorillonite.

Organic particulates in natural waters include organic matter associated with soil particles, organic particles from the effluents of sewage and industrial treatment plants, plant and animal debris, organic colloids, and microorganisms. The greatest number of particulates suspended in most natural waters are colloidal in nature. Colloids suspended in water may be divided into two groups — hydrophilic and hydrophobic — based on their affinity to water. Clays and many inorganic particles behave as hydrophobic colloids, whereas most organic particles behave as hydrophilic colloids. Recent studies by Murray and Parks[1] provided direct evidence that viruses suspended in water behave like hydrophilic colloids. Their results established a basis for predicting virus adsorption to solids and the influence of the compositions of these solids on the magnitude of their association. They indicated that poliovirus adsorption to a wide range of materials is as follows: metals (strong adsorption) > sulfides > transition metals > oxides > SiO_2 > organics (weak adsorption). These investigators also cited experimental evidence to indicate that virus adsorption to surfaces is influenced by ionic strength, pH, and dissolved components such as organics.

Most of the early research on the occurrence of enteric viruses in natural waters used virus concentration techniques that did not provide for estimating the number of viruses associated with suspended solids. As a result, much of the published data on virus numbers in environmental waters may constitute a very small fraction of the total amount of virus actually present. Recent research indicates that virus association with solids appears to be of the utmost importance in viral ecology.

This chapter evaluates the information on the nature and extent of enterovirus association with solids in different types of waters and their survival, infectivity, and transport from polluted to nonpolluted waters.

II. VIRUS ADSORPTION TO SOLIDS AND FACTORS AFFECTING THEIR INTERACTION

Human enteric viruses transmitted by water or food have emanated from human intestines, where they are either embedded in or adsorbed on fecal solids. This association with solids persists in raw sewage, settled sludge, and treated effluents. Discharge of effluents into coastal estuarine waters may result in elution of virus from sewage solids, but the viruses readily readsorb in the same estuarine water to naturally occurring solids and sediments.

Wellings et al.[2] demonstrated in 50% of the tests conducted that influents at a sewage treatment plant showed at least four- to fivefold higher virus concentration in the solids portion than in the supernatant. Influent, effluent, and chlorinated effluent samples showed from 16 to 100% of the total virus demonstrated in the samples to be solids-associated. The percentage of enteric viruses associated with solids in secondarily treated sewage discharges ranged from 3 to 100%.[3] Seventy-two percent of enteroviruses and 50% of rotaviruses recovered from polluted coastal water in Galveston Bay were solids-associated.[4] The association of viruses with solids has a protective effect, resulting in enhanced virus survival in natural waters and resistance to inactivation by chlorine.[5-9] It is especially important that adsorbed viruses can be eluted from solids under certain conditions[10] and that adsorbed viruses may be as infectious as free viruses.[11]

Factors that play an important role in the adsorption of virus on suspended solids include cation concentration, pH, and organic matter.

A. Cation Concentration

Schaub and Sagik[11] studied the adsorption of Columbia SK (an encephalomyocarditis [EMC] virus, genus *Cardiovirus*) to both organic and inorganic solids suspended in deionized water over a wide range of pH levels and with various concentrations and species of metal cations. Organic solids (dry dog food, 56 mg/ℓ) suspended in deionized water (pH, 5.5 to 6.5) adsorbed only 30% of added virus. Virus adsorption increased to 78.5% with the addition of 10^{-1} M MgCl$_2$. A further improvement of virus adsorption (98.8%) was obtained at a 10^{-2} M MgCl$_2$ concentration. Addition of 10^{-2} M CaCl$_2$ was equally effective in facilitating adsorption of 98.8% of the virus. Montmorillonite clay (inorganic suspended solid) at a concentration of 36 mg/ℓ in the presence of 10^{-3} M CaCl$_2$ adsorbed 98.4% of the virus. Natural suspended solids from storm runoff and discharge of some domestic water into a lake at Austin, Tex. also were examined for EMC virus adsorption. Between 15 and 35% of virus became solids-associated in lake water at pH 8.1. The addition of calcium ion (10^{-2} M) to lake water at pH 8.1 increased virus adsorption to 97.4%. Lowering the pH of lake water to 6.0 followed by addition of 10^{-3} M calcium chloride resulted in 72% adsorption of the virus.

The characteristics of virus adsorption to bentonite were studied by Schaub and colleagues.[12] A concentration of 10^4 pfu/mℓ of virus was allowed a 30-min adsorption period to 100 mg/ℓ bentonite plus 10^{-2} M CaCl$_2$ in deionized water. Poliovirus 1, EMC virus (Columbia SK) and coxsackievirus B readily adsorbed to clay. In all instances >93% of the virus was adsorbed. Virus adsorption to bentonite suspended in primary sewage, tap water, and deionized water at various cation concentrations indicated that the amount of virus adsorption with the divalent cation began to level off above 0.001 M and was maximal at 0.005 M. Sodium chloride, representing a monovalent cation, required approximately 100 times greater molar concentration for equivalent virus adsorption. Virus adsorption to montmorillonite clay over a pH range from 3.5 to 9.5, which encompasses the pH of most natural waters, was very similar throughout the entire pH range. These results confirm the findings of Carlson et al.,[6] who reported that on a molar basis divalent cations provided greater virus adsorption than monovalent cations on clay.

The investigators also examined the adsorption of Columbia SK virus to suspended solids occurring in natural waters. The waters selected were of diverse nature: a river with hard water, a freshwater reservoir, and a gulf coast estuary. Physical and chemical characteristics of these three waters, along with results of virus adsorption to solids, are given in Table 1.

Table 1
VIRUS ADSORPTION TO NATURALLY OCCURRING
SUSPENDED SOLIDS IN WATERS FROM DIFFERENT SOURCES[12]

| Source | pH | Conductivity (μmhos) | Total hardness (as CaCO₃) | Suspended solids | | Virus adsorbed (%) |
				Total (mg/ℓ)	Volatile (%)	
River water	7.9	13,250	2,090	230	12	98
Gulf coast estuary	8.4	16,750	1,830	102	20	34
Reservoir water	8.2	423	209	55	77	32

The river water contained a high level of suspended solids, mainly clay. Of the 230 mg/ℓ of total suspended solids, volatile suspended solids (phytoplankton, zooplankton, and detritus) constituted only 12%; the remaining 88% was inorganic. The reservoir water had alow conductivity, low hardness, and low suspended solids content. Seventy-seven percent of 55 mg/ℓ of suspended solids was volatile. Enterovirus adsorption in river water was 98%, whereas in reservoir water virus adsorption was only 32%. These results indicate that high conductivity and a high concentration of inorganic solids promote better virus adsorption. This point was further substantiated when natural solids (containing a high level of volatile solids) were replaced by montmorillonite clay (inorganic solids) in reservoir waters. Virus adsorption improved to 93% even though the water was low in hardness and conductivity. Surprisingly, the estuary water with its high conductivity and hardness did not adsorb virus to the montmorillonite clay. This is contrary to results of extensive adsorption of poliovirus on silt and suspended solids in seawater.[4,13]

B. pH

Sproul[14] has demonstrated that the extent of adsorption of bacteriophages to clay minerals generally increases as the pH of the suspending medium is lowered below 7.0.

Schaub and Sagik[11] studied the effects of pH on Columbia SK virus adsorption to montmorillonite clay in the presence of 10^{-3} M CaCl$_2$. The greatest degree of adsorption (99.2%) occurred at pH 5.5. At the two pH extremes of 3.5 and 9.5, virus adsorption was 92.7%and 96.7%, respectively. It was noted that pH was not an important factor as long as a cation was present in the test sample. This is not surprising, as metal cations lower the net electronegative ion atmosphere by ion incorporation, an effect that is duplicated by lowering the pH. The investigators anticipated that in natural waters virus adsorption to clays would occur throughout the normal pH range of water if cations were present in adequate concentration.

The influence of pH on virus adsorption to solids in raw wastewater and effluent appears to be different. Moore and co-workers[15] demonstrated that poliovirus and bacteriophages T7, T2, and f2 showed greater affinity to raw waste water solids at pH 10. In the final effluent nearly all of the seeded poliovirus and the three phages became associated withsolids at pH 10. The authors noted that divalent cations were not added to the samples and that a good floc was formed at pH 10. It appears that flocculation of virus and solids at this pH may have caused maximum virus removal from the effluent. Quantitative differences in the nature of the suspended solids, the presence or absence of divalent cations, and competition from organic matter in wastewater may influence virus adsorption at different pH levels.

C. Organic Matter

Organic matter is an integral part of the aquatic and terrestrial systems, and its effect on interactions between particulates and viruses exerts a profound influence on the ecology of viruses in these environments. The effect of organic matter on the adsorption of viruses to particulates in soils and estuarine waters has been the subject of a number of studies,[16] and conflicting results have been reported. For example, secondary sewage and humic acid had no effect on the adsorption of poliovirus or echovirus to estuarine sediments,[17] and the accumulation of organic matter near the surface of soil columns did not affect adsorption of poliovirus to soil.[18] However, bovine albumin and egg albumin reduced the adsorption of poliovirus to kaolinite.[6] In these experiments Carlson and associates[6] suspended kaolinite (50 mg/ℓ) in deionized water (pH, 7) containing $CaCl_2$ (0.1 or 0.01 M) and introduced thespecific protein. Virus adsorption to clay in the absence of albumin was 92%. Addition of as little as 1 mg/ℓ egg albumin to the sample resulted in lowering bacterial virus (T2) adsorption to 53%. An increase in the egg albumin concentration to 2 mg/ℓ resulted in a further reduction of phage adsorption to 27%. Results obtained with bovine albumin were identical. Addition of 1 mℓ of raw domestic wastewater to 1 ℓ of a clay-salt-distilled water suspension had little interference on virus adsorption. It was further demonstrated that natural clays from the Missouri River interfered with virus adsorption in a manner similar to that of the pure clay minerals. Addition of serum protein at a final concentration of 0.6 mg/mℓ to montmorillonite clay (200 mg/ℓ) plus 10^{-3} M $CaCl_2$ permitted only 16% of Columbia SK virus to adsorb to clay. The suspension of soil in secondary sewage effluent reducedthe adsorption of coliphage f2 and poliovirus to soil particles, suggesting that organic components in the wastewater may have competed for adsorption sites on the soil particulates.[19] In contrast, adsorption of coliphages T7, T2, and f2 by several soil types increased with increasing organic carbon content.[20]

The mechanisms by which organic matter influences virus adsorption to particulates are poorly defined, probably because most studies have used heterogeneous organics (e.g.,sewage) or organics with similar biophysical properties (e.g., similar isoelectric points [pI] and molecular weights). Lipson and Stotzky[21] investigated the mechanisms whereby organic matter, in the form of defined proteins, affects the adsorption of reovirus to the clay minerals kaolinite and montmorillonite. The specific organics included chymotrypsin, ovalbumin, and lysozyme (ranging in molecular weight from 17,500 to 54,000 and in isoelectric pH from 4.6 to 11.2). Clay-protein complexes were prepared in such a manner that montmorillonite contained 9 mg lysozyme, 14 mg chymotrypsin, and 19 mg ovalbumin, whereaskaolinite contained 0.4 mg lysozyme and 2.6 mg ovalbumin. Both clay types were homoionic to sodium. Chymotrypsin and ovalbumin reduced the adsorption of reovirus to both the clays examined, probably because of blockage of negatively charged sites on the clays by each protein. The importance of negatively charged (i.e., cation exchange) sites in the adsorption of reovirus to clay minerals has been reported by the same investigators.[22] Lysozyme did not reduce the adsorption of the virus to kaolinite, but it did reduce adsorption to montmorillonite. Lysozyme was probably bound to kaolinite by mechanisms other than cation exchange, as changing the bulk pH from 3.13 to 9.75 had little or no effect on adsorption of these proteins. The inability of lysozyme to block reovirus adsorption to kaolinite concurred, in part, with the proposed mechanism that adsorption of reovirus was primarily to negatively charged sites on kaolinite as a result of protonation of the virus;[22] competition between the virus and lysozyme for negatively charged sites on kaolinite apparently did not occur. Maximum adsorption of chymotrypsin to kaolinite occurred below the isoelectric pH of the protein (8.1 to 8.6),[23] indicating that the protein adsorbed to negatively charged sites on the clay. Consequently, the decreased adsorption of reovirus to kaolinite complexed with chymotrypsin may have resulted from blockage of negatively charged sites.

D. Virus-Solids Interaction in Estuarine Water

Virus-solids interactions, using kaolin and silt loam (0.7 to 0.8 μm in size) mixed at 1:4

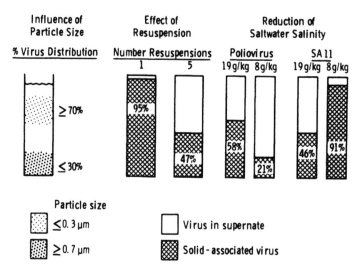

FIGURE 1.   Factors influencing virus-solids interactions. (From Metcalf, T. G., Rao, V. C., and Melnick, J. L., *Monogr. Virol.*, 15, 97, 1984. With permission.)

ratio (0.5% kaolin, 2% silt loam) and suspended in 0.01 *M* phosphate-buffered saline, were studied in Plexiglas cylinders of 5 cm diameter and 36 cm length by Metcalf et al.[9] In addition, suspended solids (clays and silts, 0.3 μm in size) collected from Galveston Bay (Texas) also were used to simulate conditions existing in virus-polluted bay waters. Salinities of 15 to 19 ppt (g/kg) were created by addition of Instant Ocean® (Aquarium Systems, Mentor, Oh.). Poliovirus 1, echovirus 1, and rotavirus SA11 were adsorbed to suspended-solids during 30 to 60 min of mixing. The factors studied included influence of particle size, effect of resuspension of virus, and salinity changes. These factors are illustrated in Figure 1. Addition of either 0.3- or 0.7-μm particulates, individually or combined, to the estuarine modeling system in the presence of virus resulted in the preferential adsorption of virus to the smaller-sized particulates. An average of 70% or more of either test enterovirus or rotavirus was recovered from 0.3-μm particulates.

The behavior of sediment-associated virus after resuspension of sediments was examined by subjecting solids-associated virus to a number of resuspensions. Kaolin and silt loam with particulates of 0.7 μm and containing adsorbed virus were allowed to settle to the bottom of the modeling system cylinder and were resuspended at 12-hr intervals with con-current sampling of the middle and upper positions of the water column. Release of virus to the water column was minimal until the fourth and fifth resuspensions. As shown in Figure 1, about one half of the initial solids-associated virus was released into the water column after the fifth resuspension. The results suggest that repeated resuspension of solids-associated virus by water turbulence could cause the eventual release of virus into the water column.

The effects of salinity changes on virus-solids interactions were also studied. As shown in Figure 1, a release to the supernatant of 37% of the initial solids-associated poliovirus occurred when the initial salinity of 19 g/kg was decreased to 8 g/kg. In contrast, rotavirus adsorption to solids increased from 46% to 96%. These results were interpreted to indicate that freshwater inflows reducing estuarine salinity would favor release of solids-associated poliovirus to the water column, but cause any freely suspended rotavirus to adsorb avidly to suspended particulates.

The effect of water quality and turbidity and the composition of suspended solids of estuarine water upon rotavirus adsorption and recovery have been investigated by Rao and colleagues.[24] Suspended solids were collected on 3- and 0.45-μm Filterite filters; the filters were backwashed with phosphate-buffered saline, and the solids were separated by centrifugation. The deposit was suspended in a small volume of seawater from the same location and from which suspended solids had been removed by centrifugation. Sample suspensions (1 mℓ) were seeded with rotavirus SA11 and shaken for 30 min at 300 rpm. The mixture was centrifuged at 2200 $\times$ g for 15 min, and supernatants were assayed to determine unadsorbed virus. Virus from the solids deposit was eluted with 2 mℓ of 10% tryptose phosphate broth (pH 9.5). The relative proportions of clay and biological material in the suspended solids were determined. Salinity and turbidity of water also were measured. Results are summarized in Table 2. Rotavirus adsorption of 99, 75, and 40% to clay concentrations of 95, 50 and 25%, respectively, indicated the significance of the relative proportion of inorganic and biological solids in a water column for virus adsorption. Solids of biological origin appeared to adsorb rotavirus poorly. Recovery of adsorbed virus was proportionately less effective with declining clay concentrations.

E. Virus-Solids Interaction in Fresh Water

An investigation was made of the influence of factors such as ionic strength of the water, presence of clays, and particle size of the clay upon adsorption of virus to Lake Houston suspended solids, and results of the study are given in Table 3.[36] The data clearly show the importance of clays as solids that promote adsorption of virus. Montmorillonite, illite, and kaolinite may represent more effective clays for adsorption of virus, but clays in general probably serve to promote formation of solids-associated virus in sewage-polluted waters. The degree to which the smallest particulates adsorb virus was illustrated by the increased adsorption detectable after centrifugal speeds $\geq$ 10,000 $\times$ g (i.e., small-particulate-adsorbed virus sedimented). An increased ionic strength promotes a greater degree of virus adsorption to solids in nonsaline waters. Differences between poliovirus and rotavirus in the degree of adsorption obtained with the same solids under the same conditions reflected the probable existence of adsorptive variability among enteric viruses.

F. Mechanism of Virus Adsorption to Suspended Solids

A detailed discussion of this topic is presented in other chapters in this volume. Two general theories of adsorption of viruses on solids are likely. Both rely on the observation that in the pH range of most natural waters clay, detritus, and enteroviruses retain a net negative electric charge, thereby causing natural repulsion of all the particles; as a result, they remain stable or stay in suspension. Reduction in pH and introduction of a metal cation can destabilize this system either by lowering the net electronegativity of the particles or by neutralization of the electronegative charge by positively charged ions, respectively. Carlson et al.[6] suggested that the cation acts to form a clay-cation-virus bridge. In their studies, reduction in cation concentration broke apart this bridge and allowed complete separation of virus from clay.

Data by Singley et al.[25] on theories of coagulation in water treatment offer a different interpretation of virus adsorption to suspended solids. Assuming that viruses are not unique to the colloid-containing suspended solids, then virus adsorption is identical to the mechanism of solids coagulation. Adsorption (coagulation) is enhanced by reduction of the interaction energy of the particles (interaction energy is the net value obtained when columbic electrostatic repulsive energy and van der Waals or other attractive energies are considered together). The metal cation reduces these repulsive forces (van der Waals forces) on the solids and virus to interact, thus providing adsorption. Studies by Schaub and Sagik[11] tend to support this second theory inasmuch as they observed that a 100-fold reduction in cation concentration did not result in total reversal in adsorption. They observed that divalent cation reduction from 10^{-2} to 10^{-4} M left 45% of virus unadsorbed to clay. The attraction of virus to solids appears to be a stronger association than a simple bridge effect.[12]

Table 2

ADSORPTION AND RECOVERY OF ROTAVIRUS SA11 FROM ESTUARINE SUSPENDED SOLIDS IN DIFFERENT MONTHS[24]

| Date | Salinity (ppt) | Turbidity (NTU) | Suspended solids (mg/ℓ) | Composition (%) of | | Virus input | Virus adsorbed (%) | Virus recovered (%) |
				Clay	Biosolids (phytoplankton, zooplankton)			
7/9/81	16	10	0.5	1	99	1.0×10^6	20	12
7/28/81[a]	6	115	10.1	95	5	6.8×10^5	99	68
10/19/81	14	14	0.04	2	98	1.2×10^7	28	18
12/08/81[a]	6	58	3.5	25	75	5.5×10^6	40	28
1/19/82[a]	—	50	2.8	40	60	7.4×10^7	58	33
2/24/82	20	28	1.5	50	50	5.2×10^6	75	45

[a] Heavy rains 2 to 3 days prior to sampling.

From Rao, V. C., Metcalf, T. G., and Melnick, J. L., *Appl. Environ. Microbiol.*, in press. With permission.

Table 3
INFLUENCE OF IONIC STRENGTH, CLAYS, AND CLAY PARTICLE SIZE ON VIRUS ADSORPTION TO LAKE HOUSTON SUSPENDED SOLIDS[a] [28]

Virus	Suspended solids (SS) (+ factor studied)	Adsorption (%)	
		Large particles only[b]	Large + small particles[c]
Poliovirus 1	SS + water	12	58
	SS + water + salt	60	82
	Clays + water	83	93
	Clays + water + salt	100	100
Rotavirus SA11	SS + water	18	42
	SS + water + salt	75	88
	Clays + water	99	99
	Clays + water + salt	100	100

[a] The ionic strength of the salt added was 13 ppt (g/kg). Clays were a mixture of montmorillonite (42%), illite (46%), and kaolinite (12%). Suspended solids consisted principally of algae, with very small amounts of silt.
[b] Suspended solids in water pelleted at $1000 \times g$.
[c] Suspended solids in water pelleted at $10,000 \times g$.

III. SEPARATION AND CONCENTRATION OF VIRUSES FROM SOLIDS

A variety of eluents has been used to separate viruses from different types of solids derived from sewage, river and estuarine water, flocs from alum and ferric chloride coagulation, and a variety of clays. The eluents include distilled water, estuarine water, buffers, tryptose phosphate broth, beef extract, and casein. Results from several investigators are summarized in Table 4.

A. Distilled Water

A solution of low ionic strength such as distilled water has been used as an eluent to recover bacteriophage T2 adsorbed on kaolinite. In laboratory experiments Carlson et al.[6] adsorbed the virus to kaolinite (50 mg/ℓ) in distilled water containing 0.005 M $CaCl_2$. Of the seeded virus, 91% was adsorbed on the clay after 30 min of slow mixing. When a sample of the clay suspension was pipetted and placed in distilled water (low ionic strength), complete separation of virus from kaolinite occurred. Manwaring and co-workers[26] found that MS2 bacteriophage coagulated by $FeCl_3$ could be recovered from the flocs with deionized water with an efficiency of 7.8%. Elution of coliphages from sewage effluent solids using distilled water recovered only 12% of the virus.[3]

B. Natural Waters: Lake, Estuarine, and Tap Water

To determine whether the association of viruses with solids persists after their discharge into natural waters, various fresh and estuarine waters were tested for their ability to elute virus from sewage solids.[3] First, 7.7 ℓ of treated sewage from a trickling filter plant was passed through a stacked series of serum-treated, 142-mm diameter, 3.0-, 0.45-, and 0.25-µm pore size Filterite filters. Then, 50 mℓ of the elution fluid was passed through the multiple filter unit. Tap, lake, and estuarine waters were all found to be capable of eluting

Table 4
ELUTION OF VIRUSES FROM SOLIDS

Eluent	Nature of solids	Virus type	Virus recovery (%)	Ref.
Distilled water	Kaolinite + deionized water + 0.005 *M* CaCl₂	Bacteriophage T2	89	6
	Flocs from FeCl₃ coagulation	Bacteriophage MS2	7.8	26
	Sewage effluent solids	Natural coliphages	12	3
Natural waters				
Tap water (pH, 8.2)	Sewage solids	Natural coliphages	600 pfu	3
Lake water (pH, 7.9)	Sewage solids	Natural coliphages	550 pfu	3
Polluted coastal canal water (pH, 8.0; salinity, 16 ppt)	Sewage solids	Natural coliphages	1250 pfu	3
Gulf shore water (pH, 8.0; salinity, 20 ppt)	Sewage solids	Natural coliphages	2300 pfu	3
Filtered estuarine water	Clay + deionized water + 10⁻³ *M* CaCl₂	Columbia SK virus	3.9—31.8	12
	River water solids	Columbia SK virus	21.7—32.6	12
	Filtered river water + clay	Columbia SK virus	33.8—45.9	12
Buffers				
Tris (hydroxymethyl aminomethane)	Flocs from FeCl₃ coagulation of water	Bacteriophage MS2	21	26
Glycine (0.05 *M*, pH,11.5)	Sewage solids	Natural coliphage	75	3
Glycine (0.1 *M* + EDTA pH,11) + 1% calf serum	Bentonite + primary effluent + 0.01 *M* CaCl₂	Poliovirus 1	71	27
	Sewage solids + primary effluent + 0.01 *M* CaCl₂	Poliovirus 1	71	27
Borate saline (pH, 9)	Organic solids (dry dog food) + deionized water + 0.01 *M* CaCl₂	Poliovirus 1 / Bacteriophage T2	>20—99 / 51—99	15
	Inorganic solids			
	Bentonite + 0.01 *M* CaCl₂	Poliovirus 1	14—23	
	Kaolinite + 0.01 *M* CaCl₂	Poliovirus 1	99	
	Natural suspended solids in effluent	Poliovirus 1	77	
Proteinaceous solutions				
3% Beef extract (pH, 9.4 to 9.6)	Solids in effluent, effluent, and chlorinated effluent	Natural enteroviruses	16—100	2
3% Beef extract-0.05 *M* glycine (pH, 9.5)	Solids from Galveston Bay water	Natural enteroviruses	72	4
	Solids from Lake Houston	Poliovirus 1	100	28
		Hepatitis A virus	40	28
TPB (pH, 7.2)	Bentonite and sewage solids + 0.01 *M* CaCl₂	Poliovirus 1	48	27
	Mixed liquor suspended solids	Natural enteroviruses	—	29
10% TPB-0.05 *M* glycine(pH, 9.5)	Estuarine suspended solids	Rotavirus SA11	11—68	24
	Freshwater suspended solids	Rotavirus SA11	100	28

solid-associated coliphages (Table 4). The highest numbers of viruses (2300 pfu) were recovered from solids when eluted with estuarine water, which appeared to be related to the salinity of that water.

Schaub, Sorber, and Taylor[12] found that Columbia SK virus adsorbed to montmorillonite clay or to natural solids from river water could be eluted by high conductivity estuarine water with efficiencies ranging from 3.9 to 31.8% and from 21.7 to 32.6%, respectively. When the virus was adsorbed on clay suspended in filtered river water, elution of clay with estuarine water resulted in improved virus recoveries varying from 33.8 to 45.9%.

C. Buffers
1. Tris
Manwaring and co-workers[26] found that MS2 bacteriophage coagulated by $FeCl_3$ could be recovered from the floc with an efficiency of 21% when hydroxymethyl aminomethane (tris) buffer was used as an eluent.

2. Glycine
Investigations by Schaub and Sorber[27] demonstrated that elution of virus by glycine from bentonite can be efficient. Data presented in Table 4 illustrate that 10^4 pfu/mℓ of poliovirus 1, adsorbed for 30 min to bentonite or sewage solids in primary effluent containing 0.01 M $CaCl_2$, can be eluted by a variety of methods. After the adsorption period the bentonite was poured onto Millipore AP25 fiberglass prefilters, and 10 mℓ of eluent was allowed to soak the prefilter for several minutes, after which the eluent was collected and assayed. Of the input virus, 71% was recovered by using 0.1 M glycine (pH 11.0) supplemented with 0.01 M EDTA and 1% fetal calf serum (FCS).

Solids-associated coliphages in unchlorinated effluents of a trickling filter plant were collected on Filterite filters and eluted with 0.05 M glycine buffer (pH 11.5). Average recovery of virus was 75%.[3]

3. Borate Saline
Moore and colleagues[15] studied poliovirus and coliphage T2 adsorption to organic, inorganic, and naturally occurring suspended solids and the subsequent elution of the viruses by borate saline. Inorganic suspended solids included bentonite (95 mg/ℓ) and kaolin (16 mg/ℓ) in the presence of 0.01 M $CaCl_2$. Organic suspended solids were prepared by suspending dry dog food in deionized water containing 0.01 M $CaCl_2$ (63 mg/ℓ). Natural suspended solids were derived from raw wastewater and the final effluent from an activated sludge plant. The total suspended solids concentration of raw wastewater averaged 145 mg/ℓ, of which 75% was volatile. The final effluent contained 16 mg/ℓ total suspended solids, 70% of which was volatile. Samples were seeded with virus, which was allowed to adsorb for 30 min; aliquots were then centrifuged at 12,000 $\times$ g for 15 min to pellet the solids. Pelleted solids were suspended in borate saline (pH 9.0), eluted for 15 min, and centrifuged at 12,000 $\times g$ for 15 min to sediment the solids. Supernatants were assayed for percentage of virus recovered based on the amount of virus originally adsorbed to the solids. The results (Table 4) show that a much lower percentage (14 to 23%) of virus was recovered from inorganic suspended solids than from organic solids, except in the case of poliovirus where 99% of the adsorbed virus was eluted from the kaolin plus calcium chloride. Poliovirus associated with suspended solids in the activated sludge effluent was recovered with an efficiency of 77%.

D. Proteinaceous Solutions
1. Beef Extract
Wellings and associates[2] collected influents and effluents from sewage treatment plants, concentrated them in a dialysis bag surrounded by polyethylene glycol overnight at 4°C, collected the residue in the bag in phosphate-buffered saline, and pelleted the solids by centrifugation. The solids were suspended in 3% beef extract buffered to pH 9.4 to 9.6 with tris-saline. Elution of virus from the solids was further assisted by sonication of the samples at 100 W for 15 min, mechanical stirring at 4°C for 18 hr, or fluorocarbon (freon) extraction. The authors reported that influent, effluent and chlorinated effluent samples showed that 16 to 100% of the total enteroviruses demonstrated in samples were solids-associated.

In investigations designed to quantitate suspended solids-associated virus in a polluted estuary, Rao et al.[4] used 3% beef extract-0.05 M glycine (pH 9.5) to recover enteroviruses from solids retained on 3.0- and 0.45-μm Filterite filters. Reconcentration of beef extract eluates of solids resulted in recovering virus from 72% of the samples. In laboratory experiments, poliovirus and hepatitis A virus adsorbed on suspended solids collected from Lake Houston were recovered 100% and 40%, respectively, by elution with beef extract-glycine (pH, 9.5).[28]

2. Tryptose Phosphate Broth

Poliovirus 1 adsorbed to bentonite and sewage solids in primary effluent containing 10^{-2} M $CaCl_2$ was eluted with tryptose phosphate broth (TPB; pH, 7.2) with an efficiency of recovery of 48%.[27]. Moore and associates[29] recovered viruses from activated-sludge mixed-liquor-suspended solids (MLSS) using a variety of eluents. More efficient recovery of virus was obtained with TPB. TPB was used as an eluent to recover rotavirus SA11 adsorbed on estuarine suspended solids with an efficiency of 11 to 68%.[24] With TPB as an eluent, 100% of rotavirus adsorbed on freshwater suspended solids could be recovered.[28]

E. Concentration of Virus from Solids Eluates

Most of the published data on virus association with suspended solids was obtained by either direct plating of solids or eluates from solids. No attempts were made to concentrate virus from the eluates. Rao and colleagues[4] recognized that at least 1 ℓ of eluent was required to recover solids retained on tubular Filterite filters after passing 100 gal of estuarine or lake water. Virus released into 1-ℓ eluates had to be concentrated to a smaller volume for an economical assay. Two methods have been evaluated for concentration of virus: iron oxide-adsorption-elution and organic flocculation.

1. Iron Oxide Adsorption-Elution

Solids from fliters were eluted with 1 ℓ of 3% beef extract prepared in 0.05 M glycine (BE-Gly; pH, 9.5) to recover enteroviruses, including poliovirus and hepatitis A virus. A solution of 10% TPB-0.05 M glycine (TPB-Gly; pH, 9.5) was used to recover rotavirus. Virus from these eluates was concentrated according to the scheme outlined in Figure 2. Average recoveries were poliovirus, 58%; hepatitis A virus, 45%; and rotavirus SA11, 53% (Table 5).[28] The iron oxide method was used to concentrate rotaviruses from TPB eluates of suspended solids collected from Galveston Bay; rotaviruses were recovered from 50% of the samples.[4]

2. Organic Flocculation

Researchers derived 100-mℓ eluates of 3% BE-Gly and 10% TPB-Gly from suspended solids seeded with known amounts of poliovirus, hepatitis A virus, and rotavirus SA11 and concentrated them by organic flocculation. Samples were adjusted to pH 3.5, stirred for 15 min on a magnetic stirrer, and centrifuged at 2200 × g for 15 min. The precipitate was suspended in 5 mℓ of 0.15 M Na_2HPO_4, sonicated for 30 sec at 100 W, and assayed. Average recoveries were poliovirus, 48%; hepatitis A virus, 42%; and rotavirus, 18% (Table 5). Organic flocculation was routinely used to concentrate enteric viruses from suspended solids collected from Galveston Bay during 1981 and 1982; enteroviruses were detected in 72% of the samples.[4]

IV. INFECTIVITY OF SOLIDS-ASSOCIATED VIRUSES

Many of the studies on virus removal in treatment of water and wastewater have equated virus inactivation and virus association with solids. Recent studies, however, have demonstrated that virus association with solids does not necessarily mean inactivation, nor does it mean that the virions once associated cannot be eluted and still retain infectivity.[4,6,11,12,15]

Table 5
CONCENTRATION OF VIRUS FROM
SOLIDS ELUATES[a]

| Virus | Trial | Virus recovery (%) | |
		Magnetic iron oxide	Organic flocculation
Poliovirus 1	1	68	40
	2	45	48
	3	61	56
	Average	58	48
Rotavirus SA11	1	64	13
	2	47	18
	3	48	23
	Average	53	18
Hepatitis A virus	1	58	40
	2	35	44
	3	42	—
	Average	45	42

[a] A 100-mℓ eluate of 3% beef extract-0.5 M glycine was seeded with known numbers of poliovirus 1 or hepatitis A virus; the pH was adjusted to 7 for poliovirus and to 3.5 for hepatitis A virus. Ten percent TPB-0.5 M glycine was seeded with rotavirus SA11 and adjusted to pH 3.5. Then 0.2 g of MIO was added. After 30 min of stirring, the supernatant was discarded while magnetic iron oxide was held on a magnet. Virus was eluted from the magnetic iron oxide with 2 mℓ of 2% casein (pH, 8.5). For organic flocculation, 100-mℓ samples were adjusted to pH 3.5, stirred for 15 min, and centrifuged at 2200 × g for 15 min. The precipitate was dissolved in 0.15 M Na$_2$HPO$_4$ (Rao, Metcalf, and Melnick, unpublished data).

Schaub and Sagik[11] performed tests to determine the tissue culture infectivity of suspended solids-bound viruses. In these studies, Columbia SK cardiovirus was adsorbed on montmorillonite clay suspended in deionized water or in filtered lake water. Virus also was adsorbed on natural solids in lake, river, and bay water. The infectivity of viruses adsorbed on these solids was determined by assaying the virus-solids mixture on L-929 cell monolayers in 60-mm plates for plaque formation. Water samples containing virus but no solids served as controls. Results in Table 6 indicate that the virus adsorbed to clay or to solids in natural water was more infectious and yielded a higher titer than the controls. It appears that the clay solids may facilitate virus to establish better proximity to the cells. However, it has not been demonstrated whether the clay or the solids actually accompany viruses into susceptible cells, whether the cells strip the viruses from the solids, or whether elution into the medium surrounding the cells is required prior to attachment.

These investigators also tested the infectivity of solids-associated virus in mice via the oral route, because the oral route obviously is the most probable route of virus transmission by water to humans. Mengovirus (a cardiovirus with greater mouse infectivity than Columbia SK) was suspended with clay in deionized water containing 10^{-3} M CaCl$_2$. Suckling white mice (7 to 8 days old) were given an oral dose between 0.01 and 0.02 mℓ containing 10^5 pfu per mouse. Infectivity of the virus was measured by the percentage mortality of the mice between 2 and 10 days after administration of the virus. Virus adsorbed to clay was as infectious as the virus in the CaCl$_2$ controls.

PRIMARY ELUATE (100 mℓ)

3% beef extract–0.5 *M* glycine: 10% tryptose phosphate broth–0.5 *M* glycine:

Add poliovirus   or   Add hepatitis A virus   Add rotavirus SA11
Adjust pH to 7.0   Adjust pH to 3.5   Adjust pH to 3.5

Add 0.2 g of magnetic iron oxide
↓
Stir with a glass rod for 30 min
↓
Virus adsorbs on iron oxide
↓
Settle iron oxide by keeping a
magnet underneath the beaker

Discard supernate

Add 1 mℓ of 0.5 *M* glycine (pH 9.0) to
iron oxide and mix for 1 min
↓
Then add 3 mℓ of 2% casein (pH 8.5) and
stir for 20 min to elute virus
↓
Settle iron oxide by a magnet, collect eluate,
treat with antibiotics, and assay

FIGURE 2.   Concentration of virus from solids eluates by the iron oxide adsorption-elution method.

Moore et al.[15] studied the infectivity of virus adsorbed to bentonite in 0.01 *M* $CaCl_2$. Experiments were performed using low levels (60 to 215 pfu) of poliovirus 1 and bacteriophages T7, T2, and f2, which allowed direct plating of samples. Immediately after inoculation of virus a sample was removed and assayed (0 time). After a 30-min adsorption period a second sample was assayed. It is apparent that the three viruses retained a significant degree (76 to 93%) of their infectivity when associated with bentonite (Table 6). Whether solids-associated virus would retain infectivity over prolonged periods of time has been examined by Rao and colleagues.[4] Naturally occurring suspended solids and sediments collected from Galveston Bay were seeded with poliovirus 1 and rotavirus SA11 in the laboratory. After determining the amount of virus adsorbed, test samples were suspended in estuarine water from which natural solids were removed by filtration. Samples were assayed at 3-day intervals for 19 days. Infectivity of the solids-associated virus was demonstrated throughout the period of testing. These studies indicate a potential for transport of solids-associated virus over distances of several miles in Galveston Bay. A 19-day survival of solids-associated poliovirus and rotavirus may be a potential health hazard for recreational water users and for consumers of contaminated shellfish.

Organic matter in sewage and natural waters may enhance or inhibit the infectivity of viruses associated with suspended solids. Lipson and Stotzky[21] investigated the role of defined proteins (chymotrypsin, lysozyme, and ovalbumin) on the infectivity of reovirus adsorbed on montmorillonite and kaolinite. Free chymotrypsin reduced the infectivity of reovirus in distilled water by approximately 85%, whereas chymotrypsin complexed with montmorillonite did not affect infectivity. Virus infectivity was decreased by free lysozyme by approximately 99%, but by only 38% when lysozyme was complexed with montmorillonite. The lower inhibition of reovirus infectivity by lysozyme and chymotrypsin complexed with montmorillonite was probably the result of the binding of the proteins to the clay and the consequent alteration of the biophysical characteristics of proteins of the clay-protein-virus interface. On the other hand, electrostatic attraction between the net positively charged

Table 6
INFECTIVITY OF SOLIDS-ASSOCIATED VIRUSES

Virus type	Sample	Virus recovered	Amt. recovered (%)	Ref.
Columbia SK	Deionized water + $10^{-3}\,M$ CaCl$_2$ control	3.2×10^5		11
	Deionized water + $10^{-3}\,M$ CaCl$_2$ + clay	4.3×10^5	134	
	Filtered town lake control	3.0×10^5		
	Filtered town lake + clay	3.2×10^5	107	
	Filtered Wichita River control	6.4×10^5		
	Wichita River with natural solids	7.7×10^5	121	
	Filtered Red Fish Bay control	4.7×10^5		
	Red Fish Bay with natural solids	6.5×10^5	138	
	Filtered town lake control	8.0×10^5		
	Town lake with natural solids	7.0×10^5	87	
Poliovirus	Deionized water + 0.01 M CaCl$_2$ + bentonite + poliovirus 1 (0 time control)	2.15×10^2		15
	After 30 min adsorption of virus on solids	2.0×10^2	93	
Bacteriophage T7		1.08×10^2		
		9.3×10^1	82	
Bacteriophage T2		6.8×10^1		
		5.6×10^1	76	
Poliovirus 1	Estuarine water + natural suspended solids (0 time control)	3.3×10^7		4
	On day 19	56	—	
Rotavirus SA11	Estuarine water + natural suspended solids (0 times control)	9.4×10^7		
	On day 19	1.1×10^4	0.01	

proteins in distilled water in the absence of clay and the net negatively charged reovirus probably also prevented contact of the virus with cells in the monolayer, thereby decreasing the infectivity titer.

Ovalbumin enhanced the infectivity titer of reovirus by 150%, and this may have been related to the low isoelectric pH (4.6 to 4.7) of this protein. At neutral pH, both ovalbumin and reovirus have a net negative charge, and the electrostatic repulsive forces may have dispersed aggregates of the virus, resulting in higher infectivity of cells.

V. ENUMERATION OF SOLIDS-ASSOCIATED VIRUSES RECOVERED FROM FIELD SAMPLES

Although many investigators assume that a large proportion of viruses in natural waters will be associated with solids, the number of reports on quantitative recovery of solids-associated viruses from environmental samples is meager.

Wellings and associates[2] presented data to demonstrate the relatively high multiplicity of solids-associated virus in raw sewage and in treated effluents before and after chlorination. In an activated sludge package treatment plant, the virus content of solids varied from 23.4 to 80.8%; in the effluent, 90 to 100% of virus was solids-associated. Quantitative estimates of virus numbers recovered from solids in 500-mℓ samples of wastewater were 11 to 278 pfu from influent, 0 to 36 pfu from effluent, and 0 to 5 pfu from chlorinated effluent samples. Gerba et al.[3] quantitated solids-associated viruses in sewage effluents and reported that the

Table 7
**RECOVERY OF SOLIDS-ASSOCIATED VIRUS FROM FIELD
SAMPLES**

Nature of sample	Vol. tested (ℓ)	Quantity of virus recovered (pfu)	Ref.
Raw sewage	0.5	11—278	2
Effluent	0.5	0—36	
Chlorinated effluent	0.5	0—5	
Unchlorinated effluent	7.7	2400 (coliphage)	3
	1.0	0.1—22.6 (enterovirus)	
Raw sewage	1.0	95—510 (rotavirus)	30
Primary effluent	1.0	40—120 (rotavirus)	
Mixed liquor suspended solids	1.0	320—800 (rotavirus)	
Chlorinated effluent	1.0	0—20 (rotavirus)	
Estuarine water	378	4—40 (enterovirus)	4
	378	1800—4980 (rotavirus)	24

percentage of total coliphages recovered varied considerably, from <1% to 24%. Using 0.05 M glycine (pH, 11.5), a maximum of 2400 pfu of coliphage was recovered from 7.7 ℓ of sewage effluent from a trickling filter plant. Distribution of animal enteric viruses associated with solids in these effluents ranged from 3 to 49% of the freely suspended virus. The quantity of virus in several samples varied from 0.1 to 22.6 pfu/ℓ.

Rao and colleagues[30] recovered indigenous rotaviruses from suspended solids obtained from raw sewage, primary effluents, MLSS, and the final chlorinated effluent in different seasons (Table 7). Highest numbers (320 to 800) were detected in the MLSS, whereas numbers in raw sewage varied from 95 to 510. Even after chlorination (dose 10 mg/ℓ), the solids in the effluent contained up to 20 rotaviruses per liter.

The distribution and quantitation of enteroviruses and rotaviruses in water, suspended solids, and sediments have been studied in different seasons in Galveston Bay.[4] Virus was most often attached to suspended solids. Of these samples, 72% were positive for enteroviruses; the numbers recovered ranged from 4 to 40 pfu from solids separated from 378 ℓ. Rotaviruses were recovered in 50% of the suspended solids samples; numbers varied from 1800 to 4980 fluorescing foci (pfu) per 378 ℓ of water (Table 7).

VI. SIGNIFICANCE OF VIRUS-SOLIDS INTERACTION

Association of viruses with solids during sewage treatment is considered very useful because virus levels in the effluent are reduced. Although this is a useful property of viruses, in some ways it poses problems, in that the disposal of sludge can act as a significant source of virus contamination of the environment.

When drinking water sources are treated, viruses associated with suspended solids in raw water settle and the viruses in sludge are found in high concentrations. Thus, coagulation and sedimentation, while reducing virus levels in treated water, can produce undesirable levels of virus in sludge. Not only do the viruses survive for prolonged periods, but they also retain their infectivity. Disposal of these sludges is a potential hazard to the environment.

Virus association with suspended solids in natural waters affects the efficiency of virus recovery from waters. Techniques presently used to concentrate viruses from waste, estuarine, and potable waters require descending pore-sized filters to obtain maximum clarification of water in order to filter a large volume of the sample. In this process solids are selectively retained on the filters. When these filters are treated with media for *in situ* elution,[31] it is believed that virus adsorbed on the surface of solids may be eluted along with virus adsorbed on the filters, but the efficiency of such a recovery has not been determined. Besides, viruses embedded in solids retained on those filters are not expected to be released, and this fact was confirmed by Wellings and co-workers.[2]

Laboratory and field investigations have shown that the persistence of viruses in aquatic and terrestrial systems is enhanced when viruses are associated with naturally occurring inorganic particulates such as clay minerals and sediments.[4,6,7,9,11,12,17] Prolonged survival of viruses adsorbed on solids in a polluted estuary has several implications. Viruses associated with large-sized solids discharged into coastal waters leave the water column and settle as a loose, fluffy layer on sediments. De Flora et al.[32] and Gerba et al.[33] recognized sediments as potential reservoirs of enteric viruses that can be released into the water column as a result of agitation of the sediment by storms, dredging, and boating. Rao et al.[4] indicated that a large fraction of virus associated with small-sized suspended solids constantly floating in the water represents a newly recognized source of viral hazard in coastal waters. These findings offer a new perspective on the amount of virus that might be transported and permit an estimate of how far viruses may be transported in an infectious state. Quantitative estimates reveal that fluffy sediment and suspended solids-associated virus constitute approximately three fourths of enteroviruses recovered and that poliovirus and rotavirus SA11 survived for 19 days when adsorbed on these solids. These findings attest to the possible transportation of considerable numbers of viruses to nonpolluted areas of Galveston Bay, especially shellfish beds, depending on water circulation patterns and prevailing winds. Shellfish have been shown to accumulate more virus when it is present as a crude suspension rather than as purified virus, indicating that virus associated with particulate matter is taken up more efficiently by shellfish.[34] A recent study[35] on the accumulation of sediment-associated viruses in shellfish indicated that the influence of viral uptake was higher when sediments were resuspended in the water column and the resuspended virus remained solids-associated.

Virus association with suspended solids in natural waters is so extensive and significant that if virus monitoring of a water supply or an assessment of the virological quality of bathing beaches or shellfish harvesting waters is the goal and only free viruses are sought a false sense of security with important public health implications may be engendered.

VII. SUMMARY AND CONCLUSIONS

Viruses adsorb effectively on clays as well as on solids collected from natural waters in laboratory studies. Inorganic suspended solids allow greater adsorption of viruses when compared to volatile solids consisting of biological material. The cation content of a given water is very important to the adsorption process. Divalent cations on a molar basis provide greater virus adsorption than monovalent cations. Organic matter interferes with adsorption of virus on suspended solids by competing for adsorption sites.

Adsorption of viruses on suspended solids was proven to be reversible. Eluents capable of recovering viruses from solids include distilled water; estuarine water; buffers such as tris, glycine, and borate saline; and proteinaceous media such as beef extract and TPB.

Association with solids has no deleterious effect on a virus; on the other hand, the virus seems to be protected from natural inactivation in waters. Solids-virus association appears to be stable over a prolonged period, and only a change in the physicochemical characteristics would cause the virions to dissociate.

Solids-associated viruses retain their infectivity to cell cultures over a prolonged period of time and also to mice by the oral route. This implies that viruses adsorbed on solids, when transmitted by contaminated water, can infect humans by the oral route.

Viruses associated with solids, when introduced into coastal water, would assume the dispersal characteristics of solids and have ample opportunities of being transported from polluted waters to nonpolluted bathing beaches and shellfish beds, thereby causing a potential public health hazard.

Many of the methods used to concentrate and assay viruses in natural waters and wastewaters do not recover those viruses adsorbed on suspended solids. Viral monitoring of waters must include detection of solids-associated viruses.

REFERENCES

1. **Murray, J. P. and Parks, G. A.,** Poliovirus adsorption on oxide surfaces. Correspondence with the DLVO-Lifstitz theory of colloidal stability, in *Particulates in Water: Characterization, Fate, Effects and Removal* (Adv. Chem. Ser. 189), Kavanaugh, M. C. and Leckie, J. O., Eds., American Chemical Society, Washington, D.C., 1980, chap. 5.
2. **Wellings, F. M., Lewis, A. L., and Mountain, C. W.,** Demonstration of solids-associated virus in wastewater and sludge, *Appl. Environ. Microbiol.*, 31, 354, 1976.
3. **Gerba, C. P., Stagg, C. H., and Abadie, M. G.,** Characterization of sewage solid-associated viruses and behavior in natural waters, *Water Res.*, 12, 805, 1978.
4. **Rao, V. C., Seidel, K. M., Goyal, S. M., Metcalf, T. G., and Melnick, J. L.,** Isolation of enteroviruses from water, suspended solids, and sediments from Galveston Bay: survival of poliovirus and rotavirus adsorbed to sediments, *Appl. Environ. Microbiol.*, 48, 404, 1984.
5. **Bitton, G. and Mitchell, R.,** Effect of colloids on the survival of bacteriophages in seawater, *Water Res.*, 8, 227, 1974.
6. **Carlson, G. F., Woodward, F. E., Wentworth, D. F., and Sproul, O. J.,** Virus inactivation on clay particles in natural waters, *J. Water Pollut. Control Fed.*, 40, R89, 1968.
7. **Gerba, C. P. and Schaiberger, G. E.,** Effect of particulates on virus survival in seawater, *J. Water Pollut. Control Fed.*, 47, 93, 1975.
8. **Stagg, C. H., Wallis, C., and Ward, C. H.,** Inactivation of clay-associated bacteriophage MS-2 by chlorine, *Appl. Environ. Microbiol.*, 33, 385, 1977.
9. **Metcalf, T. G., Rao, V. C., and Melnick, J. L.,** Solid-associated viruses in a polluted estuary, *Monogr. Virol.*, 15, 97, 1984.
10. **Bitton, G.,** Adsorption of viruses onto surfaces in soil and water, *Water Res.*, 9, 473, 1975.
11. **Schaub, S. A. and Sagik, B. P.,** Association of enteroviruses with natural and artificially introduced solids in water and infectivity of solids-associated virions, *Appl. Microbiol.*, 30, 212, 1975.
12. **Schaub, S. A., Sorber, C. A., and Taylor, G. W.,** The association of enteric viruses with natural turbidity in the aquatic environment, in *Virus Survival in Water and Wastewater Systems*, Malina, J. F. and Sagik, B. P., Eds., Center for Research in Water Resources, University of Texas, Austin, 1974, 71.
13. **Hamblet, F. E., Hill, W. F., Akin, E. W., and Benton, W. H.,** Poliovirus uptake and elimination, oysters and human viruses: the effect of seawater turbidity, *Am. J. Epidemiol.*, 89, 562, 1969.
14. **Sproul, O. J.,** Adsorption of Viruses on Mineral Surfaces, Water Research Catalogue, U.S. Department of the Interior, Washington, D.C., 1969.
15. **Moore, B. E., Sagik, B. P., and Malina, J. F.,** Viral association with suspended solids, *Water Res.*, 9, 197, 1975.
16. **Dubois, S. M., Moore, B. E., Sorber, C. A., and Sagik, B. P.,** Viruses in soil systems, *Crit. Rev. Microbiol.*, 7, 245, 1980.
17. **LaBelle, R. L. and Gerba, C. P.,** Influence of pH, salinity, and organic matter on the adsorption of enteric viruses to estuarine sediments, *Appl. Environ. Microbiol.*, 38, 93, 1979.
18. **Lance, C. and Gerba, C. P.,** Poliovirus movement during high rate land filtration of sewage water, *J. Environ. Qual.*, 9, 31, 1980.
19. **Schaub, S. A., Bausum, H. T., and Taylor, G. W.,** Fate of virus in wastewater applied to slow infiltration of land treatment systems, *Appl. Environ. Microbiol.*, 44, 382, 1976.
20. **Drewry, W. A. and Eliassen, R. E.,** Virus movement in ground water, *J. Water Pollut. Control Fed.*, 40, 257, 1968.
21. **Lipson, S. M. and Stotzky, G.,** Effect of proteins on reovirus adsorption to clay minerals, *Appl. Environ. Microbiol.*, 48, 525, 1984.
22. **Lipson, S. M. and Stotzky, G.,** Adsorption of reovirus to clay minerals: effect of cation exchange capacity, cation saturation, and surface area, *Appl. Environ. Microbiol.*, 46, 673, 1983.
23. **McLaren, A. D., Peterson, G. H., and Barshad, I.,** The adsorption and reactions of enzymes and proteins on clay minerals. IV. Kaolinite and montmorillonite, *Soil Sci. Soc. Am. Proc.*, 22, 239, 1958.

24. **Rao, V. C., Metcalf, T. G., and Melnick, J. L.,** Development of a method for concentration of rotavirus and its application to recovery of rotaviruses from estaurine waters, *Appl. Environ. Microbiol.*, 52, 484, 1986.
25. **Committee Report, Singley, J. E.** Chairman, State of the art of coagulation mechanism and stoichiometry, *J. Am. Water Works Assoc.*, 63, 99, 1971.
26. **Manwaring, J. F., Chandhuri, M., and Engelbrecht, R. S.,** Removal of viruses by coagulation and flocculation, *J. Am. Water Works Assoc.*, 63, 298, 1970.
27. **Schaub, S. A. and Sorber, C. A.,** Viruses on solids in water, in *Viruses in Water*, Berg, G., Bodily, H., Lennette, E. H., Melnick, J. L., and Metcalf, T. G., Eds., American Public Health Association, Washington, D.C., 1976, 128.
28. **Rao, V. C., Metcalf, T. G., and Melnick, J. L.,** unpublished data.
29. **Moore, B. E., Funderburg, L., Sagik, B. P., and Malina, J. F.,** The application of viral concentration techniques to field sampling, in *Virus Survival in Water and Wastewater Systems*, Malina, J. F. and Sagik, B. P., Eds., Center for Research in Water Resources, University of Texas, Austin, 1974, 71.
30. **Rao, V. C., Metcalf, T. G., and Melnick, J. L.,** Removal of indigenous rotaviruses during primary settling and activated-sludge treatment of raw sewage, *Water Res.* 21, 171, 1987.
31. **Wallis, C., Homma, A., and Melnick, J. L.,** A portable virus concentrator for testing water in the field, *Water Res.*, 6, 1249, 1972.
32. **DeFlora, S., De Renzi, G. P., and Badolati, G.,** Detection of animal viruses in coastal seawater and sediments, *Appl. Microbiol.*, 30, 472, 1975.
33. **Gerba, C. P., Smith, E. M., Schaiberger, G. E., and Edmond, T. D.,** Field evaluation of methods for the detection of enteric viruses in marine sediments, in *Methodology for Biomass Determinations and Microbial Activities in Sediments*, Litchfield, C. D., and Seyfried, P. L., Eds., American Society for Testing and Materials, Philadelphia, 1979, 64.
34. **Hoff, J. C. and Becker, R. C.,** The accumulation and elimination of crude and clarified poliovirus suspension by shellfish, *Am. J. Epidemiol.*, 90, 53, 1969.
35. **Landry, E. F., Vaughn, J. M., Vicale, T. J., and Mann, R.,** Accumulation of sediment-associated viruses in shellfish, *Appl. Environ. Microbiol.*, 45, 238, 1983.
36. **Rao, V. C., et al.,** unpublished data.

Chapter 6

BIOACCUMULATION AND DISPOSITION OF SOLIDS-ASSOCIATED VIRUS BY SHELLFISH

T. G. Metcalf

TABLE OF CONTENTS

I. INTRODUCTION

The bioaccumulation and disposition of virus is only one aspect of the general phenomenon of shellfish uptake and disposition of particulates in their quest of food. It has special significance, however, because potential health hazards are associated with eating raw or inadequately cooked, virus-polluted shellfish. Biological and virological aspects of the uptake and disposition processes, and the influence of the solids-associated state upon these activities are discussed.

Shellfish selected for discussion are oysters, clams, and mussels, the bivalve forms most often involved in transmission of virus-caused diseases.

These shellfish are set apart from other mollusks by their use of a filter-feeding process to obtain needed food materials. Anatomic and physiologic differences, and ecological modes of existence affecting the likelihood of shellfish exposure to viruses are considered, and the relationship of these factors to shellfish virus transmission potential are discussed.

The potential health hazards of virus etiology associated with the consumption of shellfish are identified, and the effectiveness of purification methods as virus depletion procedures is discussed.

II. FILTER-FEEDING SHELLFISH

Filter-feeding shellfish are members of the Pelecypoda, a class within the phylum Mollusca characterized by possession of two external calcified valves hinged together by an elastic, noncalcified ligament. Beneath the calcified valves, or shells, a mantle structure encloses body parts, including a mantle cavity. In oysters and mussels the cavity is divided into inhalant and exhalant chambers, with semicircular extension of gill structures from the junction between right and left mantle folds to the mouthparts. Clams have fused mantle lobes with formation of siphon structures, divided into inhalant and exhalant tubules.

The mantle is composed of connective tissue which contains blood vessels, nerves, and muscles. Blood cells are found in mantle tissue and congregate on its outer surface. Apart from its principal role of helping in the formation of shell structures, the mantle secretes mucus, stores glycogen and lipids, and participates in the disposal of blood cells containing waste products. Cilia at the mantle border are ennervated and impart a sensory function to the mouth.

Filter feeding is initiated during the pumping of water through gill slits, or osta. Particulate matter is removed by mucus sheets, with transport to the mouthparts, where particulates accepted as food pass first into an esophagus and then into a stomach. Phagocytic cells originating in the hemolymph actively ingest small-sized particulates in contact with gill surfaces. The outside gill surface is covered by cilia tracts that provide propulsion of seawater, sort out particulates, or transport food particles to mouthparts. Particulates rejected as food sources are eliminated from the mantle cavity in the form of pseudofeces. Secretion of mucus sheets stops when feeding is interrupted, even though the gills continue to pump water, an indication that feeding activity is not always synonymous with pumping activity.

Arrangement of adductor muscles seems to be related to pumping potential. Oysters possessing one adductor muscle may have a pumping rate of up to 113 ℓ/hr/oz of tissue, while clams and mussels that have two muscles have pumping rates of only up to 19 ℓ.

Bivalve feeding is influenced by several factors. Water temperature ranges for feeding vary from 2 to 20°C for the softshell clam, *Mya arenaria*, to 10 to 20°C for the hardshell clam, *Mercenaria mercenaria*, to 10 to 25°C for the eastern oyster, *Crassostrea virginica*, and the mussel, *Mytilus edulis*. Water salinity, pH, turbidity, and dissolved oxygen are among the more important factors influencing feeding activities.

Shellfish feed on microscopic phyto- and zooplankton. Algae, protozoa, and bacterial species indigenous to shellfish waters are the usual ingredients of shellfish diets. The concept that shellfish serve a sort of vacuum cleaner function in estuarine waters needs to be qualified. Although stomach contents may reflect particulate composition of the water column in the immediate vicinity of feeding shellfish, a considerable degree of selectivity in particulates accepted as food is actually exercised. Uptake of particulates 4 μm or less in size is usually favored. Positively charged polyvalent ions are bioaccumulated, while monovalent or negatively charged ions tend to be rejected. A particular species of diatom or bacterium may be favored over other forms present. A preference for organic detritus over living microorganisms has been observed.[1] Outright rejection of undesirable plankton from an otherwise acceptable mixture of plankton has been shown repeatedly. The existence of chemotactic influences affecting which particulates are bioaccumulated has also been observed.[2,3]

The process of digestion and assimilation of food materials begins in the stomach. Oysters, mussels, and softshell clams have a crystalline style that protrudes into the stomach from a style sac. The style helps to mix and pulverize food particles. It contains glycogenases and amylases that are released and help to begin the digestive process. Undigested substances are directed into the intestinal tract, while dissolved and partially digested materials are swept by ciliary action over the surface of ducts leading to digestive diverticula. The diverticula represent the so-called digestion gland that is made up of structures completely surrounding the stomach. Due largely to its liver-like color and appearance, the digestive gland is frequently called hepatopancreatic tissue. The digestive diverticula are composed of a number of interlinked dead-end tubules from which excreted or nonabsorbed materials are removed by passage into the large ducts that connect with the stomach.

Shellfish digestion of materials accepted as food is largely a case of intracellular digestion carried out within phagocyte cells. Lining cells of the digestive diverticula tubules serve a phagocytic function, and wandering granulocytes from the blood migrate throughout the digestive diverticula, actively phagocytosing particulate material encountered. The digestive process is carried out within phagosomes and lysosomes with the help of a number of enzymes. The enzyme complement of shellfish varies among the different species, but proteases, lipases, lysozyme, acid and alkaline phosphatases, β-glucuronidase, and enzymes acting upon lactose, maltose, and other carbohydrates are found in several bivalve forms in addition to the style-associated glycogenases and amylases mentioned previously.[4]

Products of digestion needed for energy and growth requirements are stored largely in the form of glycogen and fats in diverticula tissues. Waste products and unutilized materials pass into the hindgut region of the intestinal tract. With the help of mucus secreted into the lumen of the gut, fecal boluses are formed and moved by cilial activity to the anal opening, where expulsion in the form of fecal ribbons follows. Materials passing from stomach into the hind- and midgut sections of the intestinal tract usually require about 1.5 to 3 hr on the average to pass through the gut and be released from the anus. Passage of waste materials through the intestine within an average period of time depends upon continuation of shellfish pumping and feeding activities over a period of hours. Interruption of feeding interrupts the rate of waste passage and increases the time required for ultimate excretion of intestinal contents.

Fecal ribbons contain both living and dead cellular material. Viruses, bacteria, algae, yeasts, and protozoa may pass through the intestinal tract without damage and appear in a culturable or active state. The presence of a living bacterium or other microorganism in fecal ribbons does not mean the shellfish is incapable of digesting it. It is simply an indication of the inefficiency of the shellfish digestive process.

The bivalves have circulatory systems and a colorless blood called hemolymph. The latter bathes various tissues and helps to carry the phagocytic granulocytes throughout body tissues. The amount of hemolymph present can be considerable; it may make up half the weight of bivalve tissues. Hemolymph-carried phagocytes may migrate onto gill surfaces, where they phagocytize particulate materials and carry them within a cell, or to mouthpart surfaces, where they are ingested and pass into the stomach.

Bivalves possess excretory systems for disposal of waste products. Nephridia are the prime structures that serve to remove nitrogenous waste products. Excretion is also carried out by pericardial glands, which are special areas of epithelial cells making up the pericardial wall. Hemolymph cells also participate in the disposal of wastes by transport of undissolved or undigested particulate matter into the intestinal tract. These same cells may use another route to dispose of waste products. They are known to migrate through the outer mantle wall, carrying particulate matter to the outside.

III. SHELLFISH HABITAT

The saltwater bivalve forms are estuarine in habitat. They are found in brackish waters defining the aquatic interface that exists between fresh- and saltwater domains and can withstand wide variations of salinity, temperature, and other environmental conditions. The location of estuaries places them at risk from both treated and untreated sewage that is discharged into coastal waters at a rate of billions of gallons daily. Because enteric viruses may survive secondary sewage treatment and chlorination, the number and quantity of viruses released into estuary waters may be large. An average of about 380 virus plaque-forming units (pfu) per gallon of sewage is estimated for the U.S. Estimates of virus pfu per gallon of sewage in other parts of the world range as high as 100 times the U.S. numbers.[5]

The bivalve forms have structures that have evolved from differing modes of existence. Differences in structures and functions among the four species representative of the bivalve forms most commonly involved in virus transmission can be illustrated by their ecological modes of existence. Clams burrow into sand or mud and draw upon the interface between bottom sediments and water for their food source. In the case of the softshell clam, siphon structures are long because these clams burrow deeply into sand or mud. The hardshell clam has a short siphon structure, since it is more shallowly located. Softshell clams are smaller in size and have a more fragile shell structure than hardshell clams. Oysters and mussels usually are found attached to rocks, reefs, and a variety of underwater structures. These forms will not adapt to soft mud estuary bottoms unless such bottoms contain rocks or similar hard objects, or are altered by introduction of solid structures to which these shellfish can attach. Oysters and mussels located on rocks, reefs, or piling structures draw upon water columns above the bottom sediment interface for their food source.

The location of a bivalve and its feeding source is important to an understanding of its potential for uptake of viruses. Mussels and oysters more often collect particulates suspended in water columns, while clams are more likely to take up particulates settled onto the uppermost fluffy layer of bottom sediments.[6,7] These particulate collection sources offer differing opportunities of exposure to viruses among the four bivalve forms.

Estuaries are traps that filter out fluvial sediments from rivers draining farmlands or receive waste treatment plant effluents or discharges from activities that cause substantial loss of solids such as mining. Virtually all fluvial sediments introduced into Long Island Sound and Chesapeake Bay on the Atlantic coast of the U.S. settle out in the upper regions of these estuaries.[8,9] At the same time inner continental shelf and shoreline sediments are introduced into the lower regions of the same estuaries.[9-11]

Sand, silt, and clay particulates are the most common sedimentary materials filtered out in estuaries. These are fine-grained materials which in their smallest size are ≤ 4 μm. Clays are the smallest and probably most significant potential virus-adsorbing particulates. The clay composition of sediments in U.S. estuaries varies according to geographic location. The differing composition can be illustrated by comparison of four selected clays. Illite is found in greater proportion in west coast sediments, while illite and chlorite make up the greater proportion of east coast sediments. Gulf coast sediments consist of greater proportions of kaolinite and montmorillonite.

The amount and composition of estuarine sediments is in part the direct result of a series of events involving fine-grained particulates like clays and silts, and suspension and deposit feeders. A biological processing in which zooplankton ingest particulates in suspension in a water column and egest the same particulates along with waste products in fecal pellets converts suspended particulate matter to sedimentary particulate matter. Deposit-feeding benthic forms cause a mixing and, in the case of mobile feeders, lateral translocation of sediment particulate matter. Bottom-feeding worms may cause translocation of sediment from lower compact layers to the uppermost fluffy sediment layer.

The interactions involving feeding bivalves and particulate matter in estuarine water columns are multifaceted, with much yet to be learned. As indicated previously, it is believed that oysters and mussels are more apt to ingest fine-grained particulates in suspension in a water column, leading to possible uptake of substances adsorbed to these particulates. Bacteria, viruses, dissolved or colloidally dispersed proteinaceous and carbohydrate substances, metals, and toxic chemicals may be ingested. It is also believed that clams are more apt to ingest particulate matter at the sediment-water column interface. The nature of the particulate materials ingested can be similar to that of materials bioaccumulated from water columns as a result of particulate biological processing or different if sediment particulate matter is more a result of physicochemical forces differing in type or degree from those associated with biological processing.

Since bivalves are filter feeders, they play an important role in estuarine sedimentation processes. It has been suggested that the Netherlands Walden Sea mussels are capable of an annual deposition of 150,000 t of sediment.[12] It has also been claimed that about 1.6 g of sediment per week can be deposited per small Chesapeake Bay oyster.[13]

IV. BIOACCUMULATION AND DISPOSITION OF VIRUSES

Bioaccumulation of viruses can occur in any of the filter-feeding species representative of the four bivalve forms described. The broad aspects of the bioaccumulation process and, to a certain extent, the fate of viruses taken up are known, but several details important to a fuller understanding of these phenomena remain to be determined.

Any of the enteric viruses excreted by man or animals that are introduced into estuary waters may be bioaccumulated. If we restrict our attention to only those viruses that may be excreted by humans, more than 120 possible candidates for bioaccumulation can be identified. Relatively speaking, only a handful of these viruses have been detected in shellfish. The reasons for failure to detect a broader spectrum of viruses can only be presumed, but it seems likely that detection methods probably discriminate in one way or another against many viruses.

Bioaccumulation of virus begins with the introduction of virus into shellfish-growing waters. Experimentally, a virus can either be taken up by a shellfish as a freely suspended particle or adsorbed to suspended solids. Under laboratory conditions the number of virus particles taken up is directly related to the dispersal state. The number of particles of monodispersed virus bioaccumulated is considerably less than the number of particles taken up when virus is in an aggregated or solid-associated form. Generally speaking, the greater the number of virus particles added to aquarium or tank waters, the greater the number of particles one can expect to be bioaccumulated. Eventually an upper limit to the number of virus particles bioaccumulated is reached and no further increase is obtained regardless of the number or form of aggregation of virus particles used.

The author has studied virus uptake by oysters and clams for a number of years. Several virus types and virtually every virus dispersal form possible to obtain have been used.

Experiments were performed for the most part in running seawater tanks, with salinity, temperature, and dissolved oxygen conditions optimized for each bivalve form being studied. The results of these studies are summarized as follows. No more than 23 pfu was taken up per shellfish following exposures of 2 to 24 hr to about 1000 pfu of freely suspended, monodispersed virus completely devoid of culture cells or debris. When monodispersed virus is adsorbed to cornstarch granules, about twice as many plaque-forming units are taken up under the same exposure conditions. Half again the number of cornstarch-associated plaque-forming units is taken up if a clay like kaolinite or bentonite replaces the cornstarch. The greatest bioaccumulation of virus is obtained — about a twofold increase over the uptakes with clay-associated virus — following the use of infant stools collected after administration of Sabin-type poliovirus vaccine.

Feces-associated progeny virus has resulted in greater numbers of virus taken up than with comparable numbers of stock virus incorporated into nonvaccinated infant stools. It perhaps should be pointed out that the actual number of virus plaque-forming units taken up per shellfish in the different experiments could have been somewhat greater since bioaccumulation capability was determined by recovery of plaque-forming virus, and the recovery method used was not 100% effective.

Before bioaccumulation of virus can begin, virus obviously must be in the immediate environment of a feeding shellfish. The definition of immediate environment has been somewhat vague, but presumably means within the limits of water intake currents established by shellfish pumping activity. Recent studies in which the author has used a specially designed modeling tank to determine how close solids-associated virus must be to oysters to ensure bioaccumulation have had somewhat surprising results. Under static hydrodynamic conditions, uptake of montmorillonite-associated hepatitis A virus occurred when clay-associated virus was 22 but not 29 cm distant from the outer shell margin of the inhalant area of the mantle cavity. Under flowing water conditions, addition of clay-associated virus to water almost 8 ft away from oysters resulted in eventual uptake of either solids-associated or freely suspended virus.

Studies in which three clay-associated virus preparations (illite, kaolinite, and montmorillonite) were used to determine virus transport potential indicated montmorillonite (Na-Ca form) adsorption could be expected to carry the virus over greater distances. Although more virus adsorbed to montmorillonite — and this could be of significance to the virus uptake potential of shellfish in waters containing this clay — the smaller particle size of the clay is the factor most likely to increase the probability of virus uptake through the greater transport potential represented as well as shellfish preferential screening of smaller particulate material as food sources.

The import of study results described for virus bioaccumulation phenomena is interpreted to indicate that the solids-associated state is crucial to virus uptake by shellfish. Where shellfish beds are distant to sources of pollution, it is plausible to assume that the solids-associated state plays a primary role in transport of virus from inshore to offshore shellfish bed areas. The solids-associated state undoubtedly is one of the important mechanisms governing sedimentation of virus in water columns. It appears that the combination of these solids-associated effects is a satisfactory explanation for the introduction of virus into the immediate environment of shellfish, whether in currents passing shellfish located to take advantage of suspended particulates in a water column or shellfish located so as to feed at a bottom sediment-water interface.

The adsorption of virus to finely particulate clays and especially the feces-associated virus state are probably responsible not just for qualitative aspects of virus presentation to shellfish but, possibly more importantly, for presentation of large numbers of virus due to the presence of virus aggregates.

Shellfish acceptance and processing of virus-associated particulates begins with envelopment in mucus sheets. Studies in which freely suspended virus was used showed that this initial contact of virus with mucus was not a nonspecific envelopment process, but the result of ionic bonding of virus to sulfate radicals of mucopolysaccharide.[14] Other studies have shown that oysters possess large but finite numbers of virus adsorption sites and that nonenveloped viruses are bioaccumulated more rapidly than lipoprotein-enveloped viruses.[15] One has to infer that solids-associated virus is processed as described for freely suspended virus. It is not known whether virus is separated from solids during mucus sheet binding and processed separately, or whether virus and particulates are processed together as a unit. Whichever situation may exist, it is clear that virus gains access to the stomach, where it is exposed to the initiation of shellfish digestive processes. The proportion of virus bound in mucus sheets that is cast off in the pseudofeces to virus passing into the stomach is estimated to be small from calculations based upon (1) virus presented to shellfish, (2) virus bioaccumulated, and (3) virus remaining in seawater.

The fate of bioaccumulated virus during the process of shellfish digestion and assimilation of food is not fully understood. Some facts are known, but much remains to be determined. Beginning with the extracellular phase of digestion in the stomach, secretion of enzymes, chiefly glycogenases and amylases, appears to have minimal effect upon virus, either because this phase is inefficient or because virus remains solids-associated and protected from enzyme action. Recovery of up to 90% of fully active virus in shellfish fecal ribbons is indicative of the almost complete lack of effect upon virus of extracellular digestion carried out in the stomach.

Some bioaccumulated virus is either diverted to the digestive diverticula tissue or transported from the stomach to this tissue by actively phagocytic granulocytes from the hemolymph. Phagocytosis originally was demonstrated experimentally by oyster granulocyte envelopment of a 60-nm virus less than 2 hr after exposure of virus to the granulocytes. The deciding factor that determines whether virus is transported to the digestive diverticula or remains within the intestinal tract may be whether it remains solids-associated or is separated from solids. Virus deposited within tubules found throughout the diverticula tissue may also gain entry to diverticula tissue cells via phagocytosis by cells lining the blind end parts of these tubules.

Intracellular digestion of food by shellfish is carried out with the help of enzymes that break down food materials. Since diverticula tissue cells are a primary site of digestion and assimilation of food, introduction of virus into these cells presumably exposes them to enzyme activity. It would seem that virus capsid structures exposed to a number of active enzymes like proteases, lipases, lysozyme, and others would be degraded and virion integrity lost. The net result would be destruction of virus.

A number of observations contradict the premise that virus within diverticula cells will be destroyed. Survival of virus within diverticula tissues of active shellfish for several days has been found repeatedly. It is possible that virion capsid structures are resistant to enzymatic degradation because they lack either vulnerable structures or some other key factor necessary for enzyme activity. Virus may also be protected from enzyme action in some unrecognized way. Whether a solids-associated state is involved at this point is unknown. This is an area of shellfish virology that needs more attention if we are to fully understand virus survival capability.

The distribution of bioaccumulated virus in bivalves has been studied by analyses of fluids and tissue parts following careful dissections and separations. The greater amount of virus is found in stomach-intestine and diverticula tissue samples followed by mantle fluid, mouth-esophagus, and gill samples, respectively. The distribution pattern and virus quantities found reflect the passage of virus into mantle fluid followed by its appearance in the mouth, esophagus, and stomach, with concentration in stomach and digestive gland samples. Under experimental conditions in the laboratory where virus numbers added to seawater are adequate for making accurate assays, concentration of virus within shellfish has been shown, with numbers detected exceeding those found in the seawater. Concentrations of virus within shellfish 50-fold or greater are not uncommon.

Elimination of bioaccumulated virus is closely related to the degree of physiologic vigor shown by a bivalve. Wherever virus may be distributed within a shellfish following bioaccumulation, maintenance of pumping and feeding activities over the course of several days will usually lead to elimination or reduction of virus numbers below detectable levels. The author showed conclusively how closely survival and elimination of bioaccumulated virus is related to shellfish physiologic state. In experiments conducted in a North Atlantic estuary during winter months, oysters with bioaccumulated virus were submerged in water at 1°C. Oysters recovered and assayed for virus over a 4-month period of time showed virtually no loss of virus as long as water temperature remained below 4.3°C. When water temperatures began to rise during spring, bioaccumulated virus began to disappear from the oysters as the physiologic shutdown of "hibernation" months eased. No virus could be detected at about $4^1/_2$ months, at which time water temperatures rose above 7°C and the oysters were observed to be actively pumping water and feeding. Long-term carriage of virus in virtually undiminished numbers during the winter months, with subsequent release upon termination of the carriage state in the springtime, represents "overwintering" within shellfish, a means of prolonging the survival of infectious virus.

Bioaccumulation of virus is not followed by virus replication within shellfish. Whatever the number of bioaccumulated viruses may be, each virion detected is taken up from the surrounding seawater and does not originate within a shellfish. The term *virus carriage* should be used to define bioaccumulated virus; the *passive carriage state* accurately describes the relationship that exists.

Elimination of virus for the most part seems to be via the intestinal tract, with virus firmly enclosed within a fecal bolus consisting of waste products, undigested materials, and mucus. As indicated previously, most studies concerned with the elimination of virus in the feces have found most of the virus to be unaffected by shellfish digestive processes; i.e., virus fully infectious for cell cultures can be recovered.

V. PUBLIC HEALTH ASPECTS OF VIRUS CARRIAGE

Transmission of bacterial pathogens by shellfish was known to occur before the turn of the 20th century. Typhoid fever bacteria were shown to be bioaccumulated by oysters, confirming the etiologic role of shellfish in outbreaks of typhoid fever among oyster-eating groups.[16] Uptake and concentration of bacteria by oysters and clams was given quantitative definition by Kelly and Arcisz, who showed concentrations in shellfish that were greater than those existing in the surrounding seawater.[17,18]

Recognition of an outbreak of 30 cases of hepatitis attributed to shellfish was made in the U.S. in 1953, and a similar but much larger outbreak occurred in Sweden in 1955. Both outbreaks were the result of ingestion of raw oysters containing type A hepatitis virus. In the case of the Swedish outbreak, details of the virus pollution of oysters and subsequent virus transmission events were determined by epidemiologic studies.[19] Since then a number of outbreaks of shellfish-transmitted hepatitis have been recognized by epidemiologic studies. Reliance upon epidemiologic methods for recognition of an outbreak has been necessary up to now because no means of recovering or detecting hepatitis A virus in shellfish has been available. Although epidemiological studies have been most helpful in delineating the source and course of transmission in some outbreaks of hepatitis, limitations of epidemiologic methods involving disease recognition and reporting are known to exist. Those limitations have been cited as reasons for the belief that a number of outbreaks of shellfish-transmitted illness may have gone unrecognized.[20] Demonstration that type A hepatitis would be transmitted by ingestion of hepatitis A virus-containing shellfish was responsible for the original impression that shellfish carriage of almost any enteric virus was an automatic indication of the existence of a possible health hazard. This opinion has had to be revised as a result of the accumulated findings of three decades of public health research on shellfish-transmitted disease.

A number of the routinely culturable enteroviruses, some reoviruses, and type B hepatitis virus antigen have been recovered from shellfish. Enteroviruses recovered have included all 3 poliovirus types, 2 coxsackie A types, 3 coxsackie B types, and 11 echovirus types — about 30% of the known enteroviruses. With the exception of hepatitis A virus, no enterovirus, reovirus, or type B hepatitis virus has been responsible for shellfish-transmitted illness, although all, with the exception of reoviruses, cause identifiable disease in humans when transmitted by other routes.

The only viruses capable of causing shellfish-transmitted illness in addition to type A hepatitis virus are the nonculturable gastroenteritis viruses. Chief among this group is the Norwalk virus, which has been implicated as the cause of one of the largest outbreaks ever recorded.[21] More than 7000 cases of acute gastroenteritis were traced to the consumption of Norwalk virus-polluted shellfish in Australia, and an opinion voiced by Australian public health officials indicated that the actual number of cases undoubtedly was considerably greater. Other gastroenteritis viruses, including rotavirus and some unidentifiable viruses, are suspected of being shellfish-transmitted pathogens. There have been more than 1800 cases of acute nonbacterial gastroenteritis associated with the consumption of shellfish for which no causative agent has been identified.

The virus carriage state in shellfish was found by the author to be of potential public health significance during the course of an investigation of an experimental aquaculture system proposed for the production of shellfish on a commercial scale.[22] The system under discussion was developed with coastal communities in mind. It was to serve both biological tertiary sewage treatment and shellfish production purposes. The sewage treatment aspect involved removal of nutrients and pathogenic microorganisms from municipal wastewater treatment plant effluents prior to their discharge into marine waters. Shellfish production was but one objective of a polyculture system that contemplated secondary production of polychaete worms (for sale as fish bait), fin fish, lobsters and abalone (for human consumption), and carrageenan-producing seaweeds (carrageenan is a source of polysaccharides of commercial value).[23] The system began with cultivation of unicellular marine algae, utilizing the nutrients present in sewage effluents. The algae were released in raceways containing seed oysters and clams, and the shellfish fed upon the algae. Shellfish feces deposits were fed upon by polychaete worms and other small invertebrates in the seawater. The worms and invertebrates served as food for fin fish and lobsters. Soluble wastes from shellfish and sewage sources were assimilated by seaweeds which could be used as a source of carrageenan or food for browsing abalone.

The system was shown to provide opportunities for transmission of virus within a hypothetical food chain. Viruses present in sewage effluents passed into shellfish raceways, adsorbed to either algae or sewage particulates, and were shown to be bioaccumulated by the shellfish. The inability of the shellfish to inactivate virus resulted in excretion of infectious virus in the shellfish feces. Polychaete worms feeding upon these feces were shown to pick up infectious virus which was later recovered from flounder that fed upon the worms. Virus was recovered from the alimentary tract and from visceral and body flesh tissues of flounder 48 hr post-feeding time.

The sequential virus transmission process through invertebrates to a vertebrate host used as a food source by man was of interest because it represented a natural transmission process. No significant public health threat was considered to exist because the virus carriage state in each respective host was passive and involved only small numbers of virus. The only potential health hazard that might exist would be virus-carrying shellfish. Introduction of such shellfish into the marketplace without institution of purification procedures could have hazardous consequences.

With the exception of hepatitis A virus, the failure to demonstrate shellfish-transmitted enterovirus illnesses has been difficult to understand. The oral-anal cycle of enterovirus transmission involving food, for example, is one of the common routes of infection. The essential differences, important to infection, that exist between ingestion of enterovirus-polluted shellfish and introduction into the mouth of an enterovirus-contaminated finger, toy, or food has been difficult to answer satisfactorily. What contributes to the unique ability of type A hepatitis virus, among all of the other enteroviruses, to cause water- and shellfish-transmitted illness? Do type A hepatitis virus and the gastroenteritis-causing viruses have unique portals of entry or special receptor sites located within the gastrointestinal tract, or do they possess some special virulence factors? Are the differences in shellfish transmissibility due to the number of infectious virions ingested? Studies directed toward the solution of these questions are needed.

VI. VIRUS DEPLETION

The health hazard potential posed for those ingesting virus-polluted shellfish in either raw or inadequately cooked form has led to virus depletion procedures calculated to eliminate any existing virus carriage state. Two procedures for purification of shellfish may be used. One is called relaying and, as its name indicates, involves moving of shellfish from waters not classified as approved to waters that are approved for natural purification. The other procedure is a process of controlled purification called depuration. Relaying may involve moving shellfish a short distance in an estuary, or it could involve transport from waters of one state to approved waters in another state. Depuration involves removal of shellfish to a plant where they are submerged in flowing seawater free from pollutants and allowed to purge themselves of bioaccumulated bacterial or viral pathogens. Relaying is usually carried out for 14 days, while depuration usually is carried out for 2 to 3 days.

A degree of control exercised over depuration cannot be similarly exercised over relaying. In depuration, important parameters controlled for which the degree of variance is known include pathogen-free water, turbidity, salinity, dissolved oxygen, and temperature. In relaying, unnoticed events may occur which may either result in the uptake of pathogens in the case of the introduction of fecal pollution or reduce the effectiveness of the natural purification activity of shellfish. Although depuration in a plant has the advantage of greater control over shellfish self-purification, the 2- or 3-day depuration period requires more rigid monitoring, and the costs involved in plant operation and maintenance usually exceed relay costs.

Virus depletion effectiveness is evaluated by the results of bacterial indicator tests. Use of bacterial-indicated sanitary quality for evaluating the effectiveness of virus depletion may be anathema to a virologist, but it is regarded pragmatically as the only useful biological parameter by virtually everyone else. There is general consensus among elements of the shellfish industry that relaying is an effective method of eliminating fecal coliform bacteria from at least oysters. As far as bacteria are concerned, this may be true; unfortunately, since bacterial-indicated microbiological quality has been shown repeatedly to be an unreliable monitoring criterion for predicting the presence of virus, the possible presence of virus cannot be excluded. A similar evaluation has been made for the fecal coliform removal effectiveness of controlled depuration with shellfish, and again it is necessary to point out that bacterial indicator removal effectiveness is not a reliable indication of the virus depletion effectiveness of controlled depuration.

Shellfish sanitarians and public health authorities recognize that health hazard risks are inevitably associated with the eating of raw shellfish. Their policy has been and remains one that features reduction of health hazards to the lowest minimum possible. The less than 100% guarantee of virus elimination in either relaying or controlled depuration procedures is accepted in return for the ability of these processes to reduce potential health hazards to the lowest possible level.

Both relaying and controlled depuration procedures have effectively reduced the incidence of shellfish-transmitted disease of viral etiology since their introduction. The commercial mussel purification plant built at Conway, North Wales in 1928 was the forerunner of depuration plants constructed in the U.S. In addition to the mussel, *M. edulis*, the British have successfully applied controlled depuration to the purification of three species of oysters, *C. gigas* (Japanese oyster), *Ostrea edulis* (European oyster), and *C. angulata* (Portugese oyster); and the hardshell clam, *M. mercenaria*. In the U.S. most plants have been constructed for purification of the softshell clam, *M. arenaria*.

Commercial depuration plants usually are constructed for the purification of a particular type of shellfish. This is because each species has its own limiting conditions or parameters most suitable for effective purification. These include water temperature, salinity, and dissolved oxygen; the number of shellfish it is possible to place in a tank and the water volume per unit mass of shellfish; depth of the water over shellfish, etc.

The author conducted a 3-year study of the effectiveness of commercial depuration plant practices for reduction of bioaccumulated feces-associated enteroviruses in oysters, and hard- and softshell clams. Depuration of virus-carrying shellfish was carried out in one tank set aside for the study at the same time that shellfish in other tanks were being depurated prior to entry into the marketplace. As far as is known, it was the first study in which virus-carrying shellfish were depurated side by side at the same time and under the same conditions as shellfish to be sold in the marketplace. The plant was designed and operated for the depuration of softshell clams. It was of single-pass design, i.e., seawater entering the plant flowed through depuration tanks once and was then returned to adjacent estuary waters. Plant water was UV-light sterilized. The study featured virus assays of individual shellfish, and the design of the experiments as well as the interpretation of data and expression of depuration effectiveness was based upon statistical methods.

Study results showed effective depletion of virus in each of three species. Virus reduction ratio assessments of depuration effectiveness varied from 87 to 100%. Although plant design and operating conditions addressed softshell clam depuration, oysters and hardshell clams were depurated with equal effectiveness within 48 hr. Experiments in which control (non-virus-carrying) shellfish were depurated side by side in the same basket with virus-carrying shellfish did not result in transmission of virus from test to control shellfish even once during the 3-year study. Controls also included shellfish carrying both bioaccumulated virus and *Escherichia coli*. Analyses showed differing rates of reduction of *E. coli* and virus to nondetectable levels, with detectable virus remaining upon several occasions in the absence of detectable *E. coli*.

Expression of the health risks expected for each species of depurated shellfish was made on a basis of the effectiveness of virus depletion and arbitrary assumptions that a human infective dose might consist of one, five, or ten infectious virions. This statistically based expression suggested that if ten virions were required for infection, no likely health risks would be encountered 95% of the time (i.e., at risk 5% of the time) with either oysters or hard- or softshell clams. If five virions were needed for infection, oysters would not be a health risk 95% of the time, but both clam species might be a health risk. Put another way, the lowest health risk would be associated with oysters, and hardshell clams would have a lower risk than softshell clams.

VII. SUMMARY AND CONCLUSIONS

Virus pollution of shellfish frequently, if not always, is the result of the association of virus with solids. Bioaccumulation of virus is associated with the uptake of particulates in the quest for food. Type A hepatitis and gastroenteritis illnesses are associated with the consumption of raw or inadequately cooked virus-polluted oysters, clams, and mussels. On a worldwide basis a number of species of these shellfish forms are involved in the transmission of virus with the likelihood that the actual incidence of shellfish-transmitted disease is much greater than the observed incidence.

The solids-associated state is believed to be the vehicle responsible for virus survival and transport within an estuary and the means by which virus is brought to either water column or bottom sediment-water interface-feeding shellfish. Clays already present or part of fluvial sediments introduced into an estuary are among the more important inorganic substances with which virus associates. Fecal pellets from estuarine forms of life represent organic particulates with which virus may interact at all levels of a water column. The presence of inorganic and organic particulate matter offers opportunities not only for a variety of inter-actions with virus but also may dictate whether virus uptake by suspension- or bottom-feeding shellfish is favored.

Persistence of virus in shellfish and the effectiveness of depletion procedures is believed to depend upon factors that contribute to optimization of pumping and feeding functions plus location of virus shellfish. Virus carriage with subsequent release of infectious virus provides opportunities for transmission of virus to a number of potential life forms that live in the shellfish habitat. A carrier state may persist for months if the shellfish are dormant, representing a virus reservoir in nature from which infectious virus can be released weeks or months later. Actively pumping and feeding shellfish tend to eliminate alimentary tract-located virus within 48 to 72 hr. More time may be required for elimination of tissue-sequestered virus. Relaying and controlled purification are effective bacterial or viral de-pletion procedures which reduce potential health risks but cannot guarantee pathogen-free shellfish.

REFERENCES

1. **Savage, R. C.,** The food of the oyster, *Investigations Ser. IV*, 8(1), 1925.
2. **Loosanoff, V. L.,** On the food selectivity of oysters, *Science*, 110, 122, 1949.
3. **Cheng, T. C. and Rudo, B. C.,** Chemotactic attractions of *Crassostrea virginica* hemolymph cells to *Staphylococcus lactus*, *J. Invertebr. Pathol.*, 27, 137, 1976.
4. **Cheng, T. C. and Roderick, G. E.,** Lysosomal and other enzymes in the hemolymph of *Crassostrea virginica* and *Mercenaria mercenaria*, *Comp. Biochem. Physiol.*, 52, 443, 1975.
5. **Melnick, J. L. and Metcalf, T. G.,** Distribution of viruses in the water environment, in *Banbury Report 22: Genetically Altered Viruses and the Environment*, Cold Spring Harbor Laboratory, Cold Spring Harbor, N.Y., 1985, 95.
6. **Metcalf, T. G.,** Indicators of viruses in shellfish, in *Indicators of Viruses in Water and Food*, Berg, G., Ed., Ann Arbor Science, Ann Arbor, Mich., 1978, 383.
7. **Landry, E. F., Vaughn, J. M., Vicale, T. J., and Mann, R.,** Accumulation of sediment-associated viruses in shellfish, *Appl. Environ. Microbiol.*, 45, 238, 1983.
8. **Bokuniewicz, H. J., Gebert, J., and Gordon, R. B.,** Sediment mass balance in a large estuary, *Estuarine Coastal Mar. Sci.*, 4, 523, 1976.
9. **Schubel, J. R. and Carter, H. H.,** Suspended sediment budget for Chesapeake Bay, in *Estuarine Processes*, Vol. 2, *Circulation, Sediments and Transfer of Material in the Estuary*, Wiley, M., Ed., Academic Press, New York, 1977, 48.
10. **Meade, R. H.,** Landward transport of bottom sediments in estuaries of the Atlantic coastal plain, *J. Sediment. Petrol.*, 39, 222, 1969.
11. **Hathaway, J. C.,** Regional clay minerals facies in the estuaries and continental margin of the United States east coast, in *Environmental Framework of Coastal Plain Estuaries*, Nelson, B. W., Ed., U.S. Geological Survey, Boulder, 1972, 293.

12. **Verwey, J.,** On the ecology of distribution of cockle and mussel in the Dutch Walden Sea: their role in sedimentation and the source of their food supply, with a short review of the feeding behavior of the bivalve mollusks, *Arch. Neerl. Zool.*, 10, 171, 1952.

13. **Haven, D. S. and Moraleo-Alamo, R.,** Aspects of biode position by oysters and other invertebrate filter-feeders, *Limnol. Oceanogr.*, 11, 487, 1966.

14. **DiGirolamo, R., Liston, J., and Matches, J. R.,** Ionic bonding, the mechanism of viral uptake by shellfish mucus, *Appl. Environ. Microbiol.*, 33, 19, 1977.

15. **Bedford, A. J., Williams, G., and Bellamy, A. R.,** Virus accumulation by the rock oysters, *Crassostrea glomerata*, *Appl. Environ. Microbiol.*, 35, 1012, 1978.

16. **Foote, C. J.,** A bacteriological study of oysters, with a special reference to them as a cause of typhoid infection, *Med. News*, 66, 320, 1895.

17. **Kelly, C. B. and Arcisz, W.,** Survival of enteric organisms in shellfish, *Public Health Rep.*, 69, 1205, 1954.

18. **Arcisz, W. and Kelly, C. B.,** Self purification of the soft shell clam, *Mya arenaria*, *Public Health Rep.*, 70, 605, 1955.

19. **Roos, R.,** Hepatitis epidemic conveyed by oysters, *Svenska Lakartidningen*, 53, 989, 1956.

20. **Goldfield, M.,** Epidemiological indicators for transmission of viruses by water, in *Viruses in Water*, Berg, G., Bodily, H. L., Lennette, E. H., Melnick, J. L., and Metcalf, T. G., Eds., American Public Health Association, Washington, D.C., 1976, 70.

21. **Murphy, A. M., Grohmann, G. S., Christopher, P. J., Lopes, W. A., Davey, G. R., and Millsom, R. H.,** An Australia-wide outbreak of gastroenteritis from oysters caused by Norwalk virus, *Med. J. Aust.*, 2, 329, 1979.

22. **Metcalf, T. G., Comeau, R., Mooney, R., and Ryther, J. H.,** Opportunities for virus transport within aquatic and terrestrial environments, in *Risk Assessment and Health Effects of Land Application of Municipal Wastewater and Sludges*, Sagik, B. P. and Sorber, C. A., Eds., Center for Applied Research and Technology, University of Texas, San Antonio, 1978, 77.

23. **Ryther, J. H., Dunstan, W. M., Tenore, K. R., and Huguenin, J. E.,** Controlled entrophication, increasing food production from the sea by recycling human wastes, *BioScience*, 22, 144, 1972.

Chapter 7

METHODS FOR CONCENTRATION AND RECOVERY OF VIRUSES FROM WASTEWATER SLUDGES

Samuel R. Farrah

TABLE OF CONTENTS

I. ASSOCIATION OF VIRUSES WITH WASTEWATER SLUDGES

Early studies demonstrated that poliovirus and other enteric viruses were removed from wastewater during activated sludge treatment.[1-5] Subsequent work has shown that much of this observed removal can be attributed to adsorption of the viruses by the sludge flocs.[6-8] Following adsorption to the sludge flocs, inactivation of viruses may occur during aerobic treatment of the sludge.[8,9] However, not all viruses are inactivated by aerobic treatment, and a large number of viruses may be present in the sludge removed from activated sludge units.[10-14] Further treatment is required to stabilize the sludge and to reduce the number of pathogenic organisms. Two common means of treating waste-activated sludge are anaerobic and aerobic digestion. Both anaerobic and aerobic digestion are considered processes that significantly remove pathogens (PSRP).[15] Digestion carried out under mesophilic temperatures has been found to reduce but not eliminate viruses and other pathogens.[9-12,16-19] Some of the viruses present in the digested sludge may represent contamination of the digested sludge with fresh undigested sludge.[9,11] Short-circuiting or removal of digested sludge from digesters during or after addition of fresh sludge may contribute to the viral load of the digested sludge. Thermophilic digestion under anaerobic or aerobic conditions has been found to result in greater inactivation of viruses than digestion at mesophilic temperatures.[10,11,18,19]

Since aerobic treatment and subsequent digestion may not suffice to produce a final sludge that is free of viruses, procedures have been developed to detect viruses in sludge solids. This has made it possible to determine the fate of the sludge-associated viruses during treatment processes and following disposal of sludge on land. Sludge-associated viruses may retain their infectivity[22,23] and can be assayed by direct inoculation of cell cultures with ether-treated sludge samples.[12,13] Toxicity of the sludge samples to cell cultures and the generally low numbers of viruses in sludge limit the use of this method. Therefore, procedures for recovering viruses associated with sludge flocs have been developed. These procedures have been reviewed previously.[20,21]

II. RECOVERY OF VIRUSES ASSOCIATED WITH WASTEWATER SLUDGES

A. Initial Treatment of Sludges

In order to conveniently detect viruses in relatively large volumes of liquid sludge, several investigators have started their recovery procedures with a centrifugation step to separate the sludge solids from the supernatant. The supernatant fraction may be discarded[14,24,25] or assayed for viruses separately from the solids.[26,27] Although most of the viruses have been found associated with the solids following centrifugation at ambient pH, several investigators have taken steps to ensure that the viruses in the sludge are associated with the flocs before centrifugation. These steps include adjusting the sludge to pH 3.5 with and without the addition of 0.0005 M aluminum chloride or 0.05 M magnesium chloride[28,29] and the use of a cationic polyelectrolyte[30] to promote flocculation (Table 1). Separating the solids by centrifugation is an effective concentration step and decreases the volume of sample to be processed.

B. Solutions Used to Elute Viruses Associated with Sludge Flocs

The association of viruses with sludge flocs is generally stable at ambient pH values and in the absence of exogenous organic compounds. Therefore, most procedures for recovering viruses from sludge flocs have used a solution of proteinaceous material at neutrality or high pH or a defined solution at high pH to elute viruses adsorbed to the sludge flocs. Some of the solutions used are listed in Table 2. Solutions containing proteins or peptides such as

Table 1
INITIAL TREATMENTS USED TO REDUCE SLUDGE VOLUME

Treatment	Ref.
Centrifugation at ambient pH (supernatant discarded)	14,24,25
Centrifugation at ambient pH (supernatant assayed for viruses independently of sludge flocs)	26,27
Adjustment to pH 3.5, centrifugation[a]	28
Adjustment to pH 3.5, addition of either 0.005 M aluminum chloride or 0.05 M magnesium chloride, and centrifugation[a]	29
Precipitation with soluble polyelectrolyte (Zetag)	30

[a] Since most viruses adsorb to sludge flocs under these conditions, the supernatant fraction is discarded and only the flocs are examined for the presence of viruses.

Table 2
SOLUTIONS USED TO ELUTE VIRUSES ADSORBED TO SLUDGE FLOCS

Solution	Sludge type	Indigenous (I) or seeded (S) viruses recovered	Ref.
10% beef extract	Anaerobic	I	13,30
3% beef extract	Anaerobic	I	24,31,32,33
10% beef extract + 1.34% $Na_2HPO_47H_2O$ + 0.12% citric acid	Primary Activated Anaerobic	I	10,29
1% skim milk	Anaerobic	I	11,31,36
Tryptose phosphate broth	Mixed liquor suspended solids	S	37
5% FCS + 3% beef extract + gelatin	Anaerobic	S	17
0.05 M glycine (pH, 11.5)	Aerobic Anaerobic	I, S	25,26
Freon	Raw	I	38,39
4 M urea + 0.1 M lysine (pH, 9)	Aerobic	I, S	27
1 M sodium trichloroacetate	Aerobic Anaerobic	I, S	28
Base sufficient to produce pH 9.0	Primary Activated	I	12

beef extract,[10,13,24,29-33] calf serum,[17,34,35] skim milk,[11,31,36] or tryptose phosphate broth[37] have been used by several investigators to separate viruses from sludge flocs. Viruses have also been eluted from sludge flocs by using defined solutions containing glycine,[25,26] freon,[38,39] urea,[27] or sodium trichloroacetate,[28] and by simply adjusting the pH of the sludge to 9.0.[12]

C. Physical Treatments Used to Treat Sludge Flocs

Since viruses may be embedded in the sludge flocs as well as adsorbed to the surface, mechanical procedures to disrupt the flocs in the presence of an eluting solution have been used to maximize virus recovery from sludge. Some of the procedures, including sonication,[17,24,33] shaking,[11-14,31,37,40] stirring,[10,25-28,30,33] and blending,[34,38,39] are listed in Table 3.

Following mechanical treatment of the sludge flocs in the presence of a suitable eluting solution, the solids are usually removed by centrifugation. The remaining supernatant containing the eluted viruses can be neutralized (if necessary) and assayed directly if the number is sufficiently high. It may also be necessary to remove bacterial contaminants using filtration,[29] chloroform,[24] or ether treatment.[13]

Table 3
PHYSICAL TREATMENTS USED TO RELEASE VIRUSES FROM SLUDGE FLOCS

Physical treatment	Ref.
Shaking	11—14,31, 37, 40
Blending	34,38,39
Sonication	17,24,33
Stirring	10,25—28,30,33

Table 4
METHODS USED FOR CONCENTRATING VIRUSES ELUTED FROM SLUDGE FLOCS

Conc. procedure	Ref.
Organic flocculation of beef extract	10,28,29,38
Organic flocculation of skim milk	11,31
Organic flocculation of organic compounds eluted from sludge flocs	25,26
Inorganic flocculation	27,28
Adsorption to membrane filters	25,26
Ultracentrifugation	26,38
Hydroextraction	30,33
Two-phase concentration	12
Lyphogel	37

D. Concentration of Viruses Eluted from Sludge Flocs (Table 4)

Samples that contain too few viruses for detection must be further concentrated. Several methods of concentrating eluted viruses have been used by different investigators. These include organic flocculation of beef extract,[10,28,29,38] skim milk,[11,31] or the organic compounds eluted from sludge flocs at high pH.[25,26] In addition to organic flocculation, hydroextraction,[30,33] inorganic flocculation,[27,28] adsorption to membrane filters,[25,26] ultracentrifugation,[26,38] two-phase separation,[12] and concentration with lyphogel[37] have all been used to further concentrate viruses eluted from sludge flocs. In some cases, two concentration steps have been used in sequence to achieve small final volumes of sample. Viruses eluted by a defined solution such as sodium trichloroacetate may be concentrated first by inorganic flocculation and then by organic flocculation.[28]

E. Procedures for Recovering Viruses from Wastewater Sludges

Different combinations of methods of mechanically treating sludge, eluting solutions, and concentration have been used by different investigators. Some of the more commonly used ones are listed in Table 5. All of the reported methods have been shown to be capable of recovering viruses from sludge. When seeded viruses have been used to study survival of viruses in sludge or when high levels of indigenous viruses have been found, concentration steps have not been necessary.

III. COMPARATIVE STUDIES

A. Comparative Studies of Eluents

As part of their efforts to develop procedures for recovering viruses from sludge, several investigators have compared different eluents for their ability to elute viruses adsorbed to sludge flocs. Some of the eluents tested are included in Table 6. In general, complex solutions of proteins and peptides such as beef extract, skim milk, and fetal calf serum (FCS) have been found to be better at eluting viruses than hydrolyzed proteins such as casein or lactalbumin hydrolysate.[35] Defined solutions containing individual amino acids, detergents, or buffers have not been effective eluents for adsorbed viruses.[27,36,37] A solution of 4 M urea buffered at pH 9 was found to be almost as effective in eluting viruses as was beef extract.[27]

B. Comparative Studies of Procedures for Recovering Viruses from Wastewater Sludges

In some cases, different procedures have been compared in the same laboratory (Table 7). Wellings et al.[33] compared sonication of sludge samples in the presence of beef extract to stirring. Both methods were able to separate viruses, but stirring required more time. Brashear and Ward[38] compared several methods for their ability to recover viruses from raw sludge.

Blending sludge samples in the presence of freon, centrifugation to remove the solids, and ultracentrifugation to concentrate viruses from the supernatant was the most effective procedure. In a cooperative study, eight laboratories compared procedures for recovering viruses in sludge according to the American Society for Testing and Materials (ASTM) procedures.[32] Initially, several procedures were compared in each laboratory using sludge available to the individual laboratory. Based on these preliminary studies, two procedures were selected for round-robin testing. These procedures are one proposed by Dr. Glass and one developed at the Environmental Protection Agency (EPA) by Dr. Berman. Outlines of these procedures are shown in Figure 1. In order to test the procedures, sludge samples were sent from one laboratory to the other participating laboratories. Each laboratory analyzed each sludge sample in triplicate using both methods. A portion of the final sample obtained with each method was analyzed in the individual laboratory, and a portion was sent to an EPA reference laboratory where samples from all of the individual laboratories were analyzed under the same conditions. Statistical analysis of the results showed that the EPA method was slightly more sensitive than the Glass method for recovering viruses from four of five sludge samples. For one sludge, no statistical difference was found. In general, the EPA method is less complicated than the Glass method and requires less time. The final volume with the Glass method is smaller than with the EPA method. Bacterial and fungal contamination was more of a problem with the EPA method. Since the difference in sensitivity was slight and both methods have advantages and disadvantages, both methods were recommended as tentative standard methods.

One problem with using beef extract to elute viruses is the resulting toxicity to cell cultures of the final sample obtained. Several procedures have been tested for their ability to reduce this toxicity, including treating samples with dithizone in chloroform, freon, and polyelectrolyte precipitation, and washing of cell monolayers with FCS in saline before agar overlay.[24,41] Washing cell cultures is effective in reducing toxicity, but it also results in a significant reduction in the virus titer.

IV. SUMMARY

The results obtained in comparative and individual testing of procedures have shown that procedures for recovering viruses from sludges are effective in recovering indigenous viruses in wastewater sludges. These procedures can be used to determine the fate of sludge-associated viruses during wastewater treatment and after sludge disposal. However, none of the procedures was evaluated for recovering rotavirus and hepatitis A virus from sludge. Improvements in procedures should be capable of (1) recovering viruses from different sludge types, since sludge type may influence recovery of viruses;[42] (2) producing a final sample that is small enough to permit economical assay of the entire sample; and (3) producing a final sample that is free of bacterial and fungal contaminants and not toxic to cell cultures. Specialized equipment or highly trained personnel should not be required.

Table 5
PROCEDURES FOR RECOVERING VIRUSES ASSOCIATED WITH WASTEWATER SLUDGE FLOCS

Sludge type	Eluting solution	Mixing procedure	Concentration procedure	Conc. factor (initial vol/final vol)	Viruses recovered		Ref.
					Added viruses (% recovered)	Indigenous viruses	
Anaerobically digested	10% beef extract + 1.34% sodium phosphate + 0.12% citric acid	Stirring	Organic flocculation	10	—a	77—345 pfu/100 mℓ	10,29
Anaerobically digested	5% FCS + 3% beef extract + 1% gelatin	Sonication	None	0.1	P1, CA9, CB4,b E11 (80—120%)	—	17
Anaerobically digested	10% FCS in phosphate-buffered saline	Blending	None	—	—	Enteroviruses, 0—2 pfu/mℓ; reoviruses, 2—8 pfu/mℓ	34
Anaerobically digested	BSS	Shaking	None	—	CB3	—	40
Lagooned	0.05 M glycine	Stirring	Membrane filtration, organic flocculation	—	P1 (55%)	P1, P2, P3, CB4, E14, E7, E15; 0—100 TCID$_{sv}$/g	26
Anaerobically digested	1% skim milk or 3% beef extract	Shaking	Organic flocculation	4	—	33—65 pfu/g	31
Raw, anaerobically digested	10% beef extract	Stirring	Hydroextraction	20	—	E3, E11, E15, E27, CB3, CA18	30
Anaerobically digested	3% beef extract	Sonication	Organic flocculation	16—100	P1 (49%)	—	24
Anaerobically digested	10% beef extract or Tris buffer	Shaking	None	1	—	P3, CB5; 0—0.6 TCID$_{sv}$/mg	13
Anaerobically digested, lagoon-dried	10% FCS	Shaking	None	1	—	11/28 samples of lagooned sludge were positive for reovirus, E1, and P1	14

Aerobically digested	1 M sodium trichloroacetate	Stirring	Inorganic flocculation	80	P1 (82%)	60—3000 pfu/ℓ	28
Digested, dried	3% beef extract	Sonication	Hydroextraction	—	—	Reovirus, P1, P3, E22/23	33
Aerobically digested, dried	4 M urea + 0.05 M lysine	Stirring	Inorganic, organic flocculation	100	P1, CB3, E1 (39%)	P1, CB4, E1, E4, E7: 0.02—4.6 $TCID_{50}$/g	27
Mixed liquor	10% FCS	Sonication	None	5	P1	—	6
Activated, dried	0.05 M glycine	Stirring	Membrane adsorption, organic flocculation	90	P1, E7, CB3 (77%)	P1, E7, E17: 1—489 pfu/g	25
Primary, activated	None (direct inoculation)	None	None	—	—	CB3, 5	12
Activated	Distilled water	Shaking	None	—	P1	—	8
Mixed liquor	Tryptose phosphate broth	Dounce homogenization	Lyphogel	8	P1 (27%)	—	37
Raw	10% FCS	Shaking	None	1	—	Reovirus, P1, P2, CB2, CB3, CB4, CB5	14
Dried, raw	Freon	Blending	Organic flocculation, centrifugation	20	—	149—700 pfu/mℓ	38,39

a   Not done or information not available.

b   P, polio; CA, coxsackie A; CB, coxsackie B; E, echo; pfu, plaque-forming unit, $TCID_{50}$, 50% tissue culture infectious dose.

Table 6
COMPARATIVE STUDIES OF SOLUTIONS USED TO ELUTE VIRUSES ASSOCIATED WITH SLUDGE FLOCS

Viruses	Sludge type	Eluting solution	Recovery[a]	Ref.
Indigenous	Raw	3% casein	0/10	35
		3% lactalbumin hydrolysate	8/40	
		3% beef extract	49/100	
		10% FCS	75/100	
Polio 1	Activated	3% beef extract	66%	36
Echo 1		1% skim milk	17%	
Coxsackie B5		10% newborn calf serum	51%	
		0.1% sodium dodecylsulfate	49%	
		0.01% Zetag	5%	
		Distilled water	16%	
Polio 1	Activated	3% beef extract	75%	27
		0.1% isoelectric casein	58%	
		3% lysine	20%	
		3% aspartic acid	1%	
		4 M urea + 1% lysine	70%	
Polio 1	Activated	Borate saline	32%	37
		Deionized water	34%	
		Tris-EDTA	23%	
		Tryptose phosphate broth	55%	

[a] Either fraction of total samples processed that were positive for viruses or percent of added viruses recovered.

Table 7
COMPARATIVE STUDIES OF PROCEDURES USED TO RECOVER VIRUSES FROM WASTEWATER SLUDGES

Sludge type	Procedure	Comments	Ref.
Raw	Freon treatment Elution with water Elution with 10% beef extract Adjustment to pH 3.5 and 0.0005 M aluminum chloride, elution with 10% beef extract	Blending of sludge with freon, separating flocs and freon from the supernatant by centrifugation, and concentrating the viruses in the supernatant by ultracentrifugation permitted recovery of the greatest amount of viruses.	38
Digested	Elution with 3% beef extract and sonication, stirring, or freon treatment	Variable amounts and types of viruses were recovered using the different procedures; both sonication and stirring were effective in recovering viruses, but stirring required more time.	33
Aerobically, anaerobically digested	Elution by stirring with 10% beef extract, conc. by organic flocculation (EPA) Elution by sonication with 3% beef extract, conc. by organic flocculation (Glass)	The EPA method was found to be slightly more sensitive; both methods were found to have advantages and disadvantages; and both methods were proposed as tentative standard methods.	32

EPA PROCEDURE	GLASS PROCEDURE
(300 mℓ of sludge)	(800 mℓ of sludge)

Adjust to pH 3.5 and 0.0005 M aluminum chloride, stir for 30 min, and centrifuge

Add 19.2 g beef extract and blend for 3 min

→ Discard supernatant

Suspend sediment in 300 mℓ of 10% buffered beef extract, stir for 30 min, and centrifuge

Adjust to pH 9.0, stir for 25 min, sonicate for 2 min, and centrifuge

→ Discard sediment

→ Discard sediment

Filter supernatant and concentrate by organic flocculation

Concentrate by organic flocculation

FIGURE 1. Comparison of the EPA and Glass procedures for recovering viruses from wastewater sludges.

REFERENCES

1. **Carlson, H. J., Ridenour, G. M., and McKhann, C. F., Jr.,** Effect of the activated sludge process of sewage treatment on poliomyelitis virus, *Am. J. Public Health*, 33, 1083, 1943.
2. **Clarke, N. A., Stevenson, R. E., Chang, S. L., and Kabler, P. W.,** Removal of enteric viruses from sewage by activated sludge, *Am. J. Public Health*, 51, 1118, 1961.
3. **Kelly, S., Sanderson, W. W., and Neidl, C.,** Removal of enteroviruses from sewage by activated sludge, *J. Water Pollut. Control Fed.*, 33, 1057, 1961.
4. **Lund, E., Hedstrom, C. E., and Jantzen, N.,** Occurrence of enteric viruses in wastewater after activated sludge treatment, *J. Water Pollut. Control Fed.*, 41, 169, 1969.
5. **Safferman, R. S. and Morris, M. E.,** Assessment of virus removal by a multi-stage activated sludge process, *Water Res.*, 10, 413, 1976.
6. **Balluz, S. A., Jones, H. H., and Butler, M.,** The persistence of poliovirus in activated sludge treatment, *J. Hyg. Cambr.*, 78, 165, 1977.
7. **Lund, E.,** The effect of pretreatments of the virus contents of sewage samples, *Water Res.*, 7, 873, 1973.
8. **Malina, J. F., Jr., Ranaganathan, K. R., Sagik, B. P., and Moore, B. E.,** Poliovirus inactivation by activated sludge, *J. Water Pollut. Control Fed.*, 47, 2178, 1975.
9. **Eisenhardt, A., Lund, E., and Nissen, B.,** The effect of sludge digestion on virus infectivity, *Water Res.*, 11, 579, 1977.
10. **Berg, G. and Berman, D.,** Destruction by anaerobic mesophilic and thermophilic digestion of viruses and indicator bacteria indigenous to domestic sludges, *Appl. Environ. Microbiol.*, 39, 361, 1980.
11. **Goddard, M. R., Bates, J., and Butler, M.,** Recovery of indigenous enteroviruses from raw and digested sewage sludges, *Appl. Environ. Microbiol.*, 42, 1023, 1981.
12. **Lund, E. and Ronne, V.,** On the isolation of virus from sewage treatment plant sludges, *Water Res.*, 7, 863, 1973.
13. **Nielsen, A. L. and Lydholm, B.,** Methods for the isolation of viruses from raw and digested wastewater sludge, *Water Res.*, 14, 175, 1980.
14. **Sattar, S. A. and Westwood, J. C. N.,** Recovery of viruses from field samples of raw, digested, and lagoon-dried sludges, *Bull. WHO*, 57, 105, 1979.
15. *Fed. Regist.*, 40 CFR Part 257, September 13, 1979.
16. **Palfi, A.,** Survival of enteroviruses during anaerobic sludge digestion, in *Advances in Water Pollution Research*, Jenkins, S. H., Ed., Pergamon Press, Elmsford, N.Y., 1973, 99.
17. **Bertucci, J. J., Lue-Hing, C., Zenz, D., and Sedita, S. J.,** Inactivation of viruses during anaerobic sludge digestion, *J. Water Pollut. Control Fed.*, 50, 1642, 1977.
18. **Kabrick, R. M., Jewell, W. J., Salotto, B. V., and Berman, D.,** *Inactivation of Viruses, Pathogenic Bacteria and Parasites in the Autoheated Aerobic Thermophilic Digestion of Sewage Sludges*, Ann Arbor Science, Ann Arbor, Mich., 1979, 771.

19. **Sanders, D. A., Malina, J. F., Jr., Moore, B. E., Sagik, B. P., and Sorber, C. A.,** Fate of poliovirus during anaerobic digestion, *J. Water Pollut. Control Fed.,* 51, 334, 1979.

20. **Farrah, S. R.,** Isolation of viruses associated with sludge particles, in *Methods in Environmental Virology,* Gerba, C. P. and Goyal, S. M., Eds., Marcel Dekker, New York, 1982, 161.

21. **Lund, E.,** Methods for virus recovery from solids, in *Viruses and Wastewater Treatment,* Goddard, M. R. and Butler, M., Eds., Pergamon Press, Elmsford, N.Y., 1982, 189.

22. **Moore, B. E., Sagik, B. P., and Malina, J. F., Jr.,** Viral association with suspended solids, *Water Res.,* 9, 197, 1975.

23. **Schaub, S. A. and Sagik, B. P.,** Association of enteroviruses with natural and artificially introduced colloidal solids in water and infectivity of solids-associated virions, *Appl. Microbiol.,* 30, 212, 1975.

24. **Glass, S. J., Van Sluis, R. J., and Yanko, W. A.,** Practical methods for detecting poliovirus in anaerobic digester sludge, *Appl. Environ. Microbiol.,* 35, 983, 1978.

25. **Hurst, C. J., Farrah, S. R., Gerba, C. P., and Melnick, J. L.,** Development of quantitative methods for the detection of enteroviruses in sewage sludges during activation and following land disposal, *Appl. Environ. Microbiol.,* 36, 81, 1978.

26. **Farrah, S. R., Bitton, G., Hoffman, E. M., Lanni, O., Pancorbo, O. C., Lutrick, M. C., and Bertrand, J. E.,** Survival of enteroviruses and coliform bacteria in a sludge lagoon, *Appl. Environ. Microbiol.,* 41, 459, 1981.

27. **Farrah, S. R., Scheuerman, P. R., and Bitton, G.,** Urea-lysine method for recovery of enteroviruses from sludge, *Appl. Environ. Microbiol.,* 41, 455, 1981.

28. **Scheuerman, P. R., Farrah, S. R., and Bitton, G.,** Development of a method for the recovery of enteroviruses from aerobically digested wastewater sludges, *Water Res.,* in press.

29. **Berg, G., Berman, D., and Safferman, R. S.,** A method for concentrating viruses recovered from sewage sludges, *Can. J. Microbiol.,* 28, 553, 1982.

30. **Lydholm, B. and Nielsen, A. L.,** The use of soluble polyelectrolyte for the isolation of virus from sludge, in *Viruses and Wastewater Treatment,* Goddard, M. R. and Butler, M., Eds., Pergamon Press, Elmsford, N.Y., 1980, 85.

31. **Goddard, M. R. and Bates, J.,** The effect of selected sludge treatment processes on the isolation of enteroviruses from wastewater sludge, in *Viruses and Wastewater Treatment,* Goddard, M. R. and Butler, M., Eds., Pergamon Press, Elmsford, N.Y., 1980, 79.

32. **Goyal, S. M., Schaub, S. A., Wellings, F. M., Berman, D., Glass, J. S., Hurst, C. J., Brashear, D. A., Sorber, C. A., Moore, B. E., Bitton, G., Gibbs, P. H., and Farrah, S. R.,** Round robin investigation of methods for recovering human enteric viruses from sludge, *Appl. Environ. Microbiol.,* 48, 531, 1984.

33. **Wellings, F. M., Lewis, A. L., and Mountain, C. W.,** Demonstration of solids-associated virus in wastewater and sludge, *Appl. Environ. Microbiol.,* 31, 354, 1976.

34. **Cliver, D. O.,** Virus association with wastewater solids, *Environ. Lett.,* 10, 215, 1975.

35. **Sattar, S. A. and Westwood, J. C. N.,** Comparison of four eluents in the recovery of indigenous viruses from raw sludge, *Can J. Microbiol.,* 22, 1586, 1976.

36. **Bates, J. and Goddard, M. R.,** Recovery of seeded viruses from activated sludge, in *Viruses and Wastewater Treatment,* Goddard, M. R. and Butler, M., Eds., Pergamon Press, Elmsford, N.Y., 1982, 205.

37. **Moore, B. E. D., Funderberg, L., Sagik, B. P., and Malina, J. F., Jr.,** Application of viral concentration techniques to field sampling, in *Virus Survival in Water and Wastewater Systems,* Malina, J. F., Jr. and Sagik, B. P., Eds., Center for Research in Water Resources, University of Texas, Austin, 1974, 3.

38. **Brashear, D. A. and Ward, R. L.,** Comparison of methods for recovering indigenous viruses from raw wastewater sludge, *Appl. Environ. Microbiol.,* 43, 1413, 1982.

39. **Brashear, D. A. and Ward, R. L.,** Inactivation of indigenous viruses in raw sludge by air drying, *Appl. Environ. Microbiol.,* 45, 1943, 1983.

40. **Damgaard-Larsen, S., Jensen, K. O., Lund, E., and Nissen, B.,** Survival and movement of enteroviruses in connection with land disposal of sludges, *Water Res.,* 11, 503, 1977.

41. **Hurst, C. J. and Goyke, T.,** Reduction of interfering cytotoxicity associated with wastewater sludge concentrates assayed for indigenous enteric viruses, *Appl. Environ. Microbiol.,* 46, 133, 1983.

42. **Pancorbo, O. C., Scheuerman, P. R., Farrah, S. R., and Bitton, G.,** Effect of sludge type on poliovirus association with and recovery from sludge solids, *Can. J. Microbiol.,* 27, 279, 1981.

Chapter 8

MECHANISM OF VIRUS INACTIVATION IN WASTEWATER SLUDGES

Richard L. Ward

TABLE OF CONTENTS

I. INTRODUCTION

Viruses that replicate in the intestinal tract — enteric viruses — are released into the environment in association with fecal matter. Because of their natural tendencies to bind to particulates, most enteric viruses remain solids-associated during sewage treatment. Those removed during primary sedimentation become a part of raw primary sludge. Detectable enteric viruses in primary sludge often exceed 10^5 infectious units per liter.[1,2] Because of the inefficiency of the detection procedures, the actual concentrations could be much greater. The effluent recovered after primary sedimentation is normally treated by processes that result in additional sludges also containing high concentrations of infectious viruses.[3,4]

These primary and secondary sludges are normally subjected to various treatments before they are disposed of or utilized. Conventional treatment processes have been designed to reduce the potential of sludge as a food source for disease vectors; to reduce its nuisance properties, such as odor; or to reduce its volume. It is serendipitous that most of these sludge stabilization processes also cause significant amounts of viral inactivation. In some treatment plants, the sludge is subjected to other processes designed specifically for pathogen reduction.

Because enteric viruses are obligatory intracellular parasites, their numbers cannot increase outside of their hosts. A goal of environmental virologists is to reduce the population of viable enteric viruses to negligible levels between their release in fecal matter and exposure to a subsequent host. If viruses in sludges are inactivated during treatment plant operations, a considerable portion of the potential public health hazard associated with enteric viruses will have been eliminated. The purpose of this chapter is to discuss the mechanisms by which this occurs during both conventional and nonconventional sludge treatments.

Mechanisms of virus inactivation in sludges have been studied by seeding viruses into sludges which are subsequently subjected to various treatments that simulate plant operations. Natural association of indigenous viruses with sludge particulates is not expected to be identical to the associations formed between these particulates and seeded viruses. However, general mechanisms by which both indigenous and seeded viruses are inactivated in sludge can be compared. It is on this assumption that the applicability of the results reported in this chapter are based. In cases where inactivation of indigenous viruses has been studied, the results have been in general agreement with those obtained using seeded viruses.

II. CONVENTIONAL SLUDGE TREATMENT

A. Anaerobic Digestion

The most common method of sludge stabilization used in sewage treatment plants in the U.S. is anaerobic digestion. During this process, large organic molecules are biodegraded into much smaller molecules, a large percentage of which are gases if the process is carried to completion. The process itself does not generate heat, and the organisms involved in biodegradation grow much better above ambient temperatures. Therefore, a portion of the methane generated by the process is normally burned to heat the sludge. The usual temperature increases are to about 35°C, where mesophilic bacteria have optimal growth rates, or to about 50°C, where thermophilic bacteria metabolize more rapidly. Most digestors are operated in the mesophilic range because of the lesser energy requirements. Studies performed to determine the effects of mesophilic digestion on both indigenous and seeded viruses indicate that their infectivities typically decrease 90 to 99% over a 15-day period.[3,5-12]

Changes in the numbers of different pathogens in sludge during anaerobic digestion are dependent on several factors, such as chemical environment and biological activity. One of the most important, however, is temperature. The rate of decrease of infectious viruses in sludge during anaerobic digestion is temperature dependent and is more rapid at thermophilic than at mesophilic temperatures.[5,6]

The mechanism of heat inactivation of picornaviruses, the group to which polioviruses and other enteroviruses belong, has been studied by a number of investigators over the past 25 years. Woese[13] suggested that thermal inactivation of viruses occurred primarily through damage to viral nucleic acid. Later reports indicated that low-temperature inactivation involved nucleic acid damage, but that at high temperatures heat caused the denaturation of viral proteins.[14-17] The temperature at which the change in mechanism was postulated to occur is about 40°C.[17]

High-temperature heat inactivation of poliovirus causes an alteration of its antigenicity and the release of one of its four capsid proteins, VP4.[18,19] Viral RNA becomes susceptible to ribonuclease, but in the absence of this enzyme it is released intact. RNA recovered from picornaviruses inactivated by heat at high temperatures has been shown to retain full infectivity.[15,17]

Low-temperature inactivation of viruses is associated with breakdown of viral nucleic acid within antigenically unaltered viral capsids.[17] RNA recovered from picornaviruses is noninfectious and has been found in a degraded state.[17,20] Loss of infectivity of a rhinovirus at 34.5°C was attributed to the action of a nuclease that is an integral part of the virion.[21]

Indigenous components of sludge have also been identified that greatly alter the inactivation rates of different enteric viruses. The effects of these components on viral inactivation rates have also been shown to be temperature dependent. As expected, the faster rates have been observed at higher temperatures.

One component of sludge found to have a profound effect on the survival of enteroviruses is ammonia. The discovery of this agent in anaerobically digested sludge was made as a result of approaches used to measure virus inactivation rates. It was found that poliovirus inactivation was significantly more rapid when virus was seeded into digested sludge in our laboratory, while other investigators measured virus inactivation in digesting rather than previously digested sludge.[9-11] The difference in rates of virus inactivation was due to an increase in the pH of the sludge during storage following digestion.[23] This increase in pH caused indigenous ammonium ions to be partially converted into aqueous ammonia, which was subsequently shown to inactivate poliovirus and other enteroviruses.[23]

Examination of the mechanism by which ammonia inactivates enteroviruses revealed that this compound had no detectable effect on the isoelectric state of poliovirus and little effect on the ability of this virus to bind to cells.[24] It also had no effect on naked poliovirus RNA but caused RNA within virions to be inactivated by cleavage. It was suggested that ammonia activates a nuclease within the virion or can pass through hydrophobic regions of the capsid only in its nonionized state, thus raising the internal pH of the virion sufficiently to permit alkaline hydrolysis of viral RNA.

Other indigenous components of sludge shown to alter the inactivation rates of enteric viruses are ionic detergents. The effects of detergents were originally discovered when it was observed that reovirus inactivation was greatly accelerated at temperatures above 40°C in either raw or anaerobically digested sludge as compared to a buffered salt solution.[25] Rotavirus, another member of the Reoviridae family, was also found to be more rapidly inactivated in sludge.[26]

When ionic detergents were subsequently identified as the agents responsible for destabilization of these viruses in sludge,[27] the mechanism by which one such detergent, sodium dodecyl sulfate (SDS), caused rotavirus inactivation was examined. It was found that SDS-inactivated rotaviruses had decreased sedimentation values and were unable to attach to host cells.[28] These inactivated virions were also shown to have lost one outer capsid viral protein, which probably caused the observed effects on the properties of the virions.

The effects of ionic detergents on reovirus and rotavirus have been shown to be modified not only by temperature but also by pH.[29] The maximum stability of reovirus in both SDS and sludge was found near neutral pH. This supports the conclusion that SDS and similar detergents are the sludge agents that destabilize this and related viruses.

Sludge also contains agents that stabilize viruses against heat inactivation. Although ionic detergents accelerate inactivation of certain enteric viruses (Reoviridae), they protect others. Poliovirus and other enteroviruses were shown to be inactivated much more slowly in the presence of ionic detergents.[27] This protective effect, however, disappeared at both low and high pH values. It was suggested that the increased concentration of such protective agents was responsible for the observed decrease in enterovirus inactivation in dewatered sludges.[30]

Dewatering was also found to protect reovirus against heat inactivation in seeded sludge.[30] Since it had been shown that nonionic detergents could stabilize rotavirus against the potent destabilizing effects of SDS,[26] it was postulated that their increased concentration in dewatered sludge may overcome the effects of concomitant increases in the concentration of ionic detergents.

From these results it is clear that a number of competitive factors influence the inactivation rates of enteric viruses in sludge. Furthermore, these factors can often influence the rates of inactivation of different groups of viruses in different and sometimes opposite ways. The most important of these factors that have been identified in connection with anaerobic digestion are temperature, pH, ammonia, and detergents. Undoubtedly other physical and chemical factors may be involved and will eventually be identified.

B. Aerobic Digestion

This is a process conducted by agitating sludge with air or oxygen to maintain aerobic conditions.[31,32] Because most aerobic digestors are open tanks, their performance is dependent on weather conditions, especially temperature. There is little published information regarding virus inactivation during aerobic digestion of sludge. Since the process is conducted under conditions similar to activated sludge treatment of wastewater, the information available regarding the mechanism of inactivation in the latter process may be applicable.

Several investigators have shown that loss of viral infectivity occurs in activated sludge.[33-35] Evidence was presented that the cause of inactivation was due to microorganisms present in the sludge.[36] For the past 18 months our laboratory has made an intensive effort to identify the components of activated sludge that cause virus inactivation. It has been confirmed that growth of microorganisms is somehow responsible for the virucidal activity. Autoclaved primary effluent had little or no virucidal activity, but developed this activity when seeded with activated sludge (unpublished results). The activities in both the sludge and seeded effluent were sensitive to heat. Temperatures of 60°C or greater eliminated most activity within 1 hr. The active agent in both materials was also associated with the particulates, and little was found in the supernatant portion after low-speed centrifugation. Unfortunately, efforts directed at separating the active agents from other sludge particulates have been unsuccessful. Thus, the nature of the microorganism or the microbial product responsible for causing virus inactivation has still not been identified.

A study performed at the University of Florida indicated that viruses are also inactivated during aerobic sludge digestion.[37] The investigators stated that virus survival under aerobic conditions was less than during anaerobic digestion. Temperature again played a major role. Increased virus survival was observed as the temperature was reduced from 28° to 5°C.

Aerobic digestion, in contrast to anaerobic digestion, is highly exothermic. In open digestors the heat is rapidly dissipated into the atmosphere but in closed containers it can be retained. A study conducted at Cornell University showed that temperatures in excess of 45°C were attained routinely in a full-scale, autoheated aerobic digestor.[38] During this process virus inactivation was greater than in anaerobic digestors even though average retention times were only about one fifth that of the anaerobic system. Further studies are needed to determine whether microbial activity is responsible for this increased virus inactivation.

C. Composting

One of the most effective methods of stabilizing sludge is composting. Although this method of treating organic waste has been practiced for many years, composting of municipal sludges in the U.S. has only gained major attention since the early 1970s. This process involves extensive aeration of dewatered sludge in piles, windrows, or containers. During the process, microbial growth results in large releases of heat, which in turn can raise the temperature of the composting sludge above 60°C.[39]

Virus inactivation during composting is due mainly to heat. Studies have been done using the bacteriophage f2 as a model virus. This phage is more heat resistant than typical enteric viruses. However, when seeded into sludge composted by either the windrow or static pile method, its titer decreased steadily.[40,41] Reductions of 90% of virus were observed every 4 to 7 days in windrows during dry weather, and decreases occurred at about 50% of this rate in rainy weather. More rapid destruction of virus was observed in aerated piles where higher temperatures were attained.

No controlled *in situ* studies have been reported with enteric viruses in sludge during composting. However, experiments were performed to determine the stabilities of several viruses in raw and previously composted sludges as a function of moisture content and temperature.[30] It was found that raw sludge accelerated heat inactivation of reovirus and rotavirus, as previously noted, but this effect disappeared in well-composted sludge. The virucidal substances, probably ionic detergents,[27] are apparently degraded during composting.

Factors such as pH, chemical composition, and moisture content of sludge can also effect virus inactivation rates during composting. Because sludge is sometimes treated with lime before it is composted, its pH may remain above neutrality for an extended period after composting has begun. The presence of ammonia coupled with high pH and more than ambient temperature should cause rapid inactivation of enteroviruses.[23] Other types of interactions of chemical and physical factors are expected to affect virus inactivation rates during composting, but the main contributions of these factors is probably to modify the effects of heat. For this reason, the degree of virus inactivation during composting should be proportional to the temperature, which in turn depends on the efficiency of composting.

D. Lime Treatment

The purpose of this process, as with other chemical treatments of sludge, is primarily to coagulate solids for subsequent dewatering. Variable efficiencies of virus removal have been observed after chemical treatment of wastewater, and efficiencies of greater than 99.9% have been found.[42-54] Many of the viruses removed by these treatments can be recovered in the settled floc or chemical sludge.

The stabilities of viruses in chemical sludges have not been studied in detail, but most chemicals would not be expected to greatly affect virus inactivation rates. However, lime treatment is accompanied by a pH rise, the extent of which is dependent upon the quantity added to the wastewater or sludge. This pH increase can cause large reductions in viral titers. It appears that rapid inactivation of viruses in wastewater requires pH values greater than 11.[47,52,53] Although no in-depth studies have been reported, viruses have also been found to be inactivated in both raw and digested sludges at higher pH.[55,56]

Mechanistic studies on the effects of high pH have been conducted with poliovirus, but normally at above ambient temperatures. Viral inactivation rates at pH 10.6 were quite rapid at 40°C but slow at 22°C.[57] At pH 8.6, inactivation was also quite slow at 40°C. Thus, pH and heat appeared to act synergistically.

Polioviruses inactivated by high pH treatment released their RNA genomes and VP4,[58,59] the same viral protein released during heat treatment.[18,19] Empty capsids formed during this process gradually broke apart over a period of hours when maintained under these conditions.[60] Treatment of polioviruses at the same temperature but at pH 11 also resulted in the formation of empty capsids, but these lacked a second viral protein, VP2.[58] Treatment at 40°C and pH 12 caused these empty capsids to be rapidly degraded. The state of poliovirus RNA molecules released during alkaline treatment has not been studied, but even at room temperature they can be completely degraded with high alkalinity.[61]

These results suggest that disruption of the viral capsid may be the initial cause of poliovirus inactivation by high pH. This conclusion was supported by a study in which it was found that differences in the stabilities of the three serotypes of reovirus at pH 11 could be related to differences in a specific capsid protein.[62,63] Similar studies are needed with other enteric viruses to confirm that protein disruption is the mechanism of virus inactivation at high pH.

E. Air Drying

Dewatering of sludge to facilitate handling and reduce treatment or disposal costs is a common practice. Because viruses are mainly associated with sludge solids, only a small percentage should be removed with the water during treatments such as centrifugation or filter-pressing. Likewise, these same processes are neither expected nor known to cause a significant amount of viral inactivation.

A method to reduce moisture that can, in itself, cause a large amount of virus inactivation in sludge is natural evaporation. In an initial study conducted with seeded viruses in raw sludge it was established that air-drying caused a gradual decrease in viral infectivity until the solids content of the sludge reached about 73%.[64] At this point viral infectivity was reduced by 90%. Further dewatering to about 83% solids caused a much greater decrease in recoverable viral infectivity. Viral recoveries from sludge with less than 10% water were consistently less than 0.01%. These results were obtained with three separate strains of enteric viruses.

Because radioactive-labeled viruses were used in this study, it could be clearly established that loss of infectivity was due to irreversible damage to viral particles. Poliovirions recovered from air-dried sludge were disrupted, and their RNA genomes were found in a degraded state. Similar observations were subsequently made by Yeager and O'Brien with viruses seeded into moist soils which were permitted to air dry.[65,66]

The cause of inactivation appeared to be somehow due to the evaporation process itself, because similar effects occurred in distilled water when the virus particles were kept continuously in solution.[64] Perhaps this mechanism is similar to that observed in aerosols, where it was shown that virions are degraded, presumably at the air-surface interface, and release naked RNA.[67-70]

A study completed more recently compared inactivation rates of indigenous and seeded viruses in raw sludge as a function of moisture loss through evaporative drying.[91] It was found that the rate of gradual decline and point of change to rapid loss of infectivity were virtually identical for seeded and indigenous viruses. This point was very near 80% solids. The rate of rapid decline was somewhat greater with seeded than indigenous viruses, possibly a reflection of differences in their associations with sludge particulates. Similar results were found by Hurst et al.[4] during field studies following sludge application.

III. NONCONVENTIONAL SLUDGE TREATMENTS

A. Irradiation

Ionizing radiation, in the form of both γ- and β-rays, has been used in sludge treatment. γ-Rays are photons of energy released when certain unstable isotopes, such as ^{60}Co or ^{137}Ce, decay to more stable states. β-Rays used to disinfect materials are high-energy electrons emitted by an accelerator. The mechanism of damage by both types of rays is the same because γ-rays deposit their energies on electrons in the irradiated material, causing them to be accelerated and released from their atomic orbitals.[72] These energized electrons, like those emitted by an accelerator, then collide with other electrons, thus creating chemical changes within the material. Unlike β-rays, γ-rays can travel large distances before their energies are dissipated, which permits irradiation of thick samples.

Biological damage can be caused either by the direct or indirect effects of ionizing radiation. These depend on the nature of the surrounding medium.[73] Indirect effects occur in dilute aqueous environments through damage done by toxic radiation products of water and other materials in the medium.[74] Direct effects take place in the presence of high concentrations of organic substances. Damage is usually due to the interaction of energized electrons with molecules which, when altered, directly result in loss of infectivity.

Viral inactivation in sludge is due primarily to direct effects of ionizing radiation. Clearly, the doses required to cause inactivation are much greater under direct-effect than indirect-effect conditions. For example, the dose needed to inactivate an equivalent amount of poliovirus was about 14-fold greater under direct than indirect conditions.[75]

The virion component that is the primary target of irradiation differs under direct and indirect conditions. Under direct conditions, loss of infectivity is due primarily to nucleic acid damage. This has been shown with both bacterial[76-80] and human viruses.[75,81]

Most inactivated poliovirus RNA molecules obtained from viruses inactivated under direct conditions were still intact,[81] which suggested that they contained base damage or cross-link formation.[82] However, about one third of the poliovirons inactivated under direct conditions still contained infectious RNA. Although these virions had normal sedimentation values and isoelectric points, they had lost their ability to bind cells approximately in proportion to their expected loss of capsid function. Damaged capsid proteins had unaltered electrophoretic mobilities. Therefore, about two thirds of poliovirus inactivation was attributed to RNA damage and about one third to protein damage under direct-effect conditions of irradiation, but the specific types of damage to either molecular species were not identified.

Loss of poliovirus infectivity under indirect irradiation conditions was about four times faster than loss of RNA infectivity within virions.[75] This indicates that capsid damage should account for about 75% of the lost infectivity under indirect conditions, a conclusion in agreement with results found with bacterial viruses.[80,83-85] Evidence of capsid damage was shown by partial breakdown of viral particles, alterations of viral isoelectric points, loss of ability to bind to cells, and large changes in the electrophoretic mobilities of capsid proteins. Thus, irradiation under indirect conditions is much harsher on viral capsids than under direct conditions. Because of the mechanisms by which inactivation occurs under the two conditions,[74] this is the anticipated result.

B. Other Nonconventional Treatments

Sludge treatments given specifically to reduce or eliminate pathogens that do not involve irradiation have primarily consisted of inactivation by heat. These include such processes as pasteurization, thermophilic aerobic digestion, and heat drying.[32] Virus inactivation during these processes should occur by mechanisms already described for high-temperature processes.

IV. SUMMARY

Viruses in wastewater sludge are inactivated during both conventional and nonconventional treatment processes. The mechanisms of inactivation during these processes depend on factors such as pH, temperature, chemical and biological composition, and moisture content. Because viruses do not replicate in sludge, their inactivation results in reducing the potential health risks associated with sludge use or disposal.

ACKNOWLEDGMENTS

A portion of the data listed in this report was obtained as part of studies funded by the U.S.

Environmental Protection Agency under assistance agreement numbers R-810821-01-0 and R-811183-01-1.

REFERENCES

1. **Brashear, D. A. and Ward, R. L.,** Comparison of methods for recovering indigenous viruses from raw wastewater sludge, *Appl. Environ. Microbiol.*, 43, 1413, 1982.
2. **Lewis, M. A., Nath, M. W., and Johnson, J. C.,** A multiple extraction-centrifugation method for the recovery of viruses from waste water treatment plant effluents and sludges, *Can. J. Microbiol.*, 29, 1661, 1983.
3. **Lund, E. and Ronne, V.,** On the isolation of viruses from sewage treatment plant sludges, *Water Res.*, 7, 863, 1973.
4. **Hurst, C. J., Farrah, S. R., Gerba, C. P., and Melnick, J. L.,** Development of quantitative methods for the detection of enteroviruses in sewage sludges during activation and following land disposal, *Appl. Environ. Microbiol.*, 36, 81, 1978.
5. **Eisenhardt, A., Lund, E., and Nissen, B.,** The effect of sludge digestion on virus infectivity, *Water Res.*, 11, 579, 1977.
6. **Berg, G. and Berman, D.,** Destruction by anaerobic mesophilic and thermophilic digestion of viruses and indicator bacteria indigenous to domestic sludges, *Appl. Environ. Microbiol.*, 39, 361, 1980.
7. **Palfi, A.,** Survival of enteroviruses during anaerobic sludge digestion, in *Advances in Water Pollution Research*, Jenkins, S. H., Ed., Pergamon Press, Elmsford, N.Y., 1973, 99.
8. **Moore, B. E., Sagik, B. P., and Sorber, C. A.,** An assessment of potential health risks associated with land disposal of residual sludges, in *Proc. 3rd Natl. Conf. Sludge Management, Disposal, and Utilization*, Information Transfer, Rockville, Md., 1976, 108.
9. **Sanders, D. A., Malina, J. F., Jr., Moore, B. E., Sagik, B. P., and Sorber, C. A.,** Fate of poliovirus during anaerobic digestion, *J. Water Pollut. Control Fed.*, 51, 333, 1979.
10. **Bertucci, J. J., Lue-Hing, C., Zenz, D., and Sedita, S. J.,** Inactivation of viruses during anaerobic sludge digestion, *J. Water Pollut. Control Fed.*, 49, 1642, 1977.
11. **Moore, B. E., Sagik, B. P., and Sorber, C. A.,** Land application of sludges: minimizing the impact of viruses on water resources, in *Proc. Conf. Risk Assessment and Health Effects of Land Application of Municipal Wastewater and Sludges*, Sagik, B. P. and Sorber, C. A., Eds., Center for Applied Research and Technology, University of Texas, San Antonio, 1977, 154.
12. **Goddard, M. R., Bates, J., and Butler, M.,** Recovery of indigenous enteroviruses from raw and digested sewage sludges, *Appl. Environ. Microbiol.*, 42, 1023, 1981.
13. **Woese, C.,** Thermal inactivation of animal viruses, *Ann. N.Y. Acad. Sci.*, 83, 741, 1960.
14. **Ada, G. L. and Anderson, S. G.,** Yield of infective "ribonucleic acid" from impure Murray Valley encephalitis virus after different treatments, *Nature (London)*, 183, 799, 1959.
15. **Bachrach, H. L.,** Thermal degradation of foot-and-mouth disease virus into infectious ribonucleic acid, *Proc. Soc. Exp. Biol. Med.*, 17, 610, 1961.
16. **Mika, L. A., Officier, J. E., and Brown, A.,** Inactivation of two arboviruses and their associated infectious nucleic acids, *J. Infect. Dis.*, 113, 195, 1963.
17. **Dimmock, N. J.,** Differences between the thermal inactivation of picornaviruses at "high" and "low" temperatures, *Virology*, 31, 338, 1967.
18. **Breindl, M.,** The structure of heated poliovirus particles, *J. Gen. Virol.*, 11, 147, 1971.
19. **Lonberg-Holm, K., Gosser, L. B., and Kauer, J. C.,** Early alteration of poliovirus in infected cells and its specific inhibition, *J. Gen. Virol.*, 27, 329, 1975.
20. **Drees, O. and Borna, C.,** Über die Spaltung physikalisch Intakter Poliovirusteilchen in Nucleinsaure und Leere Proteinhullen duch Warmebehandlugh, *Z. Naturforsch.*, 20b, 870, 1965.
21. **Gauntt, C. J.,** Fragmentation of RNA in virus particles of rhinovirus type 14, *J. Virol.*, 13, 762, 1974.
22. **Ward, R. L. and Ashley, C. S.,** Inactivation of poliovirus in digested sludge, *Appl. Environ. Microbiol.*, 31, 921, 1976.
23. **Ward, R. L. and Ashley, C. S.,** Identification of the virucidal agent in wastewater sludge, *Appl. Environ. Microbiol.*, 33, 860, 1977.
24. **Ward, R. L.,** Mechanism of poliovirus inactivation by ammonia, *J. Virol.*, 26, 299, 1978.
25. **Ward, R. L. and Ashley, C. S.,** Discovery of an agent in wastewater sludge that reduces the heat required to inactivate reovirus, *Appl. Environ. Microbiol.*, 34, 681, 1977.
26. **Ward, R. L. and Ashley, C. S.,** Effects of wastewater sludge and its detergents on the stability of rotavirus, *Appl. Environ. Microbiol.*, 40, 1154, 1980.

27. **Ward, R. L. and Ashley, C. S.,** Identification of detergents as components of wastewater sludge that modify the thermal stability of reovirus and enteroviruses, *Appl. Environ. Microbiol.*, 36, 889, 1978.

28. **Ward, R. L. and Ashley, C. S.,** Comparative study on the mechanisms of rotavirus inactivation by sodium dodecyl sulfate and ethylenediaminetetraacetate, *Appl. Environ. Microbiol.*, 40, 1148, 1980.

29. **Ward, R. L. and Ashley, C. S.,** pH modification of the effects of detergents on the stability of enteric viruses, *Appl. Environ. Microbiol.*, 38, 314, 1979.

30. **Ward, R. L. and Ashley, C. S.,** Heat inactivation of enteric viruses in dewatered wastewater sludge, *Appl. Environ. Microbiol.*, 36, 898, 1978.

31. Process Design Manual for Sludge Treatment and Disposal, Municipal Environmental Research Laboratory, U.S. Environmental Protection Agency, Cincinnati, 1979.

32. *Fed. Regist.* 40 CFR Part 257, September 13, 1979.

33. **Malina, J. F., Jr., Ranganathan, K. R., Moore, B. E., and Sagik, B. P.,** Poliovirus inactivation by activated sludge, in *Virus Survival in Water and Wastewater Systems*, Malina, J. F., Jr. and Sagik, B. P., Eds., Center for Research in Water Resources, University of Texas, Austin, 1974, 95.

34. **Malina, J. F., Jr., Ranganathan, K. R., Sagik, B. P., and Moore, B. E.,** Poliovirus inactivation by activated sludge, *J. Water Pollut. Control Fed.*, 47, 2178, 1975.

35. **Glass, J. S. and O'Brien, R. T.,** Enterovirus and coliphage inactivation during activated sludge treatment, *Water Res.*, 14, 877, 1980.

36. **Ward, R. L.,** Evidence that microorganisms cause inactivation of viruses in activated sludge, *Appl. Environ. Microbiol.*, 43, 1221, 1982.

37. **Scheuerman, P. R., Farrah, S. R., and Bitton, G.,** Survival of virus during aerobic digestion of wastewater sludges, *Abstr. Ann. Meet. Am. Soc. Microbiol.*, Q57, 219, 1982.

38. **Kabrick, R. M. and Jewell, W. J.,** Fate of pathogens in thermophilic aerobic sludge digestion, *Water Res.*, 16, 1051, 1982.

39. **Willson, G. B., Epstein, E., and Parr, J. R.,** Recent advances in compost technology, in *Proc. 3rd Natl. Conf. Sludge Management, Disposal, and Utilization*, Information Transfer, Rockville, Md., 1976, 167.

40. **Burge, W. D., Marsh, P. B., and Millner, P. D.,** Occurrence of pathogens and microbial allergens in the sewage sludge composting environment, in *Natl. Conf. Composting of Municipal Residues and Sludges*, Information Transfer, Rockville, Md., 1977, 128.

41. **Kawata, K., Cramer, W. N., and Burge, W. D.,** Composting destroys pathogens in sewage solids, *Water Sewage Works*, 124, 76, 1977.

42. **Chang, S. L., Stevenson, R. E., Bryant, A. R., Woodward, R. L., and Kabler, P. W.,** Removal of coxsackie and bacterial viruses in water by flocculation. I. Removal of coxsackie and bacterial viruses in water of known chemical content by flocculation with aluminum sulfate or ferric chloride under various testing conditions, *Am. J. Public Health*, 48, 51, 1958.

43. **Chang, S. L., Stevenson, R. E., Bryant, A. R., Woodward, R. L., and Kabler, P. W.,** Removal of coxsackie and bacterial viruses in water by flocculation. II. Removal of coxsackie and bacterial viruses and the native bacteria in raw Ohio River water by flocculation with aluminum sulfate and ferric chloride, *Am. J. Public Health*, 48, 159, 1958.

44. **Robeck, G. G., Clarke, N. A., and Dostal, K. A.,** Effectiveness of water treatment processes in virus removal, *J. Am. Water Works Assoc.*, 54, 1275, 1962.

45. **Wentworth, D. F., Thorup, R. T., and Sproul, O. J.,** Poliovirus inactivation in water-softening precipitation processes, *J. Am. Water Work Assoc.*, 60, 939, 1968.

46. **Sproul, O. J.,** Virus removal by adsorption in treatment processes, *Water Res.*, 2, 74, 1968.

47. **Berg, G., Dean, R. B., and Dahling, D. R.,** Removal of poliovirus from secondary effluents by lime flocculation and rapid sand filtration, *J. Am. Water Works Assoc.*, 60, 193, 1968.

48. **Farrah, S. R., Goyal, S. M., Gerba, C. P., Conklin, R. H., and Smith, E. M.,** Comparison between adsorption of poliovirus and rotavirus by aluminum hydroxide and activated sludge flocs, *Appl. Environ. Microbiol.*, 35, 360, 1978.

49. **Thorup, R. J., Nixon, F. P., Wentworth, D. F., and Sproul, O. J.,** Virus removal by coagulation with polyelectrolytes, *J. Am. Water Works Assoc.*, 62, 97, 1970.

50. **Manwaring, J. F., Chaudhuri, M., and Engelbrecht, R. S.,** Removal of viruses by coagulation and flocculation, *J. Am. Water Works Assoc.*, 63, 298, 1971.

51. **York, D. W. and Drewery, W. A.,** Virus removal by chemical coagulation, *J. Am. Water Works Assoc.*, 66, 711, 1974.

52. **Sattar, S. A., Ramia, S., and Westwood, J. C. N.,** Calcium hydroxide (lime) and the elimination of human pathogenic viruses from sewage: studies with experimentally contaminated (poliovirus type 1, Sabin) and pilot plant samples, *Can. J. Public Health*, 67, 221, 1976.

53. **Grabow, W. O. K., Middendorff, I. G., and Basson, N. C.,** Role of lime treatment in the removal of bacteria, enteric viruses, and coliphages in a wastewater reclamation plant, *Appl. Environ. Microbiol.*, 35, 663, 1978.

54. **Sattar, S. A. and Ramia, S.,** Viruses in sewage: effect of phosphate removal with calcium hydroxide (lime), *Can. J. Microbiol.*, 24, 1004, 1978.
55. **Goddard, M. R., Bates, J., and Butler, M.,** Recovery of indigenous enteroviruses from raw and digested sewage sludges, *Appl. Environ. Microbiol.*, 42, 1023, 1981.
56. **Koch, K. and Strauch, D.,** Removal of poliovirus in sewage sludge by lime treatment, *Zentralbl. Bakteriol. Mikrobiol. Hyg. Abt. Orig. B*, 174, 335, 1981.
57. **Van Elsen, A. and Boeyé, A.,** Disruption of type 1 poliovirus under alkaline conditions: role of pH, temperature, and sodium dodecyl sulfate (SDS), *Virology*, 28, 481, 1966.
58. **Katagiri, S., Aikawa, S., and Hinuma, Y.,** Stepwise degradation of poliovirus capsid by alkaline treatment, *J. Gen. Virol.*, 13, 101, 1971.
59. **Maizel, J. V., Phillips, B. A., and Summers, D. F.,** Composition of artificially produced and naturally occurring empty capsids of poliovirus type 1, *Virology*, 32, 692, 1967.
60. **Boeyé, A. and Van Elsen, A.,** Alkaline disruption of poliovirus: kinetics and purification of RNA-free particles, *Virology*, 33, 335, 1967.
61. **Calvery, H. O. and Jones, W.,** The nitrogenous groups of nucleic acid, *J. Biol. Chem.*, 73, 73, 1927.
62. **Drayna, D. and Fields, B. N.,** Genetic studies on the mechanisms of chemical and physical inactivation of reovirus, *J. Gen. Virol.*, 63, 149, 1982.
63. **Drayna, D. and Fields, B. N.,** Biochemical studies on the mechanism of chemical and physical inactivation of reovirus, *J. Gen. Virol.*, 63, 161, 1982.
64. **Ward, R. L. and Ashley, C. S.,** Inactivation of enteric viruses in wastewater sludge through dewatering by evaporation, *Appl. Environ. Microbiol.*, 34, 564, 1977.
65. **Yeager, J. G. and O'Brien, R. T.,** Structural changes associated with poliovirus inactivation in soil, *Appl. Environ. Microbiol.*, 38, 702, 1979.
66. **Yeager, J. G. and O'Brien, R. T.,** Enterovirus inactivation in soil, *Appl. Environ. Microbiol.*, 38, 694, 1979.
67. **Trouwborst, T., de Jong, J. C., and Winkler, K. C.,** Mechanism of inactivation in aerosols of bacteriophage T1, *J. Gen. Virol.*, 15, 235, 1972.
68. **de Jong, J. C., Harmsen, M., and Trouwborst, T.,** The infectivity of the nucleic acid of aerosol-inactivated poliovirus, *J. Gen. Virol.*, 18, 83, 1973.
69. **de Jong, J. C., Harmsen, M., Trouwborst, T., and Winkler, K. C.,** Inactivation of encephalomyocarditis virus in aerosols: fate of virus protein and ribonucleic acid, *Appl. Microbiol.*, 27, 59, 1974.
70. **Trouwborst, T., Kuyper, S., de Jong, J. C., and Plantinga, A. D.,** Inactivation of some bacterial and animal viruses by exposure to liquid-air interfaces, *J. Gen. Virol.*, 24, 155, 1974.
71. **Brashear, D. A. and Ward, R. L.,** Inactivation of indigenous viruses in raw sludge by air drying, *Appl. Environ. Microbiol.*, 45, 1943, 1983.
72. **Dertinger, H. and Jung, H.,** *Molecular Radiation Biology*, Springer-Verlag, New York, 1970.
73. **Luria, S. E. and Exner, F. M.,** The inactivation of bacteriophages by x-rays — influence of the medium, *Proc. Natl. Acad. Sci. U.S.A.*, 27, 370, 1941.
74. **Michaels, H. B. and Hunt, J. W.,** A model for radiation damage in cells by direct effect and by indirect effect: a radiation chemistry approach, *Radiat. Res.*, 74, 23, 1978.
75. **Ward, R. L.,** Mechanisms of poliovirus inactivation by the direct and indirect effects of ionizing radiation, *Radiat. Res.*, 83, 330, 1980.
76. **Watson, J. D.,** The properties of x-ray-inactivated bacteriophage. I. Inactivation by direct effect, *J. Bacteriol.*, 60, 697, 1950.
77. **Little, C. D. and Ginoza, W.,** Intracellular development of bacteriophage X174 inactivated by gamma-ray, ultraviolet light, or nitrous acid, *Virology*, 38, 152, 1969.
78. **Coquerelle, T. and Hagen, U.,** Loss of absorption and injection abilities in gamma-irradiated phage T1, *Int. J. Radiat. Biol.*, 21, 31, 1972.
79. **Clarkson, C. E. and Dewey, D. L.,** The viability of bacteriophage T4 after irradiation of only the head component or the tail component. I. In nutrient broth projected conditions, *Radiat. Res.*, 53, 248, 1973.
80. **Becker, D., Redpath, J. L., and Grossweiner, L. I.,** Radiation inactivation of T7 phage, *Radiat. Res.*, 73, 51, 1978.
81. **Rainbow, A. J. and Mak, S.,** DNA strand breakage and biological functions of human adenovirus after gamma irradiation, *Radiat. Res.*, 50, 319, 1972.
82. **Blok, J. and Loman, A.,** The effects of gamma-radiation in DNA, *Curr. Top. Radiat. Res.*, 9, 165, 1973.
83. **Watson, J. D.,** The properties of x-ray-inactivated bacteriophage. II. Inactivation by indirect effects, *J. Bacteriol.*, 63, 473, 1952.
84. **Luthjens, L. H. and Blok, J.,** The indirect action of x-rays on the coat protein of bacteriophage particles, *Int. J. Radiat. Biol.*, 16, 101, 1969.
85. **Clarkson, C. E. and Dewey, D. L.,** The viability of bacteriophage T4 after irradiation of only the head component or the tail component. II. In aqueous suspension, *Radiat. Res.*, 54, 531, 1973.

Chapter 9

FATE OF VIRUSES DURING SLUDGE PROCESSING

Dean O. Cliver

TABLE OF CONTENTS

I. SLUDGE DISPOSAL AND VIRUS TRANSMISSION

People may be infected with viruses from the environment by ingesting contaminated food or water, and perhaps by swimming in contaminated water. Sludge produced in treatment of urban wastewater may be disposed of in several ways (Figure 1) that carry with them differing risks of contaminating food or water and perhaps leading to viral infections of humans. If viruses were present, disposal of sludge to land might lead to contamination of foods growing in the soil or to contamination of water through percolation of the virus to groundwater or runoff to surface waters. Disposal of sludge in landfills carries some risk of contamination of groundwater by viruses in leachates, whereas disposal at sea might lead to contamination of edible shellfish species or swimming beach water with viruses. Properly accomplished incineration or pyrolysis of sludge probably offers virtually no risk of virus transmission.[1] Although viral illness in humans that is traceable to sludge is virtually unknown in the U.S., one can at least conclude that disposal of liquid or dry sludge to land carries the greatest risk of such transmission. "Guidelines for the Application of Wastewater Sludge to Agricultural Land in Wisconsin" was published in 1975.[2]

The U.S. Environmental Protection Agency's (EPA) Sludge Management Program specifies that untreated sludges not be disposed of to land surfaces, whereas sludges treated by processes that significantly reduce pathogens (PSRP, defined as processes at least as effective against pathogens as is properly conducted anaerobic digestion) may be spread on land, with certain restrictions.[3] Sludges treated by processes that further reduce pathogens (PFRP, defined as processes at least as effective against pathogens as pasteurization) may be spread on land without pathogen-related restrictions. Obviously, other substances in sludge may at times be of greater concern than are pathogens. Furthermore, by no means are all pathogens present in sludge viruses.

II. VIRUSES OF CONCERN

Although transmission of viruses as a result of sludge contamination is virtually unknown in the U.S., one can identify viruses of concern on the basis of the agents that are known to be transmitted through water and food.[4] At this writing, reporting of water-borne disease in the U.S. is more up to date than that of food-borne disease; therefore, I have compiled the data for reported water-borne disease during the most recent 5-year period for which annual summaries have been issued (Table 1).[5-9] One can see that the virus of hepatitis A, which has been responsible for many water-borne outbreaks in the past,[4] has become less prevalent than the more recently recognized Norwalk-like viruses. Only a single outbreak of rotavirus gastroenteritis occurred during the period in question, nor has another in the U.S. come to my attention since that time. The situation with food-borne viruses is similar, except that rotavirus transmission has not been recorded[10] and shellfish sometimes transmit a small round featureless virus thought to resemble the parvoviruses.[11] The latter agent has not yet been shown to be transmitted by any other food. Of course, there is a great deal of acute gastrointestinal illness of unknown etiology among water- and food-borne disease outbreaks, and some of this may well be viral, but essentially no viral illness transmitted via vehicles can be attributed to agents other than those just listed.

A. Hepatitis A Virus

The virus of hepatitis A is a member of the enterovirus group: small (28 nm), round viruses containing single-stranded RNA. It replicates in many primate cell culture lines but does not cause overt cytopathic effects. Diagnosis of hepatitis A is based on demonstration of antibody of the IgM class in the blood of patients. Due to the difficulty of quantifying viral infectivity, almost no studies have yet been done on the inactivation of the hepatitis A virus in environmental samples, including sludge.

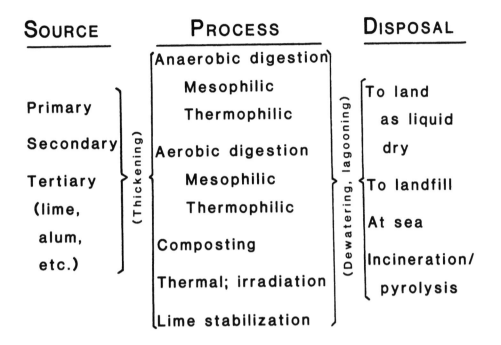

FIGURE 1.   Methods of treatment and disposal of urban sludges.

B. Norwalk Virus

The Norwalk virus (or perhaps Norwalk group) is described as having a feathery outline and a diameter of 30 to 35 nm.[11] It is not known to replicate in any cultured cells. Diagnosis is based on electron microscopy. More than one serologic type is known. Most of the studies on this group have been done in the U.K., where diagnoses are reported weekly under the acronym SRSV (small, round, structured viruses). No inactivation studies have been reported in sludge.

C. Rotavirus

The rotaviruses are approximately 70 nm in diameter and are composed of double-stranded RNA surrounded by a double-layered protein coat. Human rotaviruses have been difficult to replicate in cell cultures, so diagnosis has been based on the tests described for Norwalk virus or on enzyme-linked immunosorbent assay for the viral antigen in patients' stools. Inactivation studies in sludge have been done using a simian rotavirus (strain SA11).[12]

D. Other Human Enteric Viruses

Aside from the small, round, featureless viruses causing shellfish-associated gastroenteritis, only the polioviruses, and echovirus 4 in one instance, have been known to be transmitted through foods and perhaps water.[4] The polio- and echoviruses, together with the coxsackieviruses, the hepatitis A agent, and some serologic types that carry only numbers, comprise the human enterovirus group. Many of these replicate readily in primate cell cultures, so the enteroviruses, especially the polioviruses, have frequently served as models in inactivation studies, including studies with sludge.[13-20] The human enteroviruses cause a variety of illnesses, but seldom gastroenteritis. Several other groups of viruses have been found in human feces, sometimes in association with diarrhea; they may properly be called enteric viruses because they come from the intestines, but they have not been shown to be transmissible through food or water, so their significance, if they happen to occur in sludge, is unknown. Among these, only the inactivation of the reoviruses in sludge seems to have been studied,[15-17] except when irradiation was examined.[19]

Table 1

VIRAL AND OTHER WATER-BORNE DISEASE OUTBREAKS REPORTED IN THE U.S., 1979—1983

Year	Hepatitis A		Norwalk gastroenteritis		Rotavirus gastroenteritis		AGI[a]		All illnesses		Ref.
	Outbreaks	Cases	Outbreaks	Cases	Outbreaks	Cases	Outbreaks	Cases	Outbreaks	Cases	
1979			2	296			22	3,412	41	9,720	5
1980	1	48	5	1,914			28	13,220	50	20,008	6
1981					1	1,761	14	1,893	32	4,430	7
1982	3	103	4	750			16	1,836	40	3,456	8
1983	3	164					15	16,498	40	20,905	9
Totals	7	315	11	2,960	1	1,761	95	36,859	203	58,519	

[a] Acute gastrointestinal illness of unknown etiology.

E. Indicators of Viral Contamination

Bacteria that are regularly present in human feces have been proposed as indicators of the possible or probable presence of viruses in environmental samples such as sludge.[21] The total coliform group comprises Gram negative, nonspore-forming rods that produce acid and gas from lactose within 48 hr at 35°C. These include species of the genera *Escherichia*, *Klebsiella*, *Enterobacter*, and *Citrobacter*; though they are common in the feces of humans and other warmblooded animals, some of these organisms may also derive from nonfecal sources. They were destroyed more rapidly than viruses during mesophilic (35°C) sludge digestion and much more rapidly than viruses during thermophilic (50°C) sludge digestion.[18] Fecal coliforms are coliforms that grow at 44.5°C; these belong almost exclusively to the genera *Escherichia* and *Klebsiella*. On occasion, *Klebsiella* sp. of nonfecal origin may be detected. These were destroyed more rapidly than viruses during both mesophilic and thermophilic sludge digestion.[18] Persistence of fecal coliforms in lagooned sludge paralleled that of enteroviruses. The fecal streptococci, which in urban sludge are likely to comprise principally *Streptococcus faecalis* and *S. faecium*, are generally of fecal origin. They were found to die off at rates quite similar to the rates of inactivation of viruses in meso- and thermophilic sludge digestion, apparently because the organisms were heavily clumped rather than because the individual cells were exceptionally hardy.[18] Clumping may well explain the exceptional stability (and inactivation resembling viruses) of *S. faecalis* during γ-irradiation in sludge.[19]

Given that enteroviruses detectable in wastewater and sludge are not the types that generally cause food- or water-borne disease, the presence of these enteroviruses may be taken as indication of human fecal contamination, and the inactivation of these viruses in sludge treatment may be taken as evidence of the inactivation of the hepatitis A virus[14] (and perhaps other small, round enteric viruses) in the process. The reoviruses are closely related to the rotaviruses and are more readily cultivated in cell cultures, so their inactivation in sludge digestion may predict the inactivation of rotaviruses. Finally, coliphages (the viruses that infect the bacterium, *E. coli* have been proposed as indicators of fecal contamination, though they are not specifically of human fecal origin, and their inactivation might be of value in predicting inactivation of the human pathogenic viruses during sludge digestion.[13,17] Most of the data to be presented in the sections that follow derive from studies with human enteroviruses — most frequently the polioviruses if they have been added experimentally to sludge.

III. SLUDGES AS VIRUS SOURCES

All sludges generated during wastewater treatment contain viruses; even the grit that is removed and handled as mineral matter before treatment of the wastewater is said to have virus.[22] The presence of these viruses in sewage and in sludge is largely due to the use of the water carriage toilet to dispose of human feces. The virus tends to be associated with the fecal solids and perhaps other solids in wastewater, which presents special problems in detection and quantification of the virus and makes it likely that the majority of virus in the plant influent will separate in the sludge fractions. Methods for the recovery of viruses from sludges are not the province of this chapter; suffice to say that the problems have long been recognized and that no single recovery method is optimal for all viruses in all kinds of sludges.[23,24]

A. Primary Sludge

This is the material separated during primary wastewater treatment, in which easily sedimented solids and scum are removed; it contains 3 to 7% solids and is usually easy to thicken further.[25] Due to the difficulty of recovering viruses from wastewater solids quantitatively, one often reads that virus removal during primary wastewater treatment is negligible.[26] Nevertheless, primary sludge is consistently a virus-rich substance. Levels of enteroviruses reported in primary sludge in one study ranged from 2.4 to 15 plaque-forming units (pfu) per milliliter[22] and in another, from 87 to 17,280 pfu/ℓ.[24] Reoviruses exceeded 1 pfu/mℓ in the first report.[22]

B. Secondary Sludge

Secondary wastewater treatment is typically biological treatment intended to convert soluble or nonsedimentable materials into biomass (microbial cells and slime) that can be separated by gravity. Various activated sludge processes, rather than trickling filters as before, tend to predominate in secondary treatment of wastewater in the U.S. Secondary sludge usually has only 0.5 to 2% solids and is relatively difficult to thicken further.[25] Extended aeration is a biological treatment process for wastewater which often is not preceded by primary treatment, so that clarification at the end of the treatment would bring down a blend of solids; the resulting sludge is considered to be partially stabilized and to contain lower levels of pathogens than does raw sludge.[3]

Virus that has been through the activated sludge process may be inactivated, in that some of the virus is detectable neither in the clarified effluent nor in the sludge.[27] This inactivation appears to be a biological process. Similar inactivation was not observed when the virus studied was a coliphage (f2) rather than a human enterovirus.[28] Enterovirus levels measured in activated sludge did not exceed 5 pfu/mℓ, with the highest level having been observed in the return sludge; reovirus levels sometimes came near 5 pfu/mℓ, with the highest level observed in the waste-activated sludge.[22] Virus was not detected in sloughed solids from a trickling filter.

C. Tertiary Sludge

Tertiary treatment or advanced wastewater treatment processes are intended to remove turbidity, phosphates, and other difficult substances from wastewater prior to discharge. Addition of coagulants such as lime, salts of aluminum or iron, or polymers is common. Sludges produced with lime or polymers tend to thicken readily, whereas sludges produced with salts of aluminum or iron tend to bind water and are very difficult to thicken.[25] Though coagulants are usually used in a third stage wastewater treatment in the U.S., they can be added directly to influent raw sewage and have been found to concentrate viruses from the liquid phase.[23] Coagulants used in both drinking water and wastewater treatment were found to be more effective for removing virus (coliphage MS2) in the former.[29]

Tertiary wastewater treatment using lime should not be confused with lime stabilization of sludge, which will be discussed in Section V.F. In either case, any antiviral effect that the process may exert is probably referable principally to the high pH ordinarily attained.[30] In one study, lime treatment of secondary effluent was very effective against enteric viruses from pH 11.1 to as low as 9.6, but the lower pH range was less effective against coliphages and much less effective against bacterial indicators.[31] Lime treatment of raw sewage and of primary effluent at pH 9.5 to 10.5 produced reductions of 96 to 99% in enterovirus levels, but virus that remained infectious in the sludge was not significantly inactivated during 48 hr of storage at 28°C.[32]

Metallic coagulants added to wastewater should produce reductions of at least 90% in virus levels.[30] The pH optimum for such treatments is generally in the range of 5 to 7, where little inactivation of the virus is to be expected. Polyelectrolytes did not produce significant virus removals when used as primary coagulants and generally did not improve virus removal significantly when used as coagulant aids.

IV. ANTIVIRAL FACTORS IN SLUDGE PROCESSES

The modes by which viruses are inactivated in sludge processing are not within the scope of this chapter. However, there are only a limited number of "targets" for inactivation on a viral particle; it should be possible to identify some factors common to various sludge processes and determine the extent to which these are likely to cause virus inactivation. Ideally, one would wish to identify factors that could be measured in any sludge treatment process and used to predict the fate of viruses in that treatment.

A. Temperature

Temperature is a preeminent factor in viral inactivation in many environmental contexts, though the context also must be taken into account. Temperatures to which virus-containing sludges are subjected may range from ambient (which might be near or even below freezing in some instances) to 80°C and sometimes even higher. In general, the protein coat of enteroviruses is the most labile component at higher temperatures, whereas the nucleic acid is most affected at lower temperatures. However, pH and other environmental factors are also involved.

Inactivation of poliovirus inoculated into digested urban sludge was slightly greater in 14 days at 22°C than at 4°C.[14] Substantially fewer isolations of human enteric viruses from anaerobically digested sludge were obtained after the digestion process was changed from 21 days at 30°C to 40 days at 33°C; it was not possible to separate the effects of extended time and temperature in this instance.[33] An experimental anaerobic digestion system processed poliovirus-inoculated sludge at temperatures of 34°, 37°, and 50°C for periods of 10 to 15 days.[34] Rates of loss of poliovirus infectivity per day, for 10 days' retention, were 48% at 34°, 56% at 37°, and 99.999998% at 50°C. During up to 200 min of heating at 43°, 47°, and 51°C, raw sludge was strongly protective of inoculated poliovirus, compared to inactivation observed when the suspending medium was phosphate-buffered saline (pH not specified).[35] However, digested sludge was slightly protective to 45°C, essentially neutral at 47°C, and very detrimental to viral stability at 51°C.

Just as the direct effect of heat on viruses is influenced by the type of sludge, it also appears that heat modifies the antiviral activity of the sludge. Sludge composted at 55°C for several days was found to be less protective of poliovirus (at 39° and 43°C) and more protective of reovirus (at 45°C) than raw sludge.[36] Sludge that had been pasteurized at 60°C for 1 hr caused somewhat more rapid inactivation of coxsackievirus B3 at 32° and 35°C than did sludge from a digester operated at 35°C.[36]

B. pH

Enteric viruses are generally most stable near neutral pH but tolerate acidity better than alkalinity. The pH of most wastewater and raw sludge is near neutrality; though anaerobic digestion results in the formation of acids, these are consumed in methane production, so the pH range in anaerobic digestion is generally 6.5 to 7.5.[38] Aerobic digestion of sludges can cause the pH to fall as low as 4 or rise as high as 8.[39]

Enteroviruses ordinarily do not tolerate alkalinity above pH 10, yet storage of sludge from lime-treated wastewater for 48 hr at 28°C caused no significant reduction in the titer of experimentally added poliovirus.[32] In addition to the excess of hydroxyl ions that is likely to occur in sludge as a result of lime treatment, it appears that the high pH converts ammonium ions to the uncharged ammonia state, which has been shown to be virucidal.[15] The antiviral effect of treating sludge with lime (pH $\geq$ 12) is said to be strongly dependent on ammonia in the case of enteroviruses and even more strongly in the case of experimentally inoculated bovine parvoviruses.[40] Studies of inactivation of poliovirus 1 and coliphage f2 as a function of the levels of NH_3, NH_4^+, and OH^- in the suspensions showed that NH_3 was most effective, that NH_4^+ was no more active than Na^+, and that OH^- ions were essentially ineffective; poliovirus was much more susceptible than f2 to the effects of NH_3.[41] As will be discussed later, Ward and others have attributed many of the differences in lability and stability of viruses in sludges to the effects of detergents that are present; they have found that the effects of detergents are influenced by pH in a very complex set of relationships.[17] Rotavirus is stable over the pH range 4 to 10 but is relatively susceptible to pH 2 at 4° or 21°C; 0.1% sodium dodecyl sulfate (SDS) produces substantial or complete inactivation of the virus over the pH range 2 to 10 at 21°C and at pH 2 or 4, but not higher, at 4°C.[12]

C. Time

In that none of the viral inactivation processes that are likely to take place in sludge is instantaneous, time of exposure to a given set of conditions will determine the degree of viral inactivation that results. Time of exposure is much more readily controlled under laboratory conditions than in a sewage treatment plant, where the conduct of a process may be generally described as plug flow, completely mixed, or fill and draw. In ideal conditions, these terms mean that every drop of sludge leaves the process in exactly the same order that it entered (plug flow), that any drop of sludge is as likely as another to leave the process at a given moment (completely mixed), or that there is some minimum residence time to which a given drop of sludge will have been subjected to a process, though some may have been exposed much longer (fill and draw). Under ideal conditions, then, one can design a process so that virus is subjected to it for some minimum time or a probable minimum time. However, some digestion processes require complete mixing to keep microbiological processes in equilibrium for optimal stabilization of the sludge, and in these instances one can predict that a substantial proportion of the sludge will have been subjected to the process for far less than mean residence time. Depending on the kinetics of virus inactivation compared to the dynamics of the process, the reduced resident time — for some portion of the sludge — that results from complete mixing can be the principal determinant of the level of virus infectivity remaining in the sludge at the end of the process.[22]

D. Biological and Other Factors

Given that several sludge stabilization processes are principally biological in nature, it is of interest to note whether viruses in sludge are subject to biological degradation. Poliovirus is biologically degraded during the activated sludge process of wastewater treatment, but this occurs under aerobic conditions.[27] Anaerobic biodegradation of viruses apparently has not been recorded, and biodegradation of enteroviruses during aerobic sludge digestion seems not to have been studied.

Sludge that had been anaerobically digested was found to have developed an antiviral capacity that caused nicking of RNA inside the poliovirus particle.[42] The virucidal agent in anaerobically digested sludge was later reported to be ammonia, which was effective against enteroviruses but not reoviruses, and only above pH 8.[15] In that the presence of ammonia in anaerobically digested sludge is reasonably predictable, one might consider this effect merely to enhance the antiviral effect of alkalinity against the enteroviruses, but not the reoviruses.

The situation with detergents in sludge appears to be less straightforward. Cationic, and to a lesser extent anionic (but not nonionic), detergents have been reported to destabilize reoviruses, whereas all detergents are protective of enteroviruses.[16] However, dewatering of raw sludge was said to increase detergent concentrations and thus to enhance protection of enteroviruses. This increase of detergent concentrations led to a diminished effect against reoviruses, ostensibly due to a concurrent increase in the concentration of a hypothetical protective substance in the sludge.[36] The stabilizing and destabilizing effects of detergents in sludge have also been shown to be influenced greatly by pH.[17] A model cationic detergent had the following effects at 45° to 47°C and sometimes 4°C; at low (<5) pH, it showed little destabilizing effect against reovirus and coliphages and little protective effect toward enteroviruses; at near-neutral pH, it showed little destabilization effect against reovirus and coliphages but was protective of enteroviruses; at high (>8) pH, it strongly destabilized reovirus and coliphages and protected enteroviruses. A model anionic detergent under the same conditions strongly destabilized reovirus, coliphages, and enteroviruses at low pH; showed little destabilizing effect toward reovirus and coliphages at neutral pH but protected enteroviruses; and destabilized reovirus and coliphages but protected enteroviruses at high pH. Finally, one reads that rotavirus is destabilized by anionic detergent but is protected from this effect if nonionic detergent is present.[12] Effects also differed greatly among individual detergents of a class. It would appear extremely difficult to predict the fate of viruses in sludge processing on the basis of these detergent effects, especially as the effects of a given detergent on different viruses may be antithetical and different detergents, concurrently present, may exert opposite effects in some circumstances.

V. VIRUS INACTIVATION IN SLUDGE TREATMENT

In the preceding sections, viruses have been discussed group by group, and raw sludges have been discussed as individual classes. Most communities in the U.S. are probably practicing secondary treatment at this time, and a few are doing tertiary treatment, with more to follow. The result is that sludge to be treated in a given plant will often be a blend of highly individual primary and secondary sludges and perhaps some tertiary sludges. Furthermore, the blended sludge will ordinarily have been thickened (Figure 1), to raise the solids contents to a level not above 10 to 20%,[43] before treatment. Removal of excess water, which is ordinarily sent back through sewage treatment with the raw influent to the plant, permits treatment of smaller volumes of sludge, which saves plant area and costs.

The choice of a stabilization treatment for sludge is made at the time that a community's sewage treatment plant is designed. This choice has seldom been based on virus removals, though the requirement for PSRP and PFRP are undoubtedly changing this. Definitive statistics on the use of various processes are not available,[65] but it seems certain that over half of sludge stabilization in the U.S. is done by anaerobic digestion.

A. Anaerobic Digestion

Anaerobic digestion, conducted in the absence of air for residence times ranging from 60 days at 20°C to 15 days at 35° to 55°C, is the norm for PSRP. Mesophilic anaerobic digestion, which is more common, could be conducted anywhere in the temperature range 10° to 42°C, but is often controlled at temperatures near 35°C, especially if the yield of digester gas is to be maximized.[38] High-rate digestion seems to denote positive temperature control and continuous mixing during the process; hydraulic residence times appear generally to be at least 15 days. Thermophilic digestion takes place in the temperature range above 42°C; digester gas production is highest at 55°C. In the discussion that follows, the focus will be on the primary stage of sludge digestion, generally ignoring subsequent holding of sludge without mixing or temperature control in what may be designated secondary digestion.

1. Mesophilic Digestion

Bench-scale digestion experiments undoubtedly provide the highest degree of precision in studies of virus inactivation, though the results obtained may not be strictly indicative of what may occur in a full-scale operation. A digester inoculated once with 10^5 pfu/mℓ of porcine enterovirus was operated at 34.5°C, and daily samples were fed to germ-free pigs (dose, 10 mℓ in the first experiment and 20 mℓ thereafter).[44] Virus was detected in two of six samples after 4 days of operation and not thereafter. In another study, rates of inactivation of various types of viruses were compared during operation of a digester at 35°C.[13] The slowest inactivation observed was 74.9%/day with echovirus 11. Other rates were poliovirus 1, 90.0%/day; coxsackievirus B4, 90.2%/day; coliphage MS-2, 93.2%/day; and coxsackievirus B4, 90.3%/day. Yet another bench-scale study, using what may have been pure waste-activated sludge, showed only about 48%/day inactivation of poliovirus 1 during operation at 34°C and 60%/day at 37°C.[34] Many details of the compositions of the sludges used in these studies, which might explain some of the differences in results, are given in the papers cited. However, no component that is listed seems to play a significant role in determining viral inactivation rates.

A pilot-scale study of virus inactivation was done in a digester containing 87.5 ℓ of sludge inoculated experimentally with coxsackievirus B3.[37] Mean detention time of the sludge was 35 days. Operation at 35°C caused about 99% inactivation of the virus per day, whereas daily inactivation at 32°C averaged only about 32%.

Sampling digested sludge from community wastewater treatment plants has produced less favorable results. In an early report, entero- and reoviruses were detected in 15% of 82 samples from digesters operated at 33°C with a hydraulic retention time of 40 days (stirring only during the first 30 days); the sizes of the samples are generally not specified.[33] Digestion at another plant for 12 days at 30° to 32°C with mixing and for a further 8 days without temperature control or mixing yielded 0 pfu/mℓ of enteroviruses and 8 pfu/mℓ of reoviruses in December, 0 pfu/mℓ of enteroviruses and 3 pfu/mℓ of reoviruses in June, 0.8 pfu/mℓ of enteroviruses and 4.4 pfu/mℓ of reoviruses in August, and 2 pfu/mℓ of enteroviruses and 2.4 pfu/mℓ of reoviruses in November.[22] A large digester was fed fresh sludge once daily and later drawn back down to its working volume of somewhat less than 9×10^6 ℓ (mixing not described); average residence time was 20 days at a temperature of 35°C, during which about 90% inactivation of the indigenous viruses occurred.[18] Titers of samples ranged from 0.3 to 4.1 pfu/mℓ, almost all of which were probably enteroviruses. In studies of three plants in the U.K., digesters operated at 30° to 35°C with residence times of about 30 days (no mention of stirring) produced sludge that frequently contained viruses; inactivation rates were estimated at 6.2, 3 and 0.8%/day.[45] A comparative study of methods for testing sludges reports titers as high as 5.9 pfu/mℓ in samples from anaerobic, mesophilic digesters; the highest-titered sample came from a digester operated at 29° to 38°C, with no mention of stirring or residence time other than that the operation was "standard rate".[24] Considerably lower titers and some negative test results are also reported with samples from this and other mesophilic digesters.

The factors discussed in Section IV do not seem to account for the disparities between results in bench- and full-scale digestion. The feed-and-draw method of operation described for one community digester would appear to guarantee that some of the newly added sludge would be present in what was withdrawn, with a residence in the digester of considerably less than 24 hr, if mixing was rapid.[18] Laboratory digesters are probably fed raw sludge once a day in many instances, but feeding is more likely to be done after the day's output has been withdrawn (continuous feeding is also possible in the laboratory). Full-scale digesters receive a continuous, though not constant, influx of viruses, whereas laboratory digesters may receive only a single inoculation of virus. Finally, virus in raw sludge may be associated with solids in ways not well duplicated in the laboratory.[34] When all of these factors have been considered, they still do not seem to account for the differences in virus inactivation reported above. Therefore, it seems reasonable also to look at the flow of the sludge through the digester as a possible reason for passing some virus that has not remained in the process for nearly the designated period of time. If a digester operated at a mean residence time of 20 days were fed its day's input all at once and then quickly mixed and sampled, the virus in the added sludge should have been diluted by a factor of 20, or to 5% of the titer of the added raw sludge. If addition of raw sludge and mixing are continuous, some effluent sludge, with its solids and virus, will have been present in the digester less than 24 hrs; but in no case should the titer of what is withdrawn exceed 5% of what was added unless inactivation is much slower than was reported in most of the bench-scale studies or unless major departures from complete mixing are occurring. If one assumes that engineers who design digesters do not make such mistakes, the matter remains unresolved.

2. Thermophilic Digestion

Anaerobic, thermophilic digestion, which is less used than anaerobic, mesophilic digestion, seems also to have been less studied. One recent report describes bench-scale digestion of what may have been pure waste-activated sludge (about 1% total solids) at 50°C.[34] Inactivation of experimentally inoculated poliovirus 1 was reported to be 99.999998%/day in the sludge digester and in a control suspension of phosphate-buffered saline with 2% calf serum added. The observed inactivation was attributed entirely to the high temperature.

A digester of nearly 9×10^6 ℓ capacity, loaded on a feed-and-draw basis as described in the preceding section but operated at 49°C, produced 99 to >99.9% inactivation of indigenous viruses with an average 20-day residence time.[18] Though this is greater inactivation than would ordinarily be obtained with mesophilic digestion, it is obviously less than would be predicted from the results of the bench-scale experiments described above.

B. Aerobic Digestion

Advantages claimed for aerobic digestion are the production of better-stabilized sludge faster, more reliably, and with less capital cost.[46] As with anaerobic digestion, the mesophilic temperature range is considered to be 10° to 42°C, and the thermophilic range is that above 42°C. Some aerobic systems are designed so that thermophilic temperatures are attained with heat generated in digestion.[3] Mixing is an intrinsic part of the aeration process. Substantial acidification (pH <5) may also result.

1. Mesophilic Digestion

Virus removal by aerobic sludge digestion in the mesophilic temperature range seems not to have been studied to any great extent; nevertheless, it is designated a process that significantly reduces pathogens.[25] Bench-scale digestion, at 28°C, of sludge experimentally inoculated with poliovirus 1 produced approximately 37% inactivation per day.[66] Samples from a community digester operated at 17°C (pH, 5.7; residence time not stated) had virus titers from 0.11 to <0.007 pfu/mℓ.[24] The focus of the latter study was virus detection methods rather than inactivation by digestion.

2. Thermophilic Digestion

Aerobic digestion at 50°C with a 5-day average residence time was reported to produce about 33% inactivation of indigenous viruses per day in a pilot-scale study.[45] A full-scale, 2-year study of a community digester autoheated to 45° to 65°C showed that virus was seldom detectable in the digested samples.[48] To be regarded as a PFRP, digestion should be carried out for at least 10 days' mean residence time at 55° to 60°C.

C. Composting

Sludge to be treated by composting will often be dewatered and then mixed with a bulking agent such as wood chips before being subjected to aeration or to periodic stirring.[49] Either raw or digested sludge may be subjected to composting. One legal definition of the process specifies that a temperature of at least 60°C be sustained in all parts of the sludge mass for at least 48 hr; another specifies a temperature of at least 55°C for 3 days for a PFRP.[25]

Virus inactivation during the composting process has been very difficult to study because of the extreme heterogeneity of the material being processed. However, composted sludge has been used as a model in one virus inactivation study in which the temperatures tested were very conservative.[36] Inactivation of poliovirus 1 exceeded 99.99% in 100 min at 43°C, and inactivation of reovirus exceeded 90% in 200 min at 45°C. It would appear that if all parts of the composting sludge really are subjected to temperatures exceeding 55°C for at least 3 days virus infectivity in the product should be negligible.

D. Thermal Treatments

Many community-scale thermal treatment units for sludge are already in operation in the U.S., holding sludge at temperatures of 150° to 260° for 15 to 40 min.[50] If a temperature of 180°C is maintained for 30 min, this is a PFRP.[25] Even higher temperatures can be attained if air is added during the steam-heating phase. Temperatures of pasteurization may be as low as 70°C; if these are maintained for at least 30 min after a PSRP, this is also a PFRP.[25]

Sludge that has been pasteurized at 60°C for 1 hr was shown to be significantly less protective of experimentally inoculated coxsackievirus B3 than was anaerobically, mesophilically digested sludge, however, the temperatures of virus treatment were only 32° and 35°C.[37] In a study where pasteurization was defined as holding a temperature of at least 80°C for at least 10 min, inoculated poliovirus 1 and coxsackievirus B3 could not be detected in sludge samples taken after a temperature of 70°C had been attained.[51] The results with pasteurization of sludge suggest that virus persistence at thermal treatment temperatures is not a concern.

E. Irradiation

Ionizing radiation affords a potential means of destroying microorganisms in sludge, though viruses present smaller targets than bacteria and are therefore likely to require larger irradiation doses for any given percentage kill. Irradiation probably does not stabilize the sludge for disposal, so it is a process that further reduces pathogens if a dose of 1 Mrad is applied at room temperature to sludge that has already received a PSRP.[25] Though irradiation of sludge is not presently in use in the U.S., a good deal of research on the subject has been done.

Electrons accelerated to 1.5-mV energy level and applied in a dose of 400 krad have been shown to be effective in bacteriological disinfection of sludge.[52] Virological data from the study were not reported.

γ-Rays from ^{60}Co have greater penetrating power than accelerated electrons. A 300-krad dose was originally reported to produce a 1.5-log reduction of experimentally inoculated poliovirus 1 in sludge.[53] However, the same author and a colleague have since conducted further studies and found that the D_{10} values (doses for 90% inactivation) for four enteroviruses (poliovirus 1, echovirus 6, and coxsackieviruses A9 and B5) in raw and digested sludges were in the range of 300 to 400 krad, whereas the D_{10} value for adenovirus 1 was about 700 krad under these conditions.[54] A study by another group put the D_{10} value of ^{60}Co γ-rays for poliovirus 1 in raw sludge (3 to 5% solids) at 190 krad; D_{10} values for reovirus and a bovine parvovirus were 550 and 800 krad, respectively.[19]

γ-Rays from ^{137}Cs can also be used for irradiation of sludge.[55] The D_{10} value against poliovirus 1 in phosphate-buffered saline was shown to be 192 krad, whereas the D_{10} value was 340 krad when 2.34% solids from raw sludge were added to the suspension. Heat, as little as 10 min at 47°, was shown to act synergistically, so that the combination of heating and irradiation inactivated more virus than the sum of the inactivations achieved with each process singly.

F. Lime Stabilization

Stabilization of sludge with lime, which may be done after or in lieu of digestion, should not be confused with treatment of wastewater with lime. It was stated earlier that advanced wastewater treatment with lime at pH 9.5 to 10.5 produced a sludge in which no significant inactivation of experimentally added poliovirus 1 occurred during 48 hr at 28°C.[32]

Among the design criteria for lime stabilization processes to be applied in various situations, a central theme seems to be the attainment of a pH of 12 or above.[56] Sludge that has been stabilized with lime still contains considerable levels of volatile solids, so it is not entirely stable in some situations; nevertheless, addition of lime to attain pH 12 and maintain it for at least 2 hr is designated a process that significantly reduces pathogens.[57]

Poliovirus 1, experimentally inoculated into anaerobically digested sludge, was inactivated nearly 99.9% within 6 hr after addition of lime to bring the pH to 10.05, whereas no lime (pH, 7.45) or less lime (pH, 8.17) produced only about 90% inactivation under the same conditions.[58] The same investigators reported that addition of lime to raw sludge to attain at least pH 12 produced 99.99% inactivation of experimentally added poliovirus 1 within 30 min, but that much more lime was needed to reach pH 12 in anaerobically digested sludge, and that 99.99% inactivation of poliovirus 1 occurred within 5 days and of a bovine parvovirus, within 24 hr.[40]

G. Dewatering and Drying

The purpose of dewatering and drying of sludge is to reduce the volume that must be handled and perhaps to permit bagging and distribution of the material as a dry cake. Dewatering generally starts with digested or stabilized sludge in the U.S., but there are certainly exceptions. Chemical additives are often used, followed by filters, centrifuges, drying beds, or combinations thereof.[59] Air drying in a maximum depth of 22.5 cm for at least 3 months, during two of which daily average temperatures exceed 0°C, is a PSRP; heating the sludge to at least 80°C while reducing the moisture to 10% or lower is a PFRP.[25] These processes are not to be confused with the thickening that is often done to raw sludge before digestion or under stabilizing treatment.

Blended sludge, conditioned with lime and $FeSO_4$ and dewatered to 39% (w/v) solids, was not found to contain detectable virus, whereas high levels (about 100 pfu/g) had been present in the sludge before dewatering.[60] Sludge from another community, conditioned with polymer and press filtered to 45% (w/v) solids was also free of virus; however, sludges from three more communities, conditioned with polymer and dewatered by centrifugation to 2.7% to 21% (w/v) solids, often contained relatively high levels of virus. The authors suggest that the polymer may have dislodged virus that had been adsorbed to sludge solids, but alternative explanations seem possible.

When raw sludge was air dried at 21°C, inactivations exceeding 99.99% within 11 days were seen when solids contents exceeded 80%, but not at lower levels, with experimentally added poliovirus 1, coxsackievirus B1, and reovirus 3.[61] No inactivation was seen if virus was added to sludge that had already been dewatered to 86% solids and held for 7 days at 21°C. Inactivation was attributed to the effects of evaporation of water, which caused disintegration of the viral particle and degradation of the nucleic acid. In a further study on air drying of raw sludge at 21°C, inactivation of indigenous viruses and of experimentally inoculated poliovirus 1 was about 99% when a level of 80% solids was reached and much greater (especially for the poliovirus) with further drying.[62] Indigenous viruses in anaerobically digested sludge being dried on outdoor beds in Israel showed about 32% inactivation per week.[63]

H. Lagooning

Lagooning of sludge is regarded by some as a dewatering process and by other as a means of disposal. Lagooning can also serve the purposes of improving sludge stabilization and of storage to allow disposal to land when the season is right. With sufficiently high temperatures and long enough periods of storage, lagooning may be considered a process that further reduces pathogens.[3]

Sludge in a lagoon in northern Florida no longer contained detectable viral infectivity $1^1/_2$ months after addition of fresh sludge was halted.[20] Viruses were again detectable when addition of sludge resumed. No direct antiviral effect of lagooning of sludge is claimed: it appears that viral inactivation is largely a function of time and temperature. Although climatic data are given for the period of the study, it is unfortunate that temperatures in the lagoon itself were not reported.

VI. DISCUSSION AND SUMMARY

Virus transmission via urban sewage sludge is virtually unknown in the U.S., but it still seems worth learning as much as possible about the fate of viruses during sludge processing. Much of the information is confusing; however, one may well argue that any information is better than none.

The focus of this chapter has been on the processing of sludge for land application, including what the U.S. Environmental Protection Agency calls distribution and marketing. Agricultural and other lands are estimated to receive about 42% of the sludge generated in the U.S.[25] Land application for agricultural purposes calls for sludge treatment by a PSRP, the norm for which is anaerobic digestion. Distribution and marketing are to be done only with sludges that have received a PFRP. There can be little question of the validity of these principles, but it should be noted that viruses often are not the pathogens of greatest concern in safe sludge disposal.

Sludge processes have generally been designed to take the so-called solids from wastewater treatment and stabilize them so that they do not create a nuisance (odor or other) when disposed of where they impact on the public. Because sludge handling represents a very great portion of the cost of sewage treatment, and because sludge often contains higher levels of pathogens than wastewater in a suspension that is harder to disinfect than the fluid stream, innovation in the field of sludge treatment can be expected to continue. This chapter has not considered all the possibilities by any means; for example, vermicomposting, a process whereby urban sludges are processed by earthworms, seems not to have been studied from a virological standpoint as yet.[64]

Enteroviruses have regularly been used as indicators of the probable presence of potentially water- or food-borne viruses in sludge and to determine the extent of viral inactivation in sludge treatment. Only a few other model viruses, including coliphages, have been employed in this way.

The extent of viral inactivation seems to depend largely on the length of time the sludge is held at certain temperatures and on pH. The enteroviruses become increasingly labile with increasing temperature in the mesophilic range and are rapidly inactivated in the thermophilic range, especially at temperatures imposed by pasteurization and thermal treatments of sludge. Enteric viruses generally seem to be more tolerant of acid than alkaline conditions and to be most stable near pH 7, at which domestic wastewater and much sludge tend to be. If sludge is treated so as to render it decidedly alkaline, ammonia present may act specifically against the enteroviruses, and the alkalinity will at least favor inactivation of any other enteric viruses that may be present. The effects of detergents in sludge have been shown to be significant in some cases, but their actions are so complex as to make it very difficult to analyze a sludge for detergents in the hope of using the results to predict the fate of viruses during treatment. It must also be noted that laboratory results have been fairly consistent in predicting the fate of viruses during experimental sludge treatments, but have not always been consistent with observations in wastewater treatment plants. The mixing patterns of sludge in large digesters seem not to be completely predictable and probably contribute to the variable results.

There are still many unknowns in predicting the fate of viruses during sludge processing.

One would like to know whether biological inactivation of viruses takes place at all in anaerobic sludge digestion. More needs to be learned of the role of different classes of sludge solids in determining the stability of viruses during sludge treatment. More emphasis needs to be placed not only on means of testing sludge for viruses, but also on factors and components that govern the fate of viruses that are present. Alternatively, sludge treatments might be specified that impose one major condition (e.g., high temperature or high pH) that will essentially guarantee virus inactivation without regard for minor factors; this approach might prove costly.

It should be noted that the studies cited in this chapter were all model studies, in the sense that none was done with the virus of hepatitis A (for which the other enteroviruses used experimentally *may* be a valid model) or the Norwalk virus, the two major food- and water-borne viruses. Very limited work has been done with rotaviruses, which are rarely water-borne and perhaps never food-borne. Where doubt exists as to the continuing presence of these viruses in sludge, proper land application at sites to which the general public has little or no access, rather than distribution and marketing, will probably render public health risks negligible. If other conditions permit, the use of sludge as a soil amendment seems to be greatly preferred over methods of destruction or disposition.

ACKNOWLEDGMENTS

Contribution from the College of Agricultural and Life Sciences, University of Wisconsin-Madison. I thank Lynette F. Tenorio for her excellent work in organizing the materials cited in this chapter, and I am grateful to Dr. Joseph B. Farrell of the U.S. Environmental Protection Agency for the important information he provided.

REFERENCES

1. **Sieger, R. B. and Maroney, P. M.,** Inactivation-pyrolysis of wastewater treatment plant sludges, in Sludge Treatment and Disposal, Vol. 2, Sludge Disposal, EPA-625/4-78-012, U.S. Environmental Protection Agency, Cincinnati, 1978, 1.
2. **Keeney, D. R., Kwang, W. L., and Walsh, L. M.,** Guidelines for the Application of Wastewater Sludge to Agricultural Land in Wisconsin, Tech. Bull. No. 88, Department of Natural Resources, Madison, 1975.
3. **Farrell, J. B.,** Pathogen Reduction Studies in EPA's Sludge Management Program, EPA-600/D-84-220, U.S. Environmental Protection Agency, Cincinnati, 1984.
4. **Cliver, D. O.,** Significance of water and the environment in the transmission of virus disease, *Monogr. Virol.*, 15, 30, 1984.
5. Centers for Disease Control, Water-Related Disease Outbreaks: Annual Summary 1979 (issued September 1981), p. 3.
6. Centers for Disease Control, Water-Related Disease Outbreaks: Annual Summary 1980 (issued February 1982), p. 3.
7. Centers for Disease Control, Water-Related Disease Outbreaks: Annual Summary 1981 (issued September 1982), p. 3.
8. Centers for Disease Control, Water-Related Disease Outbreaks: Annual Summary 1982 (issued August 1983), p. 3.
9. Centers for Disease Control, Water-Related Disease Outbreaks: Annual Summary 1983 (issued September 1984), p. 3.
10. **Cliver, D. O.,** Manual on Food Virology, World Health Organization, Geneva, 1983.
11. **Appleton, H.,** Outbreaks of viral gastroenteritis associated with food, *De Ware(n)-Chemicus*, 13, 37, 1983.

12. **Ward, R. L. and Ashley, C. S.**, Effects of wastewater sludge and its detergents on the stability of rotavirus, *Appl. Environ. Microbiol.*, 39, 1154, 1980.

13. **Bertucci, J. J., Lue-Hing, C., Zenz, D., and Sedita, S. J.**, Inactivation of viruses during anaerobic sludge digestion, *J. Water Pollut. Control Fed.*, 49, 1642, 1977.

14. **Subrahmanyan, T. P.**, Persistence of enteroviruses in sewage sludge, *Bull. WHO*, 55, 431, 1977.

15. **Ward, R. L. and Ashley, C. S.**, Identification of the virucidal agent in wastewater sludge, *Appl. Environ. Microbiol.*, 33, 860, 1977.

16. **Ward, R. L. and Ashley, C. S.**, Identification of detergents as components of wastewater sludge that modify the thermal stability of reovirus and enteroviruses, *Appl. Environ. Microbiol.*, 36, 889, 1978.

17. **Ward, R. L. and Ashley, C. S.**, pH modification of the effects of detergents on the stability of enteric viruses, *Appl. Environ. Microbiol.*, 38, 314, 1979.

18. **Berg, G. and Berman, D.**, Destruction by anaerobic mesophilic and thermophilic digestion of viruses and indicator bacteria indigenous to domestic sludges, *Appl. Environ. Microbiol.*, 39, 361, 1980.

19. **Stettmund Von Brodorotti, H. and Mahnel, H.**, Inactivation of viruses and bacteria in sewage sludge by gamma radiation, *Zentralbl. Bakteriol. Parasitenkd. Infektionskr. Hyg. Abt. 1: Orig Reihe B*, 170, 71, 1980.

20. **Farrah, S. R., Bitton, G., Hoffman, E. M., Lanni, O., Pancorbo, O. C., Lutrick, M. C., and Bertrand, J. E.**, Survival of enteroviruses and coliform bacteria in a sludge lagoon, *Appl. Environ. Microbiol.*, 41, 459, 1981.

21. **Farrah, S. R. and Bitton, G.**, Bacterial survival and association with sludge flocs during aerobic and anaerobic digestion of wastewater sludge under laboratory conditions, *Appl. Environ. Microbiol.*, 45, 174, 1983.

22. **Cliver, D. O.**, Virus association with wastewater solids, *Environ. Lett.*, 10, 215, 1975.

23. **Lund, E. and Ronne, V.**, On the isolation of virus from sewage treatment plant sludges, *Water Res.*, 7, 863, 1973.

24. **Goyal, S. M., Schaub, S. A., Wellings, F. M., Berman, D., Glass, J. S., Hurst, C. J., Brashear, D. A., Sorber, C. A., Moore, B. E., Bitton, G., Gibbs, P. H., and Farrah, S. R.**, Round robin investigation of methods for recovering human enteric viruses from sludge, *Appl. Environ. Microbiol.*, 48, 531, 1984.

25. U.S. EPA Intra-Agency Sludge Task Force, Use and Disposal of Municipal Wastewater Sludge, EPA 625/10-84-003, U.S. Environmental Protection Agency, Washington, D.C., 1984.

26. **Carlson, S.**, Virusvorkommen im Rohabwasser, Abwasserschlamm und Ablauf Biologischer Kläranlagen, *Zentralbl. Bakteriol. Parasitenkd. Infektionskr. Hyg. Abt. 1: Orig.*, 212, 50, 1969.

27. **Ward, R. L.**, Evidence that microorganisms cause inactivation of viruses in activated sludge, *Appl. Environ. Microbiol.*, 43, 1221, 1982.

28. **Balluz, S. A. and Butler, M.**, The influence of operating conditions of activated-sludge treatment on the behavior of f2 coliphages, *J. Hyg. Cambr.*, 82, 285, 1979.

29. **Manwaring, J. F., Chaudhuri, M., and Englebrecht, R. S.**, Removal of viruses by coagulation and flocculation, *J. Am. Water Works Assoc.*, 63, 298, 1971.

30. **Sproul, O. J.**, Critical Review of Virus Removal by Coagulation Processes and pH Modifications, EPA-600/2-80-004, U.S. Environmental Protection Agency, Cincinnati, 1980.

31. **Grabow, W. O. K., Middendorff, I. G., and Basson, N. C.**, Role of lime treatment in the removal of bacteria, enteric viruses, and coliphages in a wastewater reclamation plant, *Appl. Environ. Microbiol.*, 35, 663, 1978.

32. **Sattar, S. A. and Ramia, S.**, Viruses in sewage: effect of phosphate removal with calcium hydroxide (lime), *Can. J. Microbiol.*, 24, 1004, 1978.

33. **Palfi, A.**, Survival of enteroviruses during anaerobic sludge digestion, in *Progress in Water Technology*, Vol. 3, *Water Quality: Management and Pollution Control Problems*, Jenkins, S. H., Ed., Pergamon Press, Oxford, 1973, 99.

34. **Sanders, D. A., Malina, J. F., Jr., Moore, B. E., Sagik, B. P., and Sorber, C. A.**, Fate of poliovirus during anaerobic digestion, *J. Water Pollut. Control Fed.*, 51, 333, 1979.

35. **Ward, R. L., Ashley, C. S., and Moseley, R. H.**, Heat inactivation of poliovirus in wastewater sludge, *Appl. Environ. Microbiol.*, 32, 339, 1976.

36. **Ward, R. L. and Ashley, C. S.**, Heat inactivation of enteric viruses in dewatered wastewater sludge, *Appl. Environ. Microbiol.*, 36, 898, 1978.

37. **Eisenhardt, A., Lund, E., and Nissen, B.**, The effect of sludge digestion on virus infectivity, *Water Res.*, 11, 579, 1977.

38. **Migone, N. A.**, Anaerobic digestion and design of municipal wastewater sludges, in Sludge Treatment and Disposal, Vol. 1, Sludge Treatment, EPA-625/4-78-012, U.S. Environmental Protection Agency, Cincinnati, 1978, 35.

39. **Jeris, J. S., Ciarcia, D., Chen, E., and Mena, M.**, Determining the Stability of Treated Municipal Wastewater Sludges, Project Summary, EPA-600/S2-85-001, U.S. Environmental Protection Agency, Cincinnati, 1985.

40. **Koch, K. and Strauch, D.,** Removal of polio- and parvovirus in sewage-sludge by lime-treatment, *Zentralbl. Bakteriol. Parasitenkd. Infektionskr. Hyg. Abt. 1: Orig. Reihe B,* 174, 335, 1981.

41. **Cramer, W. N., Burge, W. D., and Kawata, K.,** Kinetics of virus inactivation by ammonia, *Appl. Environ. Microbiol.,* 45, 760, 1983.

42. **Ward, R. L. and Ashley, C. S.,** Inactivation of poliovirus in digested sludge, *Appl. Environ. Microbiol.,* 31, 921, 1976.

43. **Noland, R. F. and Dickerson, R. B.,** Thickening of sludge, in Sludge Treatment and Disposal, Vol. 1, Sludge Treatment, EPA-625/4-78-012, U.S. Environmental Protection Agency, Cincinnati, 1978, 79.

44. **Meyer, R. C., Hinds, F. C., Isaacson, H. R., and Hinesly, T. D.,** *Porcine Enterovirus Survival and Anaerobic Sludge Digestion,* American Society for Agricultural Engineers, St. Joseph, Mich., 1971, 183.

45. **Goddard, M. R., Bates, J., and Butler, M.,** Recovery of indigenous enteroviruses from raw and digested sewage sludges, *Appl. Environ. Microbiol.,* 42, 1023, 1981.

46. **Mignone, N. A.,** Aerobic digestion and design of municipal wastewater sludges, in Sludge Treatment and Disposal, Vol. 1, Sludge Treatment, EPA-625/4-78-012, U.S. Environmental Protection Agency, Cincinnati, 1978, 57.

47. **Farrah, S. R. and Schaub, S. A.,** Viruses in wastewater sludges, in *Viral Pollution of the Environment,* Berg, G., Ed., CRC Press, Boca Raton, Fla., 1983, 147.

48. **Jewell, W. J., Kabrick, R. M., and Spada, J. A.,** Autoheated, Aerobic Thermophilic Digestion with Air Aeration, Project Summary, EPA-600/S2-82-023, U.S. Environmental Protection Agency, Cincinnati, 1982.

49. **Wesner, G. M.,** Sewage sludge composting, in Sludge Treatment and Disposal, Vol. 2, Sludge Disposal, EPA-625/4-78-012, U.S. Environmental Protection Agency, Cincinnati, 1978, 35.

50. **Wesner, G. M.,** Thermal treatment for sludge conditioning, in Sludge Treatment and Disposal, Vol. 1, Sludge Treatment, EPA-625/4-78-012, U.S. Environmental Protection Agency, Cincinnati, 1978, 69.

51. **Foliguet, J.-M. and Doncoeur, F.,** Inactivation assays of enteroviruses and *Salmonella* in fresh and digested wastewater sludges by pasteurization, *Water Res.,* 6, 1399, 1972.

52. **Trump, J. G.,** Energized electrons tackle municipal sludge, *Am. Sci.,* 69, 276, 1981.

53. **Epp, C.,** Experience with a pilot plant for the irradiation of sewage sludge: experiments on the inactivation of viruses in sewage sludge after radiation treatment, in Radiation for a Clean Environment, International Atomic Energy Agency, Vienna, 1975, 485.

54. **Epp, C. and Metz, H.,** Virological analyses of irradiated sewer sludge, *Zentralbl. Bakteriol. Parasitenkd. Infektionskr. Hyg. Abt. 1: Orig. Reihe B,* 171, 86, 1980.

55. **Ward, R. L.,** Inactivation of poliovirus in wastewater sludge with radiation and thermoradiation, *Appl. Environ. Microbiol.,* 33, 1218, 1977.

56. **Noland, R. F. and Edwards, J. D.,** Lime stabilization of wastewater treatment plant sludges, in Sludge Treatment and Disposal, Vol. 1, Sludge Treatment, EPA-625/4-78-012, U.S. Environmental Protection Agency, Cincinnati, 1978, 1.

57. **Otoski, R. M.,** Lime Stabilization and Ultimate Disposal of Municipal Wastewater Sludges, Project Summary, EPA-600/S2-81-076, U.S. Environmental Protection Agency, Cincinnati, 1981.

58. **Koch, K.,** A contribution to the recovery of poliomyelitisvirus from digested sewage sludge dependent on pH-value, *Zentralbl. Bakteriol. Parasitenkd. Infektionskr. Hyg. Abt. 1: Orig. Reihe B,* 174, 325, 1981.

59. **Harrison, J. R.,** Review of developments in dewatering wastewater sludges, in Sludge Treatment and Disposal, Vol. 1, Sludge Treatment, EPA-625/4-78-012, U.S. Environmental Protection Agency, Cincinnati, 1978, 101.

60. **Goddard, M. R., Bates, J., and Butler, M.,** Isolation of indigenous enteroviruses from chemically treated and dewatered sludge samples, *Appl. Environ. Microbiol.,* 44, 1042, 1982.

61. **Ward, R. L. and Ashley, C. S.,** Inactivation of enteric viruses in wastewater sludge through dewatering by evaporation, *Appl. Environ. Microbiol.,* 34, 564, 1977.

62. **Brashear, D. A. and Ward, R. L.,** Inactivation of indigenous viruses in raw sludge by air drying, *Appl. Environ. Microbiol.,* 45, 1943, 1983.

63. **Vasl, R. and Kott, Y.,** The fate of enteroviruses at Haifa's municipal wastewater treatment plant, in *Developments in Arid Zone Ecology and Environmental Quality,* Shuval, H., Ed., Balaban ISS, Philadelphia, 1981, 233.

64. **Donovan, J.,** Engineering Assessment of Vermicomposting Municipal Wastewater Sludges, Project Summary, EPA-600/S2-81-075, U.S. Environmental Protection Agency, Cincinnati, 1981.

65. **Farrell, J. B.,** U.S. Environmental Protection Agency, personal communication.

66. **Scheuerman, P.,** unpublished data.

Chapter 10

METHODS OF ENTEROVIRUS RECOVERY FROM DIFFERENT TYPES OF SOILS

Flora Mae Wellings

TABLE OF CONTENTS

I. INTRODUCTION

Soils have been used for the treatment of human biological wastes since man came into being, but in recent years, the practice has increased alarmingly. The need to clean up surface waters has led to land disposal of large volumes of wastewater and sludge. In addition, the ever-expanding use of septic tanks to service sprawling urban growth has added immeasurably to the wastewater load that soils must accommodate.

It has been assumed that pathogenic microorganisms would be removed by filtration through the soil and/or die off quickly under the unstable conditions of the environment. The fallacy of this assumption has been demonstrated time and again by outbreaks of cholera,[1] typhoid,[2,3] hepatitis,[4-7] and many other reported water-borne diseases.[8-10] Indisputable evidence that contamination of groundwater does occur is the data showing that over 50% of the reported water-borne disease outbreaks which have occurred in the U.S. over the past 50 years have been related to the use of untreated or chlorinated groundwater.[11] Equally striking is the fact that virus infections may account for roughly 60% of all the outbreaks.

Recent virological studies have clearly demonstrated that viruses adsorb to soil particles under certain conditions and desorb when conditions change and that some eventually reach the groundwater.[12-14] In order to scientifically determine the soil types which may enhance virus removal and the rate and route of virus movement through soils under natural conditions, methods for the recovery of virus from various types of soils are required.

II. SOIL SAMPLES

A. General Considerations

Great care must be taken to select the most representative soil sample because the result of a laboratory test is only as valid as the sample submitted. Unlike marine sediments, in which viruses are readily demonstrated because solids-associated virus tends to settle out and concentrate in the sediments on the ocean bottom,[15-18] viruses are much less frequently isolated from soils for a variety of reasons. The quantity of virus present depends upon the amount of wastewater, sludge, or septic tank leachate to which the soil had been exposed and the quantity of virus present in the virological wastes applied to the soils. However, the heterogeneous distribution of viruses in these materials would preclude a homogeneous distribution within the soil matrix. Thus, the sample selected may or may not be characteristic of the whole. Regardless of the efficiency of the methods used for recovering viruses, erroneous findings may result because of the sample per se.

Other factors involved in virus adsorption to soils, and thus the amount of virus present in a soil sample, include the type of soil, the type and strain of virus, meteorological conditions, and the hydrology and topography of the site.[12-14,19,20] Each of these currently recognized factors plays an important role in the survival and movement of virus within the soil matrix.

Laboratory-based column studies have shown that most viruses adsorb within the upper few centimeters of a soil column, depending upon the type of soil and the rate of application.[20-26] In a study in which virus seeded columns were exposed to routine dosing with distilled water to simulate rainfall over a 6-month period, bands of virus were demonstrated throughout the length of the column.[27] This resulted from the adsorption-desorption-adsorption of the virus over the life of the column without inactivation of the virus. This rainfall effect was first observed under field conditions. In a study in which secondary effluents were sprayed onto sandy soils, viruses were demonstrated in groundwater only after heavy rains.[28] Thus, based on both laboratory and field studies, it is evident that virus movement through soils does occur and directly effects the location and quantity of virus present within the soil matrix.

The most representative soil sample may or may not be that obtained from the upper few centimeters of the soil because of possible thermal inactivation, particularly in sandy soils during hot weather.[29] In order to obtain a fairly representative soil sample, it is suggested that the direction of the flow of water away from the site be determined and that samples be obtained at different depths within the area. If an aquaclude were encountered, several samples at this level may be most productive. Aquatards present in sandy soils are frequently composed of organic materials which offer excellent adsorption sites for virus and may actually serve as a virus concentration site.

B. Sample Size

The amount of soil per sample which can be processed efficiently is the limiting factor in technique development. Even though the quantity of virus expected to be found throughout the soil column exposed to virological wastes is greater than that anticipated in groundwater, recovering virus from soils is much less probable. With groundwater, sample size may be increased from a few liters to over hundreds of gallons without any real difficulty, but increasing the size of soil samples to comparable levels would be impossible from a logistical as well as processing standpoint.

Techniques presently in use are designed for samples ranging in size from 50 to 500 mℓ. Larger amounts could be processed, but the volume of the virus concentrate may be prohibitively costly to assay for virus. In addition, concentration of cytotoxic elements which are present in many of the soils would be greatly increased, which would require dilution of the concentrate and, again, increase the costs involved.

It may be desirable to obtain soil from several different sites at comparable depths and after thorough mixing to process several 50-g samples of the composite. Each investigator must evaluate the immediate situation and determine how the most representative sample may be obtained.

C. Statistical Requirements

The number of samples that are necessary for a valid virological evaluation of a waste disposal site is most difficult to determine and yet is one of the most important facets of this type of study. In determining the final number of samples to be tested, careful consideration must be given to the efficiency of the viral recovery and assay systems, the constraints of finances and personnel, and the probability of a preponderance of negative samples.

The type of statistical tests that are applied to the accumulated data must be selected with great perspicacity. The Poisson theory, which may appear to be the most applicable, may not be appropriate because of the heterogeneous distribution of virus in environmental samples and the tendency of virions to persist as clumps.[30] Consultation with a biometrician during the design of the study would be appropriate.

D. Sample Handling

Samples should be placed in individual sterile containers, either wide-mouthed, screw-capped glass or plastic jars or resealable plastic bags. It is not necessary to freeze the sample, but it should be held on wet ice while in the field and during transport to the laboratory. If a mobile unit is available with a source of electrical refrigeration, storing the samples at 4°C is ideal. At that temperature the virus is less likely to be inactivated, and bacterial growth is retarded.

Samples that are to be shipped to a distant laboratory for processing should be placed in a well-insulated container with wet ice or cold cans. They should be shipped by the most expedient carrier as soon as possible after collection so that the temperature of the sample is held as constant as possible during the entire time.

Processing of the samples should be initiated as soon as possible after their arrival at the

Table 1
ELUTING MEDIA FOR RECOVERING VIRUSES FROM SOILS

Media	pH	Media-soil ratio	Treatment (time in min)	Ref.
0.25 *M* glycine buffer with 0.05 *M* EDTA[a]	11.5	4:1	Magnetic stirrer (4)	31
10% buffered beef extract	7.0	1:1[b]	Magnetic stirrer (30) Maintain pH 7.0	32
1% casein with 0.05 *M* glycine	9.0	3:1	Overhead stirrer (1)	33
3% beef extract	10.5	4:1	Shake vigorously by hand[c] (5) Adjust pH on magnetic stirrer	34

[a]   Ethylenediaminetetraacetic acid.
[b]   Soil should be slurry. Add distilled water if necessary.
[c]   Use securely capped bottle.

laboratory. If there is to be a delay of more than 4 days between the time of collecting and processing, freezing the samples at −70°C is advised. At the time of processing, samples are thawed rapidly at 37°C.

II. VIRUS RECOVERY METHODS

A. General Considerations

Virus recovery techniques, excluding direct microfiltration, rely on the physicochemical properties of the virion per se. Because of their protein coat, virions exhibit reactions typical of macromolecular proteins, and thus techniques used for concentrating these proteins are applicable to the concentration of virus (i.e, adsorption, desorption, sedimentation, flocculation, etc.). Numerous techniques have been described for the recovery of virus from solids such as sludge, estuarine sediments, and soils. Although these techniques vary in the type of eluting medium used and the method of final concentration of the resulting virus suspension, most of the methods do include the two-phase approach: elution and concentration.

B. Virus Elution

In the absence of a standard technique, elution of virus from soil particles reflects the individual investigator's preference. Table 1 lists four elution procedures commonly used. Usually, the eluting medium is added to a beaker containing the soil sample and the beaker placed on a magnetic stirrer for 5 to 30 min. It is important to maintain the desired pH during the stirring process by the addition of an acid or base as necessary. However, in any virus recovery technique in which glycine at pH 11.5 is used, contact should be limited to 4 or 5 min and the suspension immediately adjusted to pH 9.5 because of the deleterious effect of high pH on some of the virus types which may be present. Once the viruses are eluted from solids into the aqueous phase, solids are removed by centrifugation.The supernatant fluid is decanted into a sterile container, and the sediment is disinfected before discarding. Figure 1 shows the two methods currently being considered as standards, which have been round-robin tested by eight laboratories using four different soil types. A number of problems noted with both methods will be discussed later.

C. Virus Concentration

The second phase in the recovery of virus from soils is the concentration of virus in the eluate. This may be accomplished by either organic[35] or inorganic flocculation,[31] depending

VIRUS ELUTION

Berg and Berman[32]	Goyal[34]

Add 50 mℓ buffered 10% beef extract to 50 mℓ soil slurry
↓
Mix on magnetic stirrer for 30 min; maintain pH 7.0 ± 0.1 with 5 M HCl or 5 M NaOH
↓
Centrifuge suspension at 2500 × g for 30 min at 4°C
↓
Force supernatant fluid through serum-treated, stacked filters
↓
Store at −70°C until assayed for virus
↓
Sterilize and discard sediment

Add 200 mℓ 3% beef extract (pH, 10.5) to 50 g soil in screw-capped bottle
↓
Shake vigorously by hand for 5 min; then adjust to pH 9.5
↓
Centrifuge suspension at 1500 × g for 10 min at 4°C
↓
Decant supernatant fluid into sterile beaker for concentration
↓
Sterilize and discard sediment

FIGURE 1.   Virus elution.

on the type of eluting medium used. The supernatant fluids resulting from soils eluted with beef extract, casein, or milk are readily flocculated by acidification of the suspension to pH 3.5. The flocculate is removed from suspension by centrifugation and the supernatant fluid carefully removed and discarded after disinfection. However, when glycine is used, the virus in suspension must be sedimented by inorganic flocculation.

This may be accomplished by the addition of sufficient 1 M AlCl$_3$ to achieve a final concentration of 0.06 M. When the pH is lowered to 3.5 by the addition of 1 M glycine (pH, 2.0) with continuous stirring, a flocculate will form. The suspension is clarified by centrifugation at 1500 × g for 15 min at 4°C. The supernatant fluid is removed carefully to avoid disturbing the pellet and discarded after disinfection. The flocculate is retained.

Organic flocculates may be dissolved by the addition of 0.15 M Na$_2$HPO$_4$ ·7 H$_2$O or 0.05 M glycine (pH, 11). Once the precipitates are dissolved, sufficient 1 N HCl or 1 M glycine (pH, 2.0) is added to achieve a pH of 7.5. The inorganic flocculates may be dissolved by the addition of 4 vol of fetal bovine serum and mixed by rapid pipetting. A final adjustment to pH 7.0 to 7.2 is made by the addition of 1 N NaOH. Figure 2 gives a brief outline of the concentration technique which was subjected to round-robin testing.

D. Preparation of the Concentrate for Virus Assay

The last step in the preparation of the virus suspension for inoculation onto cell monolayers is to remove interfering microorganisms and toxicity, which may destroy the cells. Soil eluates that are not concentrated may be filtered through serum-treated, stacked membranes to remove the microorganisms and then frozen and stored at −70°C until assayed for virus. Removal of interfering microorganisms in the virus concentrates may be achieved by centrifugation at 1000 × g and/or by passing the concentrate through serum-treated stacked filters and the addition of antibiotics. In the author's laboratory, concentrates are passed through a 0.45 µg serum-treated membrane in a Swinney filter. Antibiotics consisting of 2000 U penicillin, 2000 µg streptomycin, and 250 µg mycostatin per milliliter are added to each concentrate, which is then held at room temperature (21°C) for 2 hr before being frozen and stored at −85°C until assayed for virus.

Eliminating toxicity poses a greater problem than does the elimination of troublesome microorganisms. It has been noted by many investigators that some of the toxicity appears to be related to a component of the beef extract which, in concentrated form, is toxic. This

VIRUS CONCENTRATION: GOYAL'S METHOD

Adjust eluate to pH 3.5 ± 0.1 with 1 *M* glycine (pH, 2.0) or 1 *N* HCl

↓

Stir slowly until floc forms (≤30 min), adjusting pH as necessary

↓

Centrifuge at 1000 × *g* for 5 min at 4°C

↓

Carefully remove supernate with pipette; do not disturb floc

↓

Suspend pellet in 8 to 10 mℓ 0.05 *M* glycine (pH, 11.0); mix thoroughly by pipetting

↓

Adjust to pH 9.5 and centrifuge at 1000 × *g* for 10 min at 4°C

↓

Decant supernate into beaker; adjust to pH 7.5 with 1 *N* HCl or 1 *M* glycine (pH, 2.0)

↓

Add 200 U penicillin, 200 μg streptomycin, and 250 μg mycostatin mℓ

↓

Freeze and hold at −70°C until assayed for virus

FIGURE 2.   Virus concentration.

toxicity will vary extensively depending on the commercial source and type of beef extract used (i.e., paste or powder). The former produces the most cytotoxic preparations. To avoid these problems, it is advantageous to pretest new batches of beef extract for their cytotoxicity and avoid using those which are cytotoxic. Each sample concentrate should be tested for toxicity by inoculation of cell culture tubes with 0.2 mℓ of the concentrate. Careful observation of the cell cultures for 24 hr is sufficient to determine whether the inoculum is toxic. If it were, further treatment would be necessary.

There are a variety of treatments which have been used to remove toxicity. Among the most frequently used are freon[36] or dithizone extraction,[37] cationic polyelectrolyte precipitation,[38] and washing the monolayer after the virus adsorption period.[39] The efficacy of these treatments varies extensively, probably depending on the physicochemical characteristics of the cytotoxic component. Hurst and Goyke[39] evaluated three treatments (freon extraction, addition of cationic polyelectrolyte cat-floc T, and washing inoculated monolayers) for their effect on toxicity and virus titer of wastewater sludge concentrates. Although washing of the inoculated monolayer was the most effective in reducing cytotoxicity, it also resulted in the greatest loss of virus.

A relatively minor change in the Glass et al.[37] dithizone technique has been shown to remove cytotoxicity from sludge concentrates. Replacing vigorous hand shaking of the dithizone suspension with 2 min on a vortex mixer at a speed sufficient to preclude separation into the three phases for a few moments after mixing resulted in removal of the cytotoxicity in sludge samples. Companion samples shaken vigorously by hand retained their toxicity and yielded about half the number of plaque-forming units (pfu) as the vortex-mixed samples.[40] This treatment has not been tested with soils, but it may prove to be helpful.

The presence of minute solids in the final virus concentrate may result in a lower than expected number of viruses. In a recent experiment,[40] 1600 mℓ of 3% buffered beef extract eluant was inoculated with 0.5 mℓ of seed virus prepared from stools of children who had received live poliovirus vaccine. After the seed virus was well mixed, 50 mℓ was removed to determine the level of virus present before concentration by organic flocculation. After concentration, the flocculate was reconstituted to 80 mℓ, adjusted to pH 7.2, and divided into two aliquots. One was sonicated and the other was passed through a serum-treated 0.45μm membrane in a Swinney filter and stored at −85°C. On assay, 103% of input virus was demonstrated in the sample which was filtered only, whereas 180% of input virus was demonstrated in the sonicated sample. This probably reflects the freeing of solids-occluded virus present in the seed virus, as would be expected in field samples.

E. Virus Assay

It is not the purpose in this chapter to discuss virus assay in detail, but a word of caution is necessary. In counting the number of plaque-forming units, it must be remembered that a counted plaque is not a plaque-forming unit until its viral etiology has been confirmed. If 20 or fewer plaques are counted, all should be confirmed as plaque-forming units (i.e., picked and passed into a tube of cell culture). If characteristic, viral cytopathogenic changes are noted, then the plaque may be considered a plaque forming unit. When large numbers of plaques are counted, at least a 10% random sample should be confirmed.

Evidence for the need to confirm plaques was demonstrated in the recent round-robin test. Every laboratory in which plaques were counted but not confirmed had excessively high numbers of plaque-forming units compared with those laboratories in which plaques were confirmed. In one laboratory, nine plaques from one of the samples were passed, but only one was confirmed as a plaque-forming unit. Based on 11.1% (1/9) plaque-forming unit confirmation, the 231 counted plaque-forming units on another sample were reduced to 25.7 pfu. This brought the number well within the 22.9 to 44.4 pfu range reported for this soil type by the three laboratories in which all of the plaques were passed for confirmation. Thus, a counted plaque is not a plaque-forming unit until it has been confirmed.

F. Efficiency of Methodology

Evaluating the efficiency of a technique is limited because of the unknown variables involved in field experimentation as opposed to the absolute control of variables under laboratory conditions. There are gross differences between field samples and artificially seeded samples used almost routinely to determine the efficiency of a new technique. Laboratory-grown viruses are passaged frequently in the laboratory for various reasons and because of this tend to become adapted. The subtle changes produced in the physicochemical properties of these adapted strains as opposed to the wild strains may be expressed as an alteration in their adsorption capabilities,[12] their sensitivity to inactivation by chlorine,[41] and/ or other characteristics yet to be recognized.

Another serious problem associated with the use of laboratory virus strains is their lack of protection against inactivating substances which may be present in the test waters. Based on various laboratory and field studies conducted to date, it appears that the survival of viruses in the environment is directly related to their being occluded within or adsorbed to solids.[42-45]

Laboratory-grown viruses are prepared as a suspension of naked virions. Even though the virus suspension may be added to the soils and adsorption to the soil particles permitted to proceed before the technique is tested, the virus seeded sample is not the same as a field sample. Virus in the former is surficially adsorbed. It is obvious that surficially adsorbed virus may be readily desorbed, but removal of virus which is occluded within solids, a condition not achieved by the seeding process, is much more difficult. Therefore, the efficiency rating achieved with a given technique for virus recovery from laboratory seeded samples is applicable only to recovery of surficially adsorbed virus.

Enteroviruses are replicated within the cells of the human intestinal tract and are integrated within the human stool, not merely adsorbed to the surface of the solids. To simulate this natural state in seeded experiments, the ideal seed virus is that present in wastewater or in feces. Both of these have been used routinely in the author's laboratory whenever seeded specimens are required. Influent wastewater from a local treatment plant is obtained and an aliquot subjected to polyethylene glycol hydroextraction.[46] The residue is sonicated to free occluded virions and, after centrifugation and antibiotic treatment, the sample is plaque assayed to quantitate the virus present. This provides the basis for calculating the amount of the original sample required to provide a specific virus level for seeding. It is recognized that the quantity of virus seeded may be more or less than what was calculated, simply because of the presence of the solids-occluded virus. However, assay of an aliquot of the test water does provide a fairly reliable count provided the sample is taken after the seed virus has been well mixed and the solids fairly well distributed throughout the sample.

Stool specimens used for virus seeding experiments are obtained from two sources. Members of the laboratory staff and their families and friends whose children have been immunized with live poliovirus vaccine are requested to collect stools from the children during a 10-day period starting on day 3 following immunization. These are frozen in the home immediately after collection and are transported to the laboratory on dry ice by a laboratory staff member. A second source is the unused portion of stool specimens submitted to the laboratory for virus diagnostic tests. These are held in a frozen state until it has been determined that virus was present in the specimen. A test aliquot is homogenized with a mortar and pestle in the presence of sterile alundum and centrifuged, and the supernatant fluid is inoculated into cell culture and white Swiss mice. If virus is demonstrated in the cell culture host, the reserved, frozen portion of the specimen is processed as seed virus. This consists merely of partially homogenizing the specimen with a mortar and pestle. A small aliquot is well homogenized in the presence of sterile alundum, sonicated and centrifuged, and the supernatant fluid is inoculated onto cell monolayers for plaque assay quantitation. Results are used to calculate the volume of the original specimen required to provide a specified level of virus for the seeded experiments.

The two methods described of providing viruses for seeded experiments avoids the use of free, unprotected virions as well as adapted laboratory strains. This provides for a more realistic evaluation of the efficiency of the concentration methodology.

G. Problems Frequently Encountered
The types of problems which may be encountered during virus recovery from soils are related to the microorganism present in the soil, the virus concentration method, and the physicochemical characteristics of the soil per se.

1. Contaminating Microorganisms
Interfering microorganisms are of particular importance because of the propensity of some soil microorganisms to produce a very small colony which may be mistaken for a viral plaque. Usually by microscopic examination of the plaque, its bacterial origin can be recognized but, if doubt remains, further observation is required. Frequently, the bacterial colony may be identified as such within a few days, but not always. It may be necessary to pass the plaque and under liquid growth medium, bacterial contamination becomes overt.

In the technique of Berg and Berman,[32] in which the eluate is not concentrated before assay, there is a problem with the persistence of bacterial and fungal contamination in spite of filtration. The filtration of some soil eluates through the serum-treated, stacked membranes is tedious and difficult, particularly the eluate from organic muck soil because of the viscosity. Excessive force during filtration may result in membrane rupture or in accidental loss of some of the eluate. It is possible that the persistence of bacteria and fungi in these preparations is due to the difficulties encountered in the filtration procedure. It is suggested that after filtration, antibiotics be added to the eluate before storage at $-70°C$.

Sonicating the eluate or the final concentrates at 100 W for 15 min in a rosette cooling cell is helpful in eradicating some of the undesirable microbiological agents. This process may also increase the number of plaque-forming units demonstrable by producing a monodispersion of virions which may have been clustered together or may free solids-occluded virus, as alluded to previously. A final centrifugation at $20,000 \times g$ for 20 min at 4°C removes any residual debris. Filtration through a serum treated 0.45μm membrane in a Swinney filter followed by antibiotic treatment usually renders the suspension suitable for assay.

2. Problems Related to Concentration Methodology

Direct assay of the eluted virus without further concentration is costly. The volume of eluate varies between 30 and 50 mℓ. With the usual volume of inoculum per bottle being roughly between 0.5 and 2.0 mℓ, it is obvious that many bottles of cell culture are required for assay.

The final concentrate obtained from the organic muck soil with the Goyal procedure is quite gelatinous. It requires approximately twice the amount of base to dissolve the flocculate as that used for dissolving flocculates obtained from other types of soils. This increases the final volume which must be inoculated onto monolayers, thereby roughly doubling the cost of the viral assay.

3. Problems Related to Soil Type

The fine silt present in some of the sandy soils is extremely difficult to sediment. Final filtration will remove the silt in most instances, but the cytotoxicity experienced with some of the concentrates from sandy soils may be related to the passage of a portion of the very fine silt into the final concentrate.

The organic muck soil concentrate may be extremely cytotoxic even after treatment. Therefore, it may be necessary to dilute the concentrate before inoculation onto cell monolayers.

H. Cost Effectiveness

Ideally, methods used for the recovery of virus from various types of soils would be cost effective. However, it is relatively impossible to critically evaluate this criterion. An evaluation of cost effectiveness should incorporate a public health risk factor. Unfortunately, this is an unknown. It is recognized that hepatitis A disease outbreaks have been linked to ingestion of contaminated groundwater, but the number of cases is considered to be extremely small in comparison with the number of infections attributed to person-to-person contact. However, the source of infection for the index case in the person-to-person outbreak is seldom if ever identified. The epidemiologists are content to consider the source as being a subclinical case. To the environmental virologist, the source just as readily could be a water-borne infection. If such were the case, how many hepatitis A outbreaks would be avoided by action taken based on an improved understanding of virus movement within the soil matrix?

Enterovirus disease outbreaks, except for hepatitis A, have not been given the status of water-borne diseases. Yet many of the enteroviruses are so protean in their expression that a community experiencing a variety of illnesses may in fact be experiencing an unrecognized water-borne disease outbreak.

Supporting data for these suggested alternative explanations to those of the epidemiologists are beginning to be accrued.[46-51] When sufficient irrefutable data are available, decisions concerning whether a procedure is cost effective will be a valid one based on an evaluation of data rather than the invalid value judgments being made today in the absence of data.

IV. SUMMARY

The recovery of virus from different types of soils entails the elution of virus from the soil particles with direct viral assay of the eluate or concentration of the eluate before assay. Problems associated with the currently used techniques include large volumes for viral assay, persistent contamination, and toxicity of the concentrates.

At present, there is neither a standard technique nor a tentative standard technique for the recovery of virus from different types of soils. Because of this, there is no basis for a comparative evaluation of data. The initial round-robin testing of the two methods described herein indicated that each had its unique advantages and disadvantages. Perhaps within the next year or two, inprovements in either or both of the methods will be forthcoming and an efficient standard method will be available.

REFERENCES

1. **Snow, J.**, *Snow on Cholera (1854), a Reprint of Two Papers by John Snow, M.D.*, the Commonwealth Fund, Oxford University Press, London, 1936.
2. **Mallory, A., Beldin, E. A., and Brachman, P. S.**, The current status of typhoid fever in the United States and a description of an outbreak, *J. Infect. Dis.*, 119, 673, 1969.
3. **Pfeiffer, K. R.**, The Homestead typhoid outbreak, *J. Am. Water Works Assoc.*, 65, 803, 1973.
4. **Harrison, F. F.**, Infectious hepatitis: report of outbreak, apparently waterborne, *Arch. Intern. Med.*, 79, 622, 1947.
5. **Tucker, C. B., Owen, W. H., and Farrell, R. P.**, Outbreak of infectious hepatitis apparently transmitted through water, *South. Med. J.*, 47, 732, 1954.
6. **Peczenik, A., Duttweller, D. W., and Moser, R. H.**, Apparently waterborne outbreak of infectious hepatitis, *Am. J. Public Health*, 46, 1008, 1956.
7. **Mosely, J. W. and Smither, W. W.**, Infectious hepatitis: report of an outbreak probably caused by drinking water, *N. Engl. J. Med.*, 257, 590, 1957.
8. **Eliassen, R. and Cummings, R. H.**, Analysis of waterborne disease outbreaks, 1935—1945, *J. Am. Water Works Assoc.*, 40, 509, 1949.
9. **Weibel, S. R., Dixon, F. R., Weidner, R. B., and McCabe, L. J.**, Waterborne disease outbreaks, 1946—1960, *J. Am. Water Works Assoc.*, 56, 947, 1964.
10. **Craun, G. F.**, Outbreaks of waterborne disease in the United States 1971—1978, *J. Am. Water Works Assoc.*, 73, 360, 1981.
11. **Craun, G. F. and McCabe, L. J.**, Review of the causes of waterborne disease outbreaks, *J. Am. Water Works Assoc.*, 65, 1, 1973.
12. **Goyal, S. M. and Gerba, C. P.**, Comparative adsorption of human enteroviruses, simian rotavirus, and selected bacteriophages to soils, *Appl. Environ. Microbiol.*, 38, 241, 1979.
13. **Landry, E. F., Vaughn, J. M., Thomas, M. Z., and Beckwith, C. A.**, Adsorption of enteroviruses to soil cores and their subsequent elution by artificial rainwater, *Appl. Environ. Microbiol.*, 36, 544, 1979.
14. **Dizer, H., Nasser, A., and Lopez, J. M.**, Penetration of different human pathogenic viruses into sand columns percolated with distilled water, groundwater, or wastewater, *Appl. Environ. Microbiol.*, 47, 409, 1984.
15. **De Flora, S., De Renzi, G. P., and Badolati, G.**, Detection of animal viruses in coastal seawater and sediments, *Appl. Environ. Microbiol.*, 30, 472, 1975.
16. **Lo, S., Gilbert, J., and Hetrick, F.**, Stability of human enteroviruses in estuarine marine waters, *Appl. Environ. Microbiol.*, 38, 241, 1976.
17. **Smith, E. M., Gerba, C. P., and Melnick, J. L.**, Role of sediment in the persistence of enteroviruses in the estuarine environment, *Appl. Environ. Microbiol.*, 35, 685, 1978.
18. **LaBelle, R. L. and Gerba, C. P.**, Influence of estuarine sediment on virus survival under field conditions, *Appl. Environ. Microbiol.*, 39, 749, 1980.
19. **Wellings, F. M.**, Virus survival in wastewater treated soils, in *Proc. Int. Symp. Viruses and Wastewater Treatment*, Goddard, M. and Butler, M., Eds., Pergamon Press, Oxford, 1981, 117.
20. **Lance, J. C. and Gerba, C. P.**, Virus movement in soil during saturated and unsaturated flow, *Appl. Environ. Microbiol.*, 47, 335, 1984.
21. **Roebeck, G. G., Clarke, N. A., and Dostal, K. W.**, Effectiveness of water treatment processes in virus removal, *J. Am. Water Works Assoc.*, 54, 1275, 1962.
22. **Drewry, W. A. and Eliassen, R.**, Virus movement in groundwater, *J. Water Pollut. Control Fed.*, 40, 257, 1968.
23. **Wang, D., Gerba, C. P., and Lance, J. C.**, Effect of soil permeability on virus removal through soil columns, *Appl. Environ. Microbiol.*, 42, 83, 1981.
24. **Young, R. H. F. and Burbank, N. C., Jr.**, Virus removal in Hawaiian soils, *J. Am. Water Works Assoc.*, 69, 598, 1973.

25. **Duboise, S. M., Sagik, B. P., and Moore, B. E. D.,** Virus migration through soils, in *Proc. Nat. Water Resources Symp. No. 7 Virus Survival in Water and Wastewater Systems,* Malina, J. F., Jr. and Sagik, B. P., Eds., University of Texas, Austin, 1974, 233.
26. **Lefler, E. and Kott, Y.,** Virus retention and survival in sand, in *Proc. Nat. Water Resources Symp. No. 7 Virus Survival in Water and Wastewater Systems,* Malina, J. F., Jr. and Sagik, B. P., Eds., University of Texas, Austin, 1974, 84.
27. **Duboise, S. M., Moore, B. E., and Sagik, B. P.,** Poliovirus survival and movement in a sandy forest soil, *Appl. Environ. Microbiol.,* 31, 536, 1976.
28. **Wellings, F. M., Lewis, A. L., and Mountain, C. W.,** Virus survival following wastewater spray irrigation of sandy soils, in *Proc. Symp. Virus Survival in Water and Wastewater Systems,* Malina, J. F., Jr. and Sagik, B. P., Eds., Center for Research in Water Resources, Austin, 1976, 253.
29. **Yates, M. V., Gerba, C. P., and Kelley, L. M.,** Virus persistence in groundwater, *Appl. Environ. Microbiol.,* 49, 778, 1985.
30. **Sobsey, M. D.,** Field monitoring techniques and data analysis, in *Proc. Natl. Symp. Virus Aspects of Applying Municipal Waste to Land,* Baldwin, L. B., Davidson, J. M., and Gerber, J. F., Eds., Institute of Food and Agricultural Sciences, University of Florida, Gainesville, 1976, 87.
31. **Hurst, C. J. and Gerba, C. P.,** Development of a quantitative method for the detection of enteroviruses in soil, *Appl. Environ. Microbiol.,* 37, 626, 1979.
32. **Berg, G. and Berman, D.,** Test Method Submitted to ASTM Task Group Responsible for Developing Standard Methods for Virus Recovery from Soils, D19:24:04:04/05, American Society for Testing and Materials, Philadelphia, 1984.
33. **Farrah, S. R. and Bitton, G.,** Test Method Submitted to ASTM Task Group Responsible for Developing Standard Methods for Virus Recovery from Soils, D19:24:04:04/05, American Society for Testing and Materials, Philadelphia, 1984.
34. **Goyal, S. M.,** Test Method Submitted to ASTM Task Group Responsible for Developing Standard Methods for Virus Recovery from Soils, D19:24:04:04/05, American Society for Testing and Materials, Philadelphia, 1984.
35. **Katzenelson, E., Fattal, B., and Hostovesky, T.,** Organic flocculation: an efficient second-step concentration method for the detection of viruses in tap water, *Appl. Environ. Microbiol.,* 32, 638, 1976.
36. **Stark, L. M., Wellings, F. M., and Lewis, A. L.,** *Abstr. Annu. Meet. Am. Soc. Microbiol.,* Q51, 209, 1981.
37. **Glass, J. S., Van Sluis, R. J., and Yanko, W. A.,** Practical method for detecting poliovirus in anaerobic digester sludge, *Appl. Environ. Microbiol.,* 35, 983, 1978.
38. **Wait, S. A. and Sobsey, M. D.,** Method for recovery of enteric viruses from estuarine sediments with chaotropic agents, *Appl. Environ. Microbiol.,* 45, 379, 1983.
39. **Hurst, C. J. and Goyke, T.,** Reduction of interfering cytotoxicity associated with wastewater sludge concentrates assayed for indigenous enteric viruses, *Appl. Environ. Microbiol.,* 46, 133, 1983.
40. **Welllings, F. M. and Lewis, A. L.,** unpublished data, 1985.
41. **Payment, P., Tremblay, M., and Trudel, M.,** Relative resistance to chlorine of poliovirus and coxsackievirus isolates from environmental sources and drinking water, *Appl. Environ. Microbiol.,* 49, 981, 1985.
42. **Gerba, C. P. and Schaiberger, G. E.,** Effect of particulates on virus survival in seawater, *J. Water Pollut. Control Fed.,* 47, 93, 1975.
43. **Wellings, F. M., Lewis, A. L., and Mountain, C. W.,** Demonstration of solids-associated virus in wastewater and sludge, *Appl. Environ. Microbiol.,* 31, 354, 1976.
44. **Hejkal, T. W., Wellings, F. M., Lewis, A. L., and LaRock, P. A.,** Distribution of viruses associated with particles in wastewater, *Appl. Environ. Microbiol.,* 41, 628, 1981.
45. **Hejkal, T. W., Wellings, F. M., LaRock, P. A., and Lewis, A. L.,** Survival of poliovirus within organic solids during chlorination, *Appl. Environ. Microbiol.,* 38, 114, 1979.
46. **Wellings, F. M., Lewis, A. L., Mountain, C. W., and Pierce, L. V.,** Demonstration of virus in groundwater after effluent discharge onto soil, *Appl. Environ. Microbiol.,* 29, 751, 1975.
47. **Mack, W. N., Yue-Shoung, L., and Coohon, D. B.,** Isolation of poliomyelitis virus from a contaminated well, *Health Serv. Rep.,* 87, 271, 1972.
48. **Wellings, F. M., Mountain, C. W., and Lewis, A. L.,** Virus in groundwater, in *Proc. 2nd Natl. Conf. Individual On-Site Wastewater Systems,* National Sanitation Foundation, Ann Arbor, Mich., 1976, 61.
49. **Hejkal, T. W., Keswick, B., LaBelle, R. L., Gerba, C. P., Sanchez, Y., Dreesman, G., Hafkin, B., and Melnick, J. L.,** Viruses in a community water supply associated with an outbreak of gastroenteritis and infectious hepatitis, *J. Am. Water Works Assoc.,* 74, 318, 1982.
50. **Hopkins, R. S., Gaspard, G. B., Williams, F. P., Karlin, R. J., Cukor, G., and Blacklow, N. R.,** A community waterborne gastroenteritis outbreak: evidence for rotavirus as the agent, *Am. J. Public Health,* 74, 263, 1984.
51. **Bergeisen, G. H., Hinds, M. W., and Skaggs, J. W.,** A waterborne outbreak of hepatitis A in Meade County, Kentucky, *Am. J. Public Health,* 75, 161, 1985.

Chapter 11

TRANSPORT AND FATE OF VIRUSES IN SOILS: FIELD STUDIES

Charles P. Gerba

TABLE OF CONTENTS

I. INTRODUCTION

While laboratory studies are essential for an understanding of the basic mechanisms of virus fate and transport in soils, ultimate validation of actual behavior rests with field observations. Without such validation critical errors could be made in assessing virus behavior. For example, early laboratory studies on virus movement through soils suggested that viruses would be almost totally removed in the upper few centimeters of soil.[1] Field studies, though, have clearly documented that viruses can be transported significant distances in the subsurface. Further laboratory studies demonstrated that virus movement and survival in soils are highly type and strain dependent[1] and greatly influenced by water and soil characteristics, making it more difficult to actually predict virus behavior under field conditions.[2,3]

Adequate methods for the concentration and detection of human enteroviruses in large volumes of water and soil did not become available until the early 1970s,[4] and only in the last few years have methods become available for rotaviruses and hepatitis A virus (enterovirus type 72).[5,6] Microporous filter adsorption-elution methods developed for virus concentration made it feasible to sample large volumes of groundwater in the field. One of the first applications of this technology was the sampling of groundwaters beneath wastewater land application sites.[7] Thus, most of our knowledge on viruses in groundwater was derived from studies conducted within the last decade.

The prime concern of the presence of viruses in soils is their contamination of edible crops, landscaped areas used for recreation, and groundwater. In recent years in the U.S., there has also been heightened concern about groundwater contamination of our drinking water supplies. In this country approximately 25% of all water withdrawn for use is groundwater, and about 50% of the U.S. population depends on groundwater as its principal source of drinking water.[8] In rural areas in the U.S., groundwater supplies 90 to 95% of the drinking water.

From 1971 to 1982 groundwater was responsible for 51% of all waterborne outbreaks and 40% of all waterborne illness in the U.S.[9] Etiologic agents were determined in 48% of the outbreaks. Nine percent were known to be caused by enteric viruses (i.e., hepatitis A, Norwalk, and rotavirus). The most frequently reported source of contamination in outbreaks involving groundwater is overflow or seepage of sewage from septic tanks. It has been estimated that the total volume of waste disposed via septic tanks is approximately 800 billion gal/year, virtually all of which is disposed in the subsurface.[10]

In addition to septic tanks, it is estimated that there are over 3000 sites for land application of wastewater in the U.S.[10] Irrigation or spraying of sewage liquids and settled solids (sludges and sludge slurries) to agricultural lands is usually carried out on soils which only permit low-rate infiltration of associated liquids. An objective of this type of treatment is not only disposal of the wastes but also as an aid in crop production. Land treatment by overland flow also uses soils with low permeability that often has an underlying, confining clay layer to prevent percolation through the soil. Overland flow of wastewater is primarily used as a treatment method for water quality improvement and not for crop production. Rapid infiltration systems are used with soils of moderate to high permeability. Usually they are constructed as a series of basins in sandy soils. Deep-well injection of domestic wastewater, which is practiced in some regions of the U.S., is also another method by which viruses can be introduced into the subsurface. Other potential sources of viruses in the subsurface are listed in Table 1.

An understanding of virus migration through the soil is critical for judging the proper placement of septic and other forms of land disposal or treatment of wastes. In the past, estimates of safe distance between sources of sewage waste disposal and drinking water wells have been based on empirical judgment. Being able to predict virus survival and

Table 1
POTENTIAL SOURCES OF
VIRUSES IN GROUNDWATER
AND SOIL

Septic tanks
Sewer leakage
Sewage sludge
Intentional groundwater recharge with sewage
Irrigation with sewage
Direct injection of sewage
Domestic solid waste disposal (landfills)
Sewage oxidation ponds

transport can be useful criteria in developing safe distances between sewage disposal sites and drinking water supplies, especially if it can be determined from easily measured parameters and from existing data. Field studies on virus occurrence in groundwater and solids provide information about the extent of the problem, the adequacy of current guidelines, and the development of better guidelines.

II. FIELD STUDIES ON VIRUS OCCURRENCE IN SOILS AND GROUNDWATER

A. Septic Tanks

Although septic tanks appear responsible for the majority of water-borne disease outbreaks caused by contaminated groundwater, the amount of field work done on the occurrence of enteric viruses in groundwater near septic tanks is very limited. Mack et al.[11] were the first to isolate a virus from groundwater associated with a water-borne disease outbreak. The isolated poliovirus was taken in 5-gal samples from a 30.5-m deep well. The source was a septic tank drain field that allowed sewage to enter the well by passing through 5.5 m of clay, 2.5 m of shale, and 22.5 m of limestone. Although poliovirus was isolated from the well water, it was not suggested as the causative agent of the gastroenteritis outbreak.

Wellings et al.[12] reported the isolation of an echovirus 22/23 complex in 100-gal samples from a 12.2-m deep well during an outbreak of gastrointestinal illness at a migrant labor camp in Florida. The well was located 30.5 m from a solid waste landfill and was in the middle of an area bordered by septic tanks. In a study in Texas, an enterovirus was isolated from an abandoned well 25 m distance from a septic tank system serving a mobile home park.[13] Enteric viruses were isolated in groundwater on Long Island, N.Y. at distances of up to 67 m from the leaching pools of a subsurface wastewater disposal system.[14] Although a majority of the virus isolates were recovered from wells drawing from the upper part of the aquifer under study, the potential for extensive vertical entrainment of viruses within the aquifer was evidenced by their recovery from wells which had been screened at 6-, 12-, and 18-m depths. The specific limits of virus entrainment could not be determined because the wells had not been installed beyond 67 m or deeper than 18 m.

The actual isolation of naturally occurring or indigenous enteric viruses from groundwater contaminated by septic tanks is difficult to predict since viruses are not always present in such wastes. For enteric viruses to be present in septic tank effluents, an individual in the household served by the septic tank must have an infection. This is in contrast to domestic sewage effluents from communities where the population size is great enough so that at least some individual will always be infected and that enteric viruses will occur at some level year round. In one study, enteroviruses were found in only one of 78 septic tank systems examined.[15] Although the chance of finding enteric virus in an individual septic tank system may be quite variable, the concentration can be significant. Yeager and O'Brien[16] recovered approximately 2500 enteroviruses per milliliter from the contents of a single septic tank, whereas Hain and O'Brien[17] recovered 2 to 4 enteroviruses per milliliter from the contents of each of four septic tanks examined.

To overcome these limitations and gain a better understanding of virus transport from septic tank sources, several investigators have seeded septic tanks with laboratory-grown enteric viruses. Scandura and Sobsey[18] seeded four septic tank systems located in a coastal plain found in sandy soils overlaying high groundwater tables with 10^9 infectious units of bovine enterovirus and examined their migration in the subsurface for a period of 3 months. Viruses were displaced from the septic tanks at a linear rate over 5 to 6 weeks. Viruses were detected in the groundwater of all systems from as early as 3 days to 9 weeks after septic tank inoculation, in concentrations as high as 75 plaque-forming units (pfu) per milliliter. The appearance of viruses in groundwater was weakly correlated with periods of increased precipitation, pH, and total coliform concentrations.

Stramer[19] dosed several septic tank systems with single doses of vaccine poliovirus type 1 derived from cell culture or from stools of recently vaccinated infants. The viruses persisted for several months in each of the septic tank systems, and groundwater contamination was demonstrated at all the sites studied. At one site virus passed from the septic tank and traveled 50 m through silt loam and was detected in water collected from a nearby lake 43 and 71 days after septic tank dosing. On day 109, tenfold higher concentrations of poliovirus were detected in the lake sediment than had been found in the water, demonstrating that large quantities of virus must have exited the septic tank and contaminated the lake sediment. Viruses were able to travel through 0.9 to 1.6 m of unsaturated soil to reach the groundwater shortly after septic tank dosing. Viruses were observed in the groundwater for over 100 days after leaving the septic tank. The rate of poliovirus travel through silt loam (15% clay) was observed to be 200 to 250 cm/day. The occurrence of viruses in groundwater was not consistently associated with the presence of bacterial indicators or chlorides, or with increased groundwater pH. Heavy rain was more frequently associated with virus occurrence than the other parameters studied.

The results of field studies on virus migration from septic tanks clearly indicate that viruses can travel substantial distances and survive extended periods of time in the subsurface (Table 2).

B. Land Application of Sewage

The greatest amount of research effort on field studies concerned with virus fate in soils has been with the land application of domestic sewage. In many areas of the world, land application of sewage is an attractive alternative treatment and disposal method. Since the early 1970s there has been a great deal of interest in the U.S. in this area.[20,21] This interest helped stimulate research into the transport and survival of viruses in the subsurface.[22] A recent review by Keswick and Gerba[23] contains a comprehensive summary of reports of virus occurrence in groundwater near land application sites. A more recent summary of virus isolations near such sites is shown in Tables 3 and 4. A review of this information clearly demonstrates that if enteric viruses are present in the sewage being applied to the land, at least some of the viruses can be expected to penetrate the subsurface and gain entrance to the underlying groundwater.

As might be expected, viruses have been most commonly isolated at sites which practice the rapid infiltration of wastewater. Since the objective at these sites is to infiltrate as much wastewater as possible, it is not surprising to isolate viruses beneath these sites. The soils at such sites are typically of a sandy or coarse gravel nature, creating less than ideal conditions for virus removal by adsorption. Vaughn and Landry[24,25] have perhaps done the most extensive field studies on virus occurrence near sites practicing land application of sewage. They repeatedly recovered enteroviruses near several sites on Long Island, N.Y., a region

Table 2
ISOLATION OF VIRUSES FROM WELLS NEAR
SEPTIC TANKS

Location	Virus	Max distance movement (m)		Ref.
		Horizontal	Vertical	
Michigan	Polio 2	91.5	30.5	11
Texas	Enterovirus	25	NA[a]	13
New York	Enterovirus	18	67	14
New Mexico	Enteric viruses	3.5	4.6	17
North Carolina	Seeded bovine enterovirus	35	NA	18
Wisconsin	Seeded polio 1	50	1.5[b]	19

[a]   Not available.
[b]   Distance of movement through unsaturated soil.

where groundwater serves as the sole source of potable water. Viruses were observed to travel to depths of 22.8 m and distances as far as 45.7 m from the source in sandy to course gravel soils (Table 3). The basins were receiving secondary and tertiary treated wastewater. Schaub and Sorber[26] isolated viruses at depths of 30 m and at lateral distances of 183 m from basins receiving effluents. The coliphage f2 was also added to the effluent as applied to the infiltration basins at this same site. The tracer virus was found to penetrate into the groundwater at the same rate as the percolating effluent and was observed in an 18.3 m deep observation well beneath the land application site within 48 hr after application. Only about 50% of the coliphage was removed, and it was detectable for 11 days after application. The soil consisted of silty sand and gravel underlaid by bedrock. Viruses have not always been detected in groundwater near sites practicing rapid infiltration of wastewater.[27-30] Lack of virus isolation in these studies may represent soil conditions more conducive to virus retention. More probable, though, is that the techniques used for virus detection may not be sensitive enough to detect low levels of virus.[30]

In summary, it would appear from these studies that viruses can potentially travel long distances in sandy and gravel soils, although substantial removal of the viruses initially applied to the soil surface may occur.[23] Unfortunately, such studies have not revealed how far viruses move from such sites, since usually viruses were observed in the most distant or deepest wells at the study sites. Additional field work at such sites would be helpful in determining the extent of movement of naturally occurring viruses under field conditions, since they offer a continuous source of enteric viruses.

Less extensive studies have been conducted at sites practicing low-rate or crop irrigation with wastewater. Many of these studies used field lysimeters, shallow wells, and/or drains of less than 20 m.[7,31-33] Wellings et al.[7] recovered viruses from groundwater after spray irrigation of secondary sewage effluent onto sand. Of particular interest in this study was that viruses survived secondary treatment, chlorination, sunlight, spraying, and percolation through 3 to 6 m of sandy soil. Perhaps even more significant was the demonstration of a burst of viruses detected in well water which had previously been negative after a period of heavy rainfall. This study, for the first time, demonstrated that rainfall could potentially have a major effect on the migration and facilitate the appearance of viruses in groundwater.

Shaub et al.[33] demonstrated that enteroviruses could penetrate 1.7 m of silt loam in field lysimeters fed primary and secondary wastewater by testing sample volumes of only 4 ℓ. The bacteriophage f2, added as a tracer to the treated wastewater, was also found in lysimeter percolates. However, the total amount of f2 recovered in these percolates was consistently less than 0.1% of the total f2 applied, indicating a considerable degree of removal.

Table 3
**ISOLATION OF VIRUSES FROM GROUNDWATER NEAR
SITES PRACTICING RAPID INFILTRATION OF SEWAGE**

Location	Max distance of movement (m)		Nature of soil	Ref.
	Vertical	Horizontal		
East Meadow, N.Y.	11.4	3	Coarse sand and fine gravels	25
Holbrook, N.Y.	6.1	45.7	Coarse sand and fine gravels	25
Sayville, N.Y.	2.4	3	Coarse sand and fine gravels	25
Twelve Pines, N.Y.	6.4	NR[a]	Coarse sand and fine gravels	25
North Masapequa, N.Y.	9.1	NR[a]	Coarse sand and fine gravels	25
Babylon, N.Y.	22.8	408	Coarse sand and fine gravels	25
Ft. Devens, Mass.	28.9	183	Silty sand and gravel	26
Vineland, N.J.	16.8	250	Sand and coarse gravel	35
Phoenix, Ariz.	18.3	3	Loamy sand	66
Dan Region, Israel	31—67	60—270	Sand	66
England, various sites	2—18	NA[b]	Chalk	75,76

[a]   Not reported.
[b]   Not available.

Goyal et al.[34] were able to demonstrate the presence of enteric viruses in groundwater beneath three different sites where slow rate sewage irrigation was practiced (Table 4). At all sites, the sewage was secondarily treated (aeration) before land application. Soil types varied from sand to clay-loam. Viruses were detected in the deepest wells at all sites (i.e., 10 to 27.5 m). The presence of enteric viruses was also demonstrated in soil samples beneath the sites (10 to 18 cm). In contrast to these studies, Moore et al.[32] failed to detect viruses in monitoring wells at depths of 10.7 to 19.8 m at a land application site in Kerrville, Tex. even though viruses could be detected in 1.4-m deep field lysimeters. No viruses could be detected in wells from 3 to 30 m deep at a crop irrigation site in Roswell, N.M. where land application of wastewater was taking place, although in contrast to the other study sites, irrigation was seasonal and intermittent.[35]

Again, these studies demonstrate that, where even slow-rate land application of wastewater is conducted, viruses can migrate short distances from the source.

C. Land Application of Sewage Sludges and Landfills

There have been several investigations into the potential for viral contamination of groundwater resulting from sludge application to soil. Significant concentrations of enteric viruses can occur in domestic sludges even after treatment,[36] but appear to form a fairly stable association with the sludge particles and are not easily released. Although Bitton et al.[37] have pointed out that liquid sludges may contain unbound virus which could penetrate into the soil after application, several studies have failed to demonstrate the presence of viruses in wells beneath sludge application sites.[38,39] In contrast to other studies, Jorgensen and Lund[40] observed the occurrence of indigenous enteroviruses in water samples collected 3 m

Table 4
ISOLATION OF VIRUSES FROM GROUNDWATER
NEAR SITES PRACTICING IRRIGATION WITH
SEWAGE

Location	Max distance of movement (m) Vertical	Horizontal	Nature of soil	Ref.
St. Petersburg, Fla.	6	NA[a]	Sand	7
Gainesville, Fla.	3	7	Clay and sand	59
Lubbock, Tex.	27.5	NA	Sandy loam	34
Kerrville, Tex.	1.4	NA		32
Muskegon, Mich.	10	NA	Sand	34
San Angelo, Tex.	27.5	NA	Clay loam	34

[a] Not applicable.

beneath the soil surface in a forest plantation on pure diluvial sand where municipal sludge disposal was practiced. The viruses were isolated 11 weeks after the sludge had been applied to the soil.

Landfills are also potential sources of viruses because of the disposal of sewage sludges and baby diapers. Thus, Vaughn and Landry[25] reported the isolation of unidentifiable virus types from a groundwater observation well located 402 m down gradient from a sanitary landfill.

D. Tracer Studies

The previous discussion has largely focused on field studies aimed at the isolation of naturally occurring viruses from contaminated groundwater. Such studies are costly, and many parameters remain uncontrolled, such as initial virus concentrations, ability to quantitate virus removal, inability to identify all virus types present, etc. More quantitative information can be developed by using laboratory-grown virus to study virus transport under *in situ* conditions. Unfortunately, because all human enteric viruses are capable of causing disease, they cannot be used safely as tracers. To overcome this problem, vaccine strains of poliovirus have been used as tracers (Table 5). In addition, bovine enterovirus type 1 has also been used to trace the movement of viruses from a septic tank drainfield.[18]

Fletcher and Myers[41] used coliphage T4 to trace groundwater movement in the carbonate rock terrain of the Ozark region of southern Missouri. The phage was injected on the surface and was traced approximately 1600 m from the point of injection site, demonstrating a long potential migration distance for viruses in this type of subsurface terrain. The first arrival of the phages took approximately 16 hr. In a similar type of sandstone substrate a bacteriophage tracer was observed to travel 680 m over 9 days, yielding a velocity of 36 to 180 m/day.[42] At a site where wastewater was disposed on gravel soils, T4 and φX174 phages were injected at a depth of 14 m below ground level and followed over a distance of 920 m. The coliphage T4 was observed to travel 920 m in approximately 96 hr.[43]

E. Drinking Water Wells

As discussed in Section I, contaminated groundwater accounts for approximately half of the water-borne disease outbreaks every year in the U.S.[9] Although in most outbreaks no etiologic agent was identified, viruses were known to be responsible for at least 23 outbreaks of contaminated groundwater from 1971 to 1982. Hepatitis A and the Norwalk viruses were the agents identified epidemiologically as the cause of these outbreaks. Unfortunately, seldom

Table 5
FIELD STUDIES OF VIRUS MOVEMENT
THROUGH SOILS

| Virus | Max distance of movement (m) | | Ref. |
	Horizontal	Vertical	
Coliphage T4	1600	NR[a]	41
Coliphage φX174, T4	900	18	43
Coliphage f2	180	18.3	26
Coliphage f2	7.5	26	68
Type 2 phage of *Aerobacter aerogenes*	680	NR	42
Bovine enterovirus type 1	35	NR	18
Poliovirus type 1 (LSc)	50	1.5	19
Poliovirus (vaccine strain)	<40	NR	67

[a]   Not reported.

are attempts made to isolate the viruses from the water, nor is sufficient documentation provided to gain information which would be useful in promoting a better understanding of virus transport through the subsurface. Better documentation and a review of the hydrogeology of sites where such outbreaks have occurred could potentially yield useful information on the *in situ* transport of viruses in the subsurface.

Several field studies on enteric virus occurrence in groundwater have yielded information on virus penetration into the subsurface (Table 6). Perhaps the most extensive field studies were those of Marzouk et al.,[44] who collected 20- to 440-ℓ samples of groundwater from shallow wells (3 m in depth). They found that 20% of 99 samples collected were positive for enteric viruses. Numerous other investigators have documented the presence of enteric viruses in drinking water wells,[23] but unfortunately few details on the depth of the groundwater, the nature of the substrata, and the distance to the possible source are provided. This makes it difficult to use this information in understanding and predicting virus fate and transport in the subsurface. It is certain, though, that conventional indicators such as coliform and fecal coliform bacteria are not reliable indicators of the presence of enteric viruses in groundwater.[23,44-46]

The failure of conventional coliform and fecal coliform bacteria to demonstrate the presence of enteric viruses has probably been most clearly demonstrated by the recent work of Slade[45] in England. Enteroviruses were regularly isolated from well water from a chalk aquifer for over a period of 6 months in the absence of indicator bacteria. Even more surprising was the isolation of the viruses in the same well water after chlorination with 1 mg/ℓ chlorine after a contact time of 15 min. The well in question had a long history of excellent bacteriological quality. Despite an extensive investigation, no nearby source of contamination could be detected. The longer survival time of viruses in groundwater compared to that of bacteria and/or less retention by subsurface materials could account for the absence of indicator bacteria.

III. VIRUS DETECTION IN SOILS

Studies of enteric viruses in soils provide information on their persistence and penetration into the subsurface. Such data are useful in evaluating the safe use of sewage effluent for crop irrigation and the utilization of sludges on farmland. Virus persistence in soils used for food crop production presents an opportunity for virus contamination of edible portions of

Table 6
ISOLATION OF VIRUSES FROM
DRINKING WATER WELLS

Location	Virus	Ref.
Florida	Echo 22/23	12
Germany	Echo 3, 6, 30; coxsackie B1, 4, 5; U[a]	74
India	U	72
Michigan	Polio 2	11
England	Polio 2	23
Israel	Polio 1, U	73
Ghana	Polio 1, coxsackie B3	75
Mexico	Rota; coxsackie B4, 6	71
Texas	Hepatitis A, coxsackie B3	6
Georgia	Hepatitis A	70
Maryland	Hepatitis A; polio 1; echo 27, 29	69
Bolivia	Rota	46
England	Polio 1, 2, 3; coxsackie B2	45
Norfolk Island	Rota, adeno 5, polio 1	78

[a] U, unidentified.

the crop, as well as potential hazards from surface runoff and elution and further migration of viruses into the groundwater. Most studies on viruses in soils have been performed in the laboratory under controlled conditions or with laboratory-grown viruses added to soil in experimental soil plots.

From these studies, it is known that several factors are important for virus survival in soil, including soil moisture, temperature, adsorption of the virus to the soil, and nature of the virus.[47-49] Studies in which laboratory-grown viruses have been used to study these factors are discussed in Chapter 12. Field studies described in this section have demonstrated that naturally occurring enteric viruses can be isolated from soils where domestic waste disposal and irrigation are practiced.

A study in the Soviet Union was the first to report the isolation of enteroviruses from soil.[50] Soil irrigated with a mixture of household sewage and industrial waste was sampled at depths from 3 to 20 cm, and several types of enteroviruses were isolated. Development of quantitative methods for the recovery of enteroviruses from soils resulted in additional studies on their presence in soils.[51,52] Reported studies on the isolation of naturally occurring enteroviruses from soils are summarized in Table 7. It is interesting that in all studies viruses were found at the greatest depths sampled. Depending on the individual studies, these depths ranged from the soil surface to 210 cm.[54] In an analysis of 200-g samples from basins filled with sandy soil, Hurst et al.[54] found that virus concentration near the soil surface was significantly greater than that found at lower depths (Table 8), confirming observations made in laboratory studies that much of the virus removal occurs near the soil surface.[54]

IV. FIELD STUDIES ON VIRUS SURVIVAL IN GROUNDWATER AND SOIL

Actual *in situ* studies on virus survival in aquifers have not yet been directly attempted. Of necessity, such studies have generally been conducted in the laboratory or by use of dialysis chambers placed in well water,[56-58] but field results tend to substantiate prolonged virus survival in groundwater, especially at low temperatures.[19,58] Based upon time of ap-

Table 7
**STUDIES REPORTING DIRECT ISOLATION OF ENTERIC VIRUSES FROM
SOIL AFTER APPLICATION OF SEWAGE EFFLUENT**

Location	Type of wastewater	Nature of site	Depths sampled (cm)	Max depth of virus isolation (cm)	Ref.
Lubbock, Tex.	Secondary activated sludge	Crop irrigation	0—18	18	34
Muskegon, Mich.	Lagoon, chlorinated	Crop irrigation	0—10	10	34
Phoenix, Ariz.	Primary	Rapid infiltration	0—30	25	54
	Secondary	Rapid infiltration	0—25	25	54
Germany	Primary	Crop irrigation	0—210	210	55
Soviet Union	Primary	Crop irrigation	3—20	20	50

plication of secondary wastewater effluent to a cypress dome and appearance in a sampling well, Wellings et al.[59] estimated survival of indigenous enteroviruses of at least 28 days. Vaughn et al.[60] observed at 30.4 m lateral movement of indigenous viruses through a glacial aquifer. The average groundwater velocity in the study area was 15.3 cm/day, resulting in a minimal survival time of 199 days for the viruses. As previously cited, Stramer[19] observed the travel of laboratory-grown polioviruses after leaving a septic tank and traveling 50 m over a period of 105 days. Scandura and Sobsey[18] traced the travel of a bovine enterovirus in groundwater over a period of 63 days in a shallow coastal aquifer.

Quantitative studies on the survival of indigenous enteric viruses in the field are difficult because of the low numbers often observed. Studies with laboratory-grown viruses, though, indicate that enteroviruses can persist for months. Bagdasaryan[61] reported that enteroviruses added to soils in the Soviet Union survived for up to 170 days in sandy soils during the winter months. Larkin et al.[62] observed that viruses survived for periods of up to 96 days in sludge- and sewage-irrigated soils under winter conditions in Ohio. Hurst et al.[54] determined the rate of poliovirus 1 and echovirus 1 inactivation in infiltration basins receiving secondary sewage to range between 0.04 and 0.15 log/day when flooded and 0.11 to 0.52 log when dry. The increased rate of virus inactivation when the basins were not flooded was attributed to loss of soil moisture. It would not be surprising if indigenous enteroviruses also exhibited such prolonged survival in the soil.

V. SUMMARY AND CONCLUSIONS

Improvements in technology during the 1970s for virus detection in the environment provided the tools for their study in soil and groundwater. Field studies clearly demonstrated the presence and persistence of enteric viruses in these environments. Through laboratory studies, insights have been gained into factors which control the survival and transport of viruses through soil systems. Through this basic understanding, it is now possible to develop models for predicting virus behavior in the environment. Such information is needed for the siting and management of land application sites for crop irrigation and groundwater recharge with domestic wastewater, as well as septic tank placement. Estimates in the past for safe distances between sources of sewage waste disposal and drinking water wells have been largely based on empirical judgment. In recent years a number of authors have proposed the development of mathematical models for predicting virus transport and survival.[63-65] The usefulness of such models should be validated by future field studies.

Almost all of the field studies on viruses in groundwater have been aimed at studying sites where intentional land application of wastewater was being practiced or have been conducted during water-borne disease outbreaks. General field studies on the occurrence of

Table 8
ISOLATION OF
ENTEROVIRUSES FROM
SANDY SOIL FLOODED
WITH PRIMARY AND
SECONDARY
WASTEWATER[53]

	Viral pfu/200 g soil	
Soil Depth	Primary sewage	Secondary sewage
Surface layer	No data	19.5
0—2.5	101	6
2.5—10.0	6	1.5
10.0—25.0	5	0.5

enteric viruses in groundwater in the U.S. are nonexistent. It is surprising that with over half the drinking water in the U.S. originating from groundwater, we have no knowledge of the extent of virus contamination of these waters. Field studies have clearly documented that virus contamination of groundwater occurs in the absence of fecal indicator bacteria. Recent studies in England have clearly demonstrated that drinking water from a well with a long history of excellent bacteriological quality can contain significant levels of enteric viruses. More extensive field studies on viruses in drinking water wells are clearly needed to assess the exposure to potential contamination of drinking water supplies and our valuable groundwater resources.

REFERENCES

1. **Drewry, W. A. and Eliassen, R.,** Virus movement in groundwater, *J. Water Pollut. Control Fed.*, 40, R257, 1968.
2. **Gerba, C. P., Goyal, S. M., Cech, I., and Bogdan, G. F.,** Quantitative assessment of the adsorptive behavior of viruses to soils, *Environ. Sci. Technol.*, 15, 940, 1981.
3. **Duboise, S. M., Moore, B. E., Sorber, C. A., and Sagik, B. P.,** Viruses in soil systems, *Crit. Rev. Microbiol.*, 7, 245, 1979.
4. **Goyal, S. M. and Gerba, C. P.,** Concentration of viruses from water by membrane filters, in *Methods in Environmental Virology*, Gerba, C. P. and Goyal, S. M., Eds., Marcel Dekker, New York, 1982, 59.
5. **Smith, E. M. and Gerba, C. P.,** Development of a method for the detection of human rotavirus in water, *Appl. Environ. Microbiol.*, 43, 1440, 1982.
6. **Hejkal, T. W., Keswick, B., LaBelle, R. L., Gerba, C. P., Sanchez, Y., Dreesman, G., Hafkin, B., and Melnick, J. L.,** Viruses in a community water supply associated with an outbreak of gastroenteritis and infectious hepatitis, *J. Am. Water. Works Assoc.*, 74, 318, 1982.
7. **Wellings, F. M., Lewis., A. L., and Mountain, C. W.,** Virus survival following wastewater spray irrigation of sandy soils, in *Virus Survival in Water and Wastewater Systems*, Malina, J. F. and Sagik, B. P., Eds., Center for Research in Water Resources Systems, University of Texas, Austin, 1974, 253.
8. **Bitton, G. and Gerba, C. P.,** Groundwater pollution microbiology: the emerging issue, in *Groundwater Pollution Microbiology*, Bitton, G. and Gerba, C. P., Eds., John Wiley & Sons, New York, 1984, 1.
9. **Craun, G. F.,** A summary of waterborne illness transmitted through contaminated groundwater, *J. Environ. Health*, in press.

10. **Jewell, W. J. and Seabrook, B. L.,** A History of Land Application as a Treatment Alternative, EPA 430/9-79-012, U.S. Environmental Protection Agency, Washington, D.C., 1979.

11. **Mack, W. N., Lu, Y., and Coohon, D. B.,** Isolation of poliomyelitis virus from a contaminated well, *Health Serv. Rep.,* 87, 271, 1972.

12. **Wellings, F. M., Mountain, C. W., and Lewis, A. L.,** Virus in groundwater, in *Proc. 2nd Natl. Conf. Individual On-Site Wastewater Systems,* Ann Arbor Science, Ann Arbor, Mich., 1977, 61.

13. **Gerba, C. P.,** Virus occurrence in groundwater, in *Microbial Health Considerations of Soil Disposal of Domestic Wastewater,* National Center for Groundwater Research, Norman, Okla., 1981, 144.

14. **Vaughn, J. M., Landry, E. F., Chen, Y. S., and Thomas, M. Z.,** Virus entrainment in a glacial aquifer, in *Progress in Chemical Disinfection — II,* Janauer, G. E and Ghiorse, W. C., Eds., State University of New York, Binghamton, 1984, 109.

15. **Harkin, J. M., Fitzgerald, C. J., Duffy, C. P., and Kroll, D. G.,** Evaluation of Mound Systems for Purification of Septic Tank Effluent, Tech. Rep. WIS WRC 79-05, Water Resources Center, University of Wisconsin, Madison, 1979.

16. **Yeager, J. E. and O'Brien, R. T.,** Enterovirus inactivation in soil, *Appl. Environ. Microbiol.,* 38, 694, 1979.

17. **Hain, K. E. and O'Brien, R. T.,** The Survival of Enteric Viruses in Septic Tanks and Septic Tank Drainfields, New Mexico Water Resources Institute Rep. NMWRRI 108, Project No. A-052-NMEX, New Mexico State University, Las Cruces, 1979.

18. **Scandura, J. E. and Sobsey, M. D.,** Survival and fate of enteric viruses in on-site wastewater disposal systems in coastal plains soils, in *Abstr. Annu. Meet. Am. Soc. Microbiol.,* 1971, 175.

19. **Stramer, S. L.,** Fates of Poliovirus and Enteric Indicator Bacteria during Treatment in a Septic Tank System Including Septage Disinfection, Ph.D. thesis, University of Wisconsin, Madison, 1984.

20. **Sopper, W. E. and Kardos, L. T.,** *Recycling Treated Municipal Wastewater and Sludge through Forest and Cropland,* Pennsylvania State Press, University Park, 1973.

21. **Sanks, R. L. and Asano, T.,** *Land Treatment and Disposal of Municipal and Industrial Wastewater,* Ann Arbor Science, Ann Arbor, Mich., 1976.

22. **Canter, L. W., Akin, E. W., Kreissl, J. F., and McNabb, J. F.,** Microbial health considerations of soil disposal of domestic wastewaters, EPA-600/9-83-017, U.S. Environmental Protection Agency, Ada, Okla., 1983.

23. **Keswick, B. H. and Gerba, C. P.,** Viruses in groundwater, *Environ. Sci. Technol.,* 14, 1290, 1980.

24. **Vaughn, J. M., Landry, E. F., Baranosky, L. J., Beckwith, C. A., Dahl, M. C., and Delihas, W. C.,** Survey of human virus occurrence in wastewater-recharged groundwater on Long Island, *Appl. Environ. Microbiol.,* 36, 47, 1978.

25. **Vaughn, J. M. and Landry, E. F.,** An Assessment of the Occurrence of Human Viruses in Long Island Aquatic Systems, BNL 50787, Brookhaven National Laboratory, Upton, N.Y., 1977.

26. **Schaub, S. A. and Sorber, C. A.,** Virus and bacteria removal from wastewater by rapid infiltration through soil, *Appl. Environ. Microbiol.,* 33, 609, 1977.

27. **England, B., Leach, R. E., Adame, B., and Shiosaki, R.,** Virologic assessment of sewage treatment at Santee, California, in *Transmission of Viruses by the Water Route,* Berg, G., Ed., John Wiley & Sons, New York, 1965, 401.

28. **Kerfoot, W. B. and Ketchum, B. T.,** Cape Cod Wastewater Renovation and Retrieval System, a Study of Water Treatment and Conservation, Rep. WHO1-74-13, Woods Hole Oceanographic Institution, Woods Hole, Mass., 1974.

29. **Gilbert, R. G., Gerba, C. P., Rice, R. C., Bouwer, H., Wallis, C., and Melnick, J. L.,** Virus and bacteria removal from wastewater by land treatment, *Appl. Environ. Microbiol.,* 32, 333, 1976.

30. **Gerba, C. P. and Goyal, S. M.,** Pathogen removal from wastewater during groundwater recharge, in *Artificial Recharge of Groundwater,* Asano, T., Ed., Butterworths, Boston, 1985, 283.

31. **Dugan, G. L., Young, R. H. F., Lau, L. S., Ekern, P. C., and Loh, P. C. S.,** Land disposal of wastewater in Hawaii, *J. Water Pollut. Control Fed.,* 47, 2067, 1975.

32. **Moore, B. E., Sagik, B. P., and Sorber, C. A.,** Viral transport to groundwater at a wastewater land application site, *J. Water Pollut. Control Fed.,* 53, 1492, 1981.

33. **Schaub, S. A., Bausum, H. T., and Taylor, G. W.,** Fate of virus in wastewater applied to slow-infiltration land treatment systems, *Appl. Environ. Microbiol.,* 44, 383, 1982.

34. **Goyal, S. M., Keswick, B. H., and Gerba, C. P.,** Viruses in groundwater beneath sewage irrigated cropland, *Water Res.,* 18, 299, 1984.

35. **Koerner, E. L. and Haws, D. A.,** Long-Term Effects of Land Application of Domestic Wastewater: Vineland, New Jersey, Rapid Infiltration Site, EPA-600/2-79-072, U.S. Environmental Protection Agency, Washington, D. C., 1979.

36. **Farrah, S. R. and Schaub, S. A.,** Viruses in wastewater sludges, in *Viral Pollution of the Environment,* Berg, G., Ed., CRC Press, Boca Raton, Fla., 1983, 147.

37. **Bitton, G., Pancorbo, O. C., Overman, A. R., and Gifford, G. E.,** Retention of viruses during sludge application to soils, *Prog. Water Technol.*, 10, 597, 1978.

38. **Vaughn, J. M. and Landry, E. F.,** Viruses in soils and groundwater, in *Viral Pollution of the Environment*, Berg, G., Ed., CRC Press, Boca Raton, Fla., 1983, 163.

39. **Farrah, S. R., Bitton, G., Hoffman, E. M., Lanni, O., Pancorbo, O. C., Lutrick, M. C., and Bertrand, J. E.,** Survival of enteroviruses and coliform bacteria in a sludge lagoon, *Appl. Environ. Microbiol.*, 41, 459, 1981.

40. **Jorgensen, P. H. and Lund, E.,** Detection and stability of enteric viruses in sludge, soil and groundwater, *Water Sci. Technol.*, 17, 185, 1985.

41. **Fletcher, M. W. and Myers, R. L.,** Groundwater tracing in karst terrain using phage in T-4, in *Abstr. Annu. Meet. Am. Soc. Microbiol.*, 1974, 52.

42. **Martin, R. and Thomas, A.,** An example of the use of bacteriophage as a groundwater tracer, *J. Hydrol.*, 23, 73, 1974.

43. **Noonan, M. J. and McNabb, J. F.,** Contamination of Canterbury groundwater by viruses, in *The Quality and Movement of Groundwater in Alluvial Aquifers of New Zealand*, Noonan, M. J., Ed., Tech. Publ. No. 2, Lincoln College, Dept. of Agricultural Microbiology, Canterbury, New Zealand, 1979, 195.

44. **Marzouk, Y., Goyal, S. M., and Gerba, C. P.,** Prevalence of enteroviruses in groundwater in Israel, *Ground Water*, 17, 487, 1979.

45. **Slade, J. S.,** Viruses and bacteria in a chalk well, *Water Sci. Technol.*, 17, 111, 1985.

46. **Toranzos, G. A.,** Development of a Method for Concentration of Rotavirus from Water and Its Application to Field Sampling, Ph.D. thesis, University of Arizona, Tucson, 1985.

47. **Sobsey, M. D.,** Transport and fate of viruses in soils, in Microbial Health Considerations of Soil Disposal of Domestic Wastewaters, EPA-600/9-83-017, Canter, L. W., Akin, E. W., Kreissl, J. F., and McNabb, J. F., Eds., U.S. Environmental Protection Agency, Cincinnati, 1983, 175.

48. **Hurst, C. J., Gerba, C. P., and Cech, I.,** Effects of environmental variables and soil characteristics on virus survival in soil, *Appl. Environ. Microbiol.*, 40, 1067, 1980.

49. **Kowal, N. E.,** Health Effects of Land Treatment, EPA-600/1-82-007, U.S. Environmental Protection Agency, Cincinnati, 1982.

50. **Grigor'eva, L. V., Korchek, G. J., Bondarenko, V. I., and Bei, T. V.,** Sanitary characteristics (virological and bacteriological) of sewage, sludge and soil in suburbs in Kiev, *Hyg. Sanit.*, 33, 360, 1968.

51. **Hurst, C. J. and Gerba, C. P.,** Development of a quantitative method for the detection of enteroviruses in soil, *Appl. Environ. Microbiol.*, 37, 626, 1979.

52. **Bitton, G., Charles, M. J., and Farrah, S. R.,** Virus detection in soils: a comparison of four recovery methods, *Can. J. Microbiol.*, 25, 874, 1979.

53. **Lefler, E. and Kott, Y.,** Virus retention and survival in sand, in *Virus Survival in Water and Wastewater Systems*, Malina, J. F., Jr., and Sagik, B. P., Eds., University of Texas, Austin, 1974, 84.

54. **Hurst, C. J., Gerba, C. P., Lance, J. C., and Rice, R. C.,** Survival of enteroviruses in rapid-infiltration basins during the land application of wastewater, *Appl. Environ. Microbiol.*, 40, 192, 1980.

55. **Filip, Z., Seidel, K., and Dizer, H.,** Distribution of enteric viruses and microorganisms in long-term sewage-treated soil, *Water Sci. Technol.*, 15, 129, 1983.

56. **Bitton, G., Farrah, S. R., Ruskin, R. H., Butner, J., and Chou, Y. J.,** Survival of pathogenic and indicator organisms in groundwater, *Ground Water*, 21, 405, 1983.

57. **McFeters, G. A., Bissonnette, G. K., Jezeski, J. J., Thomson, C. A., and Stuart, D. G.,** Comparative survival of indicator bacteria and enteric pathogens in well water, *Appl. Microbiol.*, 27, 823, 1974.

58. **Yates, M. V. and Gerba, C. P.,** Virus persistence in groundwater, *Appl. Environ. Microbiol.*, 49, 778, 1985.

59. **Wellings, F. M., Lewis, A. L., Mountain, C. W., and Pierce, L. V.,** Demonstration of virus in groundwater after effluent discharge into soil, *Appl. Microbiol.*, 29, 751, 1975.

60. **Vaughn, J. M., Landry, E. F., and Thomas, M. Z.,** Entrainment of viruses from septic tank leach fields through a shallow, sandy soil aquifer, *Appl. Environ. Microbiol.*, 45, 1474, 1983.

61. **Bagdasaryan, G. A.,** Survival of viruses of the enterovirus group (poliomyelitis, echo, coxsackie) in soil and on vegetables, *J. Hyg. Epidemiol. Microbiol. Immunol.*, 8, 497, 1964.

62. **Larkin, E. P., Tierney, J. T., and Sullivan, R.,** Persistence of virus on sewage-irrigated vegetables, *J. Environ. Eng. Div. Proc. Am. Soc. Civ. Eng.*, 102, 29, 1976.

63. **Vilker, V. L.,** Simulating virus movement in soils, in *Modeling Wastewater Renovation*, Iskandar, I. K., Ed., John Wiley & Sons, New York, 1981, 223.

64. **Yates, M. V., Yates, S. R., Warwick, A. W., and Gerba, C. P.,** Preventing viral contamination of drinking water, in *Groundwater Contamination and Reclamation*, Schmidt, K. D., Ed., American Water Resources Association, Bethesda, Md., 1985, 117.

65. **Matthess, G. and Pekdeger, A.,** Survival and transport of pathogenic bacteria and viruses in groundwater, in *Groundwater Quality*, Ward, C. H., Giger, W., and McCarty, P. L., Eds., John Wiley & Sons, New York, 1985, 472.

66. **Gerba, C. P., Marzouk, Y., Manor, Y., Idelovitch, E., and Vaughn, J. M.,** Virus removal during land application of wastewater: comparison of three projects, in *Future of Water Reuse,* AWWA Research Foundation, Denver, Colo., 1985, 1518.

67. **Marti, F., Valle, G. D., Krech, M. K., Gees, R. A., and Baumgrat, E.,** Tracing tests in groundwater with dyes, bacteria, and viruses, *Alimenta,* 18, 135, 1979.

68. **Wilson, L. G., Gerba, C. P., Bolton, M. W., and Rose, J. B.,** Subsurface transport of urban runoff pollutants during dry-well disposal, *Groundwater Qual. Res.,* in press.

69. **Sobsey, M. D.,** personal communication, 1985.

70. **Stramer, S.,** personal communication, 1985.

71. **Deetz, T. R., Smith, E. M., Goyal, S. M., Gerba, C. P., Vollet, J. J., Tsai, L., Dupont, H. L., and Keswick, B. H.,** Occurrence of rota- and enteroviruses in drinking and environmental water in a developing nation, *Water Res.,* 18, 567, 1984.

72. **Rao, V. C., Lakhe, S. B., and Waghmare, S. V.,** Developments in environmental virology in India, *IAWPC Tech. Ann.,* 5, 1, 1978.

73. **Shuval, H. I.,** Detection and control of enteroviruses in the water environment, in *Developments in Water Quality Research,* Shuval, H. I., Ed., Ann Arbor Science, Ann Arbor, Mich., 1969, 47.

74. **Walter, R. and Rudiger, S.,** Significance of virus isolation in terms of municipal hygiene, *Z. Gesunheitstech. Staedtehyg.,* 23, 461, 1977.

75. **Addy, P. A. K. and Otatume, S.,** Ecology of enteroviruses in Ghana, *Ghana Med. J.,* 18, 102, 1976.

76. **Edworthy, K. J. and Baxter, K. M.,** Virology of wastewater recharge of the chalk aquifer. I. Hydrogeology and sampling, in *Viruses and Wastewater Treatment,* Goddard, M. and Butler, M., Eds., Pergamon Press, New York, 1981, 53.

77. **Slade, J. S. and Edworthy, K. J.,** Virology of wastewater recharge of the chalk aquifer. II. Microbiology and water quality, in *Viruses and Wastewater Treatment,* Goddard, M. and Butler, M., Eds., Pergamon Press, New York, 1981, 59.

78. **Murphy, A., Grohman, G., and Sexton, M.,** Viral gastroenteritis on Norfolk Island, in *Viral Diseases in South-East Asia and the Western Pacific,* Mackenzie, J. S., Ed., Academic Press, New York, 1982, 437.

Chapter 12

SURVIVAL AND TRANSPORT OF VIRUSES IN SOILS: MODEL STUDIES

Mark D. Sobsey and Patricia A. Shields

TABLE OF CONTENTS

I. INTRODUCTION

The transport and fate of enteric viruses in soils is an important public and environmental health consideration in the treatment and disposal of human and animal wastes. There are over 100 different human enteric viruses that can be present in feces and exudates from infected people.[1,2]

Virus concentrations in feces may be as high as 10^6 infectious units per gram for enteroviruses,[3-4b] 10^9 infectious units per gram for hepatitis A virus,[5] and 10^9 virus particles per gram for rotaviruses,[6] although not all of the latter may be infectious. Enteric virus concentrations in raw sewage in the U.S. may sometimes be as high as 10^5 infectious units per liter,[7] and conventional sewage treatment systems are not likely to reduce virus levels in raw sewage by more than 99%.[8]

Potential human health hazards from enteric viruses are associated with a variety of purposeful and inadvertent situations where virus-laden wastes may gain access to soils and groundwaters. These include land application systems for wastewaters and sludges, groundwater recharge operations, landscape and crop irrigation systems, and on-site septic waste disposal units. Additional sources of enteric viruses entering soils and groundwaters inadvertently are infiltration of fecally contaminated surface waters and surface runoff from precipitation; leachates from pervious wastewater ponds, lagoons, and basins; sewage sludge spreading, drying, and composting operations; solid waste landfills and dumps; and leakage from sewer lines.

In recent years increasing evidence for viral contamination of soils and groundwater from the sources listed above has been documented. Data available on the occurrence of water-borne disease outbreaks in the U.S. between 1971 and 1982 indicated that the use of contaminated, untreated, or inadequately treated groundwater was responsible for 51% of all water-borne outbreaks reported.[9] About 10% of these outbreaks were of viral etiology, and it is likely that many of the outbreaks of unknown etiology (52%) also may have been of viral origin. Hepatitis A virus (HAV) antigen, as well as other infectious enteroviruses, such as coxsackievirus B3, were detected in contaminated water supplies during an outbreak of gastroenteritis and infectious hepatitis in Georgetown, Tex.[10] Recently, Sobsey et al.[11] developed a method for the recovery and quantitation of HAV from water, and applied it to the isolation of this virus in an infectious form from contaminated groundwater during an outbreak of infectious hepatitis in rural Maryland. Household septic wastewater treatment systems were suspected as the source of groundwater contamination.

Direct isolation of viruses from soils and underlying groundwater supplies has been reviewed previously,[12] and more recent studies continue to illustrate the problem. Enteroviruses were recovered from wells located as much as 67 m down gradient from a domestic subsurface disposal system,[13] indicating extensive migration of human enteric viruses in soils. Goyal et al.[14] detected enteric viruses in groundwater wells beneath three different sites where slow-rate sewage irrigation of cropland was practiced. These studies illustrate that migration of viruses through soils to groundwater supplies is of considerable public health concern. As the incidence of water-borne disease outbreaks has increased over the last 35 years,[15] it is important to identify and evaluate the factors involved in viral contamination of the soil and groundwater environment.

The nature and degree of virus retention by and inactivation in soils will influence the extent of virus contamination of both the soil and the underlying groundwater. Although information on virus survival, transport, and ultimate fate in soils is still inadequate and incomplete, several controlling and influencing factors have been identified and evaluated, especially in laboratory and pilot-scale studies. This subject has previously been reviewed in considerable detail,[9,16-20] but new findings continue to be reported in the literature. Therefore, the purpose of this chapter is to highlight and update available information and identify

Table 1
FACTORS INFLUENCING VIRUS SURVIVAL OR INACTIVATION IN SOILS

Factor	Comments	Ref.
Temperature	Lower temperatures promote increased survival	21,27—31
Microbial and related chemical activity	Survival is decreased in nonsterile soils	25,27,28,32
Moisture content	Survival is decreased in drying soils	27,28,30,31,33—35
pH	Survival is decreased by pH extremes	28,36,37
Salt species and conc.	Cation type and ionic strength influence virus retention and survival	28,38,39
Virus assoc. with particulate matter	Survival often is increased when virus is in sorbed state	40,41
Virus aggregation	Aggregation protects from inactivation	42,43
Soil type	Survival is influenced by chemical, textural, and mineralogical properties, such as pH, exchangeable ion content, and water retention capacity	25,40,41
Virus type	Survival differs among virus types and strains	25,28,30
Soil organic matter	Fulvic acid masks virus infectivity	44

critical areas for further research pertaining to model systems for virus survival and transport in soils.

II. VIRUS INACTIVATION AND RETENTION BY SOILS

In considering virus retention and inactivation in soils, it is useful to distinguish between those factors and conditions in the soil environment that will play a role in virus inactivation or protection from inactivation and those that influence virus retention or movement. Some conditions or factors may influence both inactivation and retention of viruses, while others may predominantly influence only one of these two general phenomena. It is especially important to recognize that under some conditions viruses may be extensively retained in soils but not be appreciably inactivated, resulting in the possible accumulation of infectious viruses in the soil. If these accumulated viruses were subsequently released from the soil as a result of changing conditions, considerable groundwater contamination could occur. In fact, such release and migration of accumulated viruses in soils as a consequence of changing soil conditions has been demonstrated in laboratory and pilot-scale studies with soil columns[21-25] and under field conditions.[14,26] The factors and conditions influencing virus survival in and retention by soils will be discussed separately.

III. FACTORS INFLUENCING VIRUS SURVIVAL IN SOILS

The important factors influencing virus survival in soils are summarized in Table 1. For many of these factors the precise mechanisms by which they cause virus inactivation or protection have not been fully determined. Furthermore, it must be recognized that natural soils are dynamic ecosystems and that interactions between many of these factors add to the complexity of elucidating their precise roles in virus inactivation or protection. Much of what is known about the effects of these factors on virus survival in the environment is based on studies in water, wastewater, and sludge. Some of the mechanisms by which these factors affect virus survival may be similar in these different media, but the unique characteristics of soil systems may cause them to have qualitatively and quantitatively different effects in soils.

Table 2
EFFECT OF TEMPERATURE ON VIRUS INACTIVATION RATES IN SOIL MATERIALS

Virus type	Soil	Water	Approx. no. days for 99% inactivation at		Ref.
			1—8°C	20—25°C	
Poliovirus type 1	Loamy fine sand	Treated effluent	>>84	20	21
Poliovirus type 1	Sand	Distilled water	>>75	50[a],56[a]	28
	Sand	Distilled water	>>75	14[b],50[c]	
Poliovirus type 1	Sand	Septic tank effluent	416	27	29
Poliovirus type 1	Sand	Tap water	>>175[a]	42[a]	30
	Sand	Treated effluent	>>179	51	
Poliovirus type 1	?	?	>>134	77	31

[a]   Sterile soil columns.
[b]   Nonsterile, aerobic soil columns.
[c]   Nonsterile, anaerobic soil columns.

A. Temperature

Temperature has a considerable influence on virus inactivation rates in soils. As shown in Table 2, survival time of viruses in soils is shorter at higher temperatures. Thus, viruses are likely to persist longer in soils during the colder months of the year. It should be noted from Table 2 that the data of Lefler and Kott[30] and some of the data of Hurst et al.[28] were obtained using sterile soil samples, so in these cases the observed temperature effects cannot be attributed to differences in soil microbial activity at the different temperatures.

Other investigators have also reported greater virus inactivation rates at higher temperatures. Yeager and O'Brien[45] found that coxsackievirus B1 inactivation rates in sandy and sandy loam soils suspended in river water, groundwater, and septic wastewater increased as temperatures were increased from 4 to 37°C. In pilot-scale outdoor studies on poliovirus persistence on vegetables and in soils irrigated with sewage effluent in Cincinnati, Larkin et al.[46] and Tierney et al.[47] found that 99% inactivation of virus in soils took about 2 months during the winter months and only 2 to 3 days in the warm summer months of June and July. In a field study by Hurst et al.[35] on virus survival and movement in a rapid infiltration system for wastewater, the rate of inactivation of indigenous viruses was greater in the fall than in the winter, possibly due in part to the effects of higher temperatures in the fall season.

The mechanisms of virus inactivation due to thermal effects have not been adequately studied in soils, but the findings from studies in aqueous solutions may be relevant. Based on studies with poliovirus, Dimmock[48] proposed a general model for thermal inactivation in which there are different mechanisms at low (<44°C) and high (>44°C) temperatures. At low temperatures the loss of poliovirus infectivity in buffered water corresponds to the inactivation rate of viral RNA. At high temperatures, the rate of virus inactivation is greater than the rate of RNA inactivation and is associated with structural changes in virus capsids. These findings have been confirmed by other workers.[49] Because soil temperatures rarely if ever reach 44°C, it is likely that thermal inactivation in soils proceeds by the low temperature mechanism. However, there may be additional thermal mechanisms of virus inactivation in soils.

B. Soil Moisture

Soil moisture also influences virus survival in soils, although there is somewhat conflicting information in the literature on the effects of this factor. Lefler and Kott[30] observed approximately the same rates of poliovirus inactivation in dried and saturated sand, but the amount of soil moisture in the dried sand was not specified. Other workers have reported more rapid virus inactivation in dried than in wet soils. Bagdasaryan[27] reported that enteroviruses survived three to six times longer in soils with 10% moisture content than in air-dried soils. Sagik et al.[31] found that poliovirus type 1 inactivation was considerably more rapid in drying soil as the moisture content decreased from 13 to 0.6% than in the same soil type maintained at 15 or 25% moisture content. Inactivation of 99% of the initial viruses occurred within 1 week in drying soil but took 7 to 8 and 10 to 11 weeks in soils with 25 and 15% moisture content, respectively.

Yeager and O'Brien[45] compared the degree of poliovirus inactivation in eight different soils saturated with river water, groundwater, or septic wastewater and in the same soils that were allowed to dry out during the course of the experiment. Upon drying, none of the initial viruses was detectable in any of the dried soils (>99.999% inactivation), but considerable quantities were still present in the same types of saturated soils. In experiments on the rate of poliovirus inactivation at different soil moisture levels, there was a sharp increase in the inactivation rate at 1.2% soil moisture compared to that at 2.9%.

Hurst et al.[28] also observed differences in poliovirus inactivation rates at different soil moisture levels, with the greatest inactivation rate at a moisture level of 15%. Inactivation proceeded more slowly at both higher and lower moisture levels, but the slowest inactivation rates were at the two lowest moisture levels, 5 and 10%. These findings are somewhat inconsistent with previously cited studies, and the reasons for this discrepancy are unknown. It is possible that virus inactivation rates at the different moisture levels were influenced by other soil factors, such as microbial activity, or that virus inactivation rates would have been even greater at moisture levels lower than those tested.

Bitton and co-workers[33,34] found that virus survival in sludge-amended soils exposed to natural weather conditions was mainly affected by desiccation. In these experiments, lagooned sludge, seeded with echovirus 1 or poliovirus 1, was applied to columns of sand placed outside during summer and fall seasons in north central Florida. They found that in the warm, dry fall season (<0.13 cm cumulative rainfall), there was a more rapid decline in virus infectivity which took place in the top layers of the column; no virus was detected after 8 days. In contrast, virus was detected for up to 35 days in columns exposed during the wet summer months, when cumulative rainfall was over 13 cm.[34]

In a field study on virus survival in a rapid infiltration system for wastewater, Hurst et al.[35] found that virus inactivation rates were greater in more rapidly drying soils. Allowing soils in rapid infiltration systems to periodically dry and become aerated between wastewater applications enhanced virus inactivation. The effects of both drying and aerobic microbial activity may contribute to virus inactivation under these conditions.

In studies on the mechanisms of virus inactivation in soils. Yeager and O'Brien[50] found that the loss of poliovirus infectivity in moist and dried soils resulted from irreversible damage to the virus particles, including (1) dissociation of viral capsids and genomes and (2) degradation of viral RNA. In both moist and dried, nonsterile soils, viral RNA was released from capsids and found in a degraded form. In dried, sterile soils, viral RNA was released but remained largely as intact molecules. Viral capsid components were not readily recoverable from drying soils due to irreversible binding to soil material, but they were recoverable as empty capsids from moist soils. Further experiments using dried viruses showed that their capsids became isoelectrically altered. The results of these studies suggest that poliovirus and perhaps other enteric viruses are inactivated by different mechanisms in moist and in drying soils.

Table 3
VIRUS SURVIVAL IN STERILE AND NONSTERILE
SUSPENSIONS OF SOIL MATERIAL IN SECONDARY
EFFLUENT AT 5°C AND 25°C[a]

Temp. (°C)	Virus	Microbial activity	Time (weeks) for 99% virus inactivation in			
			Bentonite clay	Kaolinite clay	Corolla sand	FM loamy sand
5	HAV	Sterile	»16[b]	»16	ND[c]	ND
		Nonsterile	»16	»16	ND	ND
5	Polio 1	Sterile	»16	»16	»16	»16
		Nonsterile	»16	>16[d]	»16	»16
5	Echo 1	Sterile	»16	»16	»16	»16
		Nonsterile	»16	»16	»16	»16
25	HAV	Sterile	>16	8	8	ND
		Nonsterile	12	8	5	ND
25	Polio 1	Sterile	10	10	7	9
		Nonsterile	6	2	2	3
25	Echo 1	Sterile	10	10	8	9
		Nonsterile	4	5	5	4

[a] Soil concentrations (w/v) were 1% for clays and 5% for sands; sample pH, 7.0.
[b] », 0.0 to <1.0 $\log_{10}$ reduction at the end of 16 weeks.
[c] Not determined.
[d] >, 1.0 to <2.0 $\log 10_{10}$ reduction at the end of 16 weeks.

C. Soil Microbial Activity and Related Chemical Activity

Soil microbial activity and perhaps related chemical activity due to microbial enzymes and other chemicals appear to decrease virus survival in soils. In an early report, Bagdasaryan[27] noted greater enterovirus inactivation in nonsterile than in sterile sandy and loamy soils incubated at 3 to 10 and 18 to 23°C. In studies by Sobsey et al.[25] on rates of poliovirus and reovirus inactivation in eight different soil suspensions in settled sewage at 20°C, the time required for 99% inactivation was almost always shorter in nonsterile than in sterile suspensions. Hurst et al.[28] observed increased inactivation of poliovirus and echovirus in nonsterile sandy soil wetted with distilled water and incubated under aerobic conditions at 23 and 37°C, compared to sterile control samples. However, inactivation rates in sterile and nonsterile samples were similar at 10°C under aerobic conditions and at 1, 23 and 37°C under anaerobic conditions. Thus, appreciable virus inactivation due to microbial activity in soils appeared to occur only under aerobic conditions and at moderate to high temperatures. This temperature effect on microbially mediated antiviral activity also was reported in more recent studies by Sobsey et al.[51] on the survival of HAV and model enteroviruses in soil suspensions. At 5°C, the survival of HAV, poliovirus type 1, and echovirus type 1 in suspensions of various soils in secondary effluent was similar for both sterile and nonsterile samples (Table 3). However, at 25°C, the time required for 99% inactivation of all three viruses in the same types of soil suspensions often was found to be considerably shorter in nonsterile than in sterile suspensions.

Recently, McConnell et al.[52] evaluated the removal of radioactively labeled reovirus type 1 from drinking water by slow-rate sand filtration (SSF) using columns constructed to simulate a full scale SSF field operation. Reovirus 1 radioactivity was found throughout the length of the column, although no virus infectivity could be detected. The use of clean, unsieved sand columns, which had bacterial counts as much as 90 times lower than those for ripened, unsieved sand, resulted in similar inactivation of virus. This indicated that reovirus inactivation in the sand beds either was not due to microbial activity or occurred even at the lower microbial densities.

Although the mechanisms of microbially mediated antiviral activity in soils have not been fully elucidated, Yeager and O'Brien[50] have reported differences in poliovirus structural changes during inactivation in sterile and nonsterile soils, depending upon soil moisture levels. In both sterile and nonsterile soils under moist conditions, viral RNA was probably damaged before release from capsids. In sterile, dried soils released RNA genomes remained largely intact, but in nonsterile, dried soils the released RNA was degraded. The role of microbially produced nucleases in these findings is uncertain.

Studies on the role of microbial activity in virus inactivation in natural waters and biological sewage treatment processes may also be pertinent to microbially mediated virus inactivation in soils. Hermann et al.[32] observed enterovirus inactivation to proceed more rapidly in nonsterile lake water in dialysis bags *in situ* than in sterile-filtered lake water held under similar conditions in the laboratory. Virion degradation and probable microbial utilization of virion structural proteins as substrate also was observed. Cliver and Hermann[53] and Hermann and Cliver[54] found that laboratory cultures of *Pseudomonas aeruginosa* and *Bacillus subtilis* inactivated certain enteroviruses, possibly through the activity of microbial enzymes and other antiviral chemicals produced in the cultures. Virion structural proteins were degraded and used as substrate by the bacteria. Furthermore, purified microbial enzymes from *Streptomyces griseus* and subtilisin and neutral pronase from *B. subtilis* inactivated certain enteroviruses.

In studies on poliovirus inactivation in the activated sludge process, Malina et al.[55] reported that the viruses rapidly adsorbed to the activated sludge solids, but their loss of infectivity proceeded at a slower rate. When adsorbed virions became inactivated, their RNA was released back into the activated sludge supernatant.

Information in the literature cited above indicates that aerobic microbial activity in soils may play a role in virus inactivation, but more work is needed to determine the nature and extent of this activity and the contributing roles of other soil conditions that influence microbial activity.

D. Effects of Ionic Salts and pH

The direct effects of ionic salts and pH on virus survival in soils have been less extensively investigated than their effects on virus retention by soils. Murphy et al.[37] found that mouse encephalomyelitis virus was inactivated more rapidly in soil adjusted to pH 3.7 and 8.5 than in neutral soil. More recently, Hurst et al.[28] determined that virus inactivation in soils correlated with soil levels of resin-extractable phosphorus, exchangeable aluminum, and soil pH. Because these same factors also influence virus adsorption to soils, the observed differences in survival rates may be related to changes in the extent of virus adsorption to the soil material and therefore changes in the extent of virus protection from inactivation in the adsorbed state. As indicated in the next section, viruses adsorbed to soils may in some cases survive longer than freely suspended viruses.

Direct effects of different pH and salt levels on virus survival in aqueous solutions have been reported by Salo and Cliver,[39] and similar phenomena may occur in the soil environment. They observed that poliovirus type 1 was inactivated more rapidly at the pH extremes of 3 and 9 than at the intermediate pH levels of 5 and 7. Virus inactivation rates also increased with increasing concentration of NaCl, and the effect was most pronounced at pH 3. Under conditions of equal ionic strength, different inorganic salts produced different inactivation rates, thus indicating specific ion effects. At all pH levels tested the RNA of inactivated virions became ribonuclease sensitive, and at pH 5 and 7 the RNA was hydrolyzed in the absence of ribonuclease. These findings suggest that some alteration of the virus capsid occurs when virions are inactivated, thereby resulting in the viral RNA becoming susceptible to attack. An additional finding was that the proteolytic enzyme chymotrypsin degraded the capsids of poliovirus inactivated at pH 3, but not at the other pH levels tested.

All viruses do not respond to pH and salts in the same way; e.g., coxsackievirus A9 was inactivated more rapidly at pH 5 than at the pH extremes of 3 and 9.[39] Further evidence for different enterovirus responses to salt effects comes from studies by Cords et al.,[38] who found that a number of type A coxsackieviruses were inactivated more rapidly in solutions of low ionic strength than in those with high ionic strength. Inactivation was accompanied by the loss of capsid protein VP-4. However, such inactivation and structural alteration did not occur with poliovirus type 1 or group B coxsackieviruses.

Another example of the effects of pH on virus survival is provided by studies with adenoviruses. Studies by Jacobs[36] showed that the simian adenovirus SV11 was more rapidly inactivated at pH 5.0 and 6.0 than at pH 7.0 and 4.0. Prage et al.[56] found that the inactivation of type 2 adenovirus exposed to water buffered at pH 6.0 to 6.6 was accompanied by an increase in DNase sensitivity and the loss of structural proteins from the vertices of the virion capsids.

Although the results of these studies indicate that virus inactivation can be caused by pH and salt effects, the extent to which these factors contribute to virus inactivation in soils is uncertain and deserves further investigation.

E. Soil Type, Virus Association with Soils, and Virus Aggregation

The association of viruses with soil particles may either enhance or reduce their survival, depending on the chemical properties of the soil material. A number of investigators have observed that virus adsorption to soil particles prolongs their survival. Gerba and Schaiberger[40] found that bacteriophage MS2 survived longer in seawater when associated with kaolinite clay than in clay-free water. Although the mechanisms of protection were not established, it was speculated that kaolinite adsorption of antiviral chemicals, stabilization of virion structure in the adsorbed state, or changes in the extent of virus aggregation could be responsible. Bitton and Mitchell[57] observed that bacteriophage T7 survival was prolonged when adsorbed to montmorillonite clay or *Escherichia coli* cells. LaBelle and Gerba[58] found that estuarine sediments protected poliovirus 1 from inactivation by heat, microorganisms, or certain salts. More recently, Rao et al.[59] found that poliovirus 1 and simian rotavirus SA11 survived up to 19 days in seawater with sediments or suspended solids added, while neither virus was detected after 9 days in seawater without particulate matter.

Adsorption to soil materials does not always result in enhanced survival and in fact may in some cases enhance inactivation. Studies by Sobsey et al.[25] showed that poliovirus type 1 and reovirus type 3 in suspensions of eight different soil materials did not always survive longer than in soil-free controls. Murray and Laband[60] found that adsorption of poliovirus type 1 to oxide particles of manganese, aluminum and copper resulted in significant and rapid inactivation, while adsorption to silica and iron oxide did not result in significant inactivation.

Recent batch studies by Moore et al.[41] indicated that reovirus adsorption to several soil materials, including Ottawa sand, dolomite, montmorillonite and organic muck, resulted in considerable inactivation. Using radiolabeled viruses, it was found that viral radioactivity in the solution phase increased or stayed the same after 1 or 24 hr of adsorption, thus suggesting that adsorbed viruses may have been subsequently released from soils but were no longer infectious. Virions in solution had a normal buoyant density as determined by centrifugation, and therefore they had not undergone gross physical alterations.

Virus aggregation tends to enhance the survival of viruses exposed to various antiviral agents, and this phenomenon has been extensively studied in chemically defined solutions containing purified virions and known quantities of specific disinfectants.[42,43] Although there are no definitive reports concerning the effects of aggregation on enteric virus survival in soils, the results of studies in water suggest that virus aggregates would survive better in soils than would dispersed viruses.

F. Virus Type

Because different viruses vary in their susceptibility to inactivation by a variety of physical, chemical, and biological agents, variations in their ability to survive in soils is not unexpected. The results of an early study by Bagdasaryan[27] indicated some differences in the survival times of five enteroviruses in sandy and loamy soils. Lefler and Kott[30] noted that bacteriophage f2 survived longer than poliovirus in either dry or saturated sand. Sobsey et al.[25,51] found that the rates of inactivation of HAV, poliovirus, echovirus, and reovirus in suspensions of several soils in various wastewaters were often quite different at 25°C (Table 3).

Hurst et al.[28] reported different inactivation rates among different enteric viruses in a sandy soil wetted with sewage. Coxsackievirus A9 was inactivated most rapidly (>90% in 1 day), and coxsackievirus B3 was inactivated least rapidly (90% in about 15 days). In contrast, however, Hurst et al.[35] noted that the rates of poliovirus type 1 and echovirus type 1 inactivation were generally similar under field conditions in a rapid infiltration system.

Additional studies are needed to quantitatively determine the range of differences in enteric virus survival rates for different soils and soil conditions. Such studies are needed particularly for those viruses known to cause waterborne disease, such as HAV, Norwalk and Norwalk-like gastrointestinal viruses, and rotaviruses.

G. Soil Organic Matter

The effects of organic matter on enteric virus survival in soils have not been established, but they may play some role in virus survival. Because humic substances may influence soil properties such as pH, moisture content, or ion exchange capacity,[44] it is likely that they may influence virus survival indirectly through these factors. Additionally, recent findings suggest that fulvic and humic acids may mask virus infectivity by a reversible process. Bixby and O'Brien[61] found that fulvic acid complexation of bacteriophage MS2 caused considerable loss of infectivity and prevented adsorption to soil. The infectivity of the complexed phage could be restored by treating with 3% beef extract solution at pH 9. Further studies on the effects of soil organic matter on virus survival are needed.

IV. FACTORS INFLUENCING VIRUS TRANSPORT OR RETENTION IN SOILS

The ability of soils to retain viruses and the factors influencing virus transport or retention have been studied in some detail, and this literature has been previously reviewed.[16-20] These factors are listed in Table 4, and they will be discussed here with an emphasis on recent findings and critical needs for further research.

A. Virus Association with Particulate Matter and Virus Aggregation

Like other small particles, viruses can potentially be removed or retained in soils by three mechanisms: (1) straining or entrapment either at the soil surface or at contact junctions between subsurface soil particles; (2) sedimentation at the soil surface or on the downstream side of soil particles as a result of velocity decreases that cause suspended matter to settle out; and (3) adsorption by physicochemical interactions between virion and soil particle surfaces.[62] For free virions, that is, virions not associated with larger particles or with each other as aggregates, the first two mechanisms are likely to be of minor importance. Free virions are probably removed in soils primarily by adsorption. However, many viruses in sludges, wastewater, and water are associated either with particulate matter or with each other.[63] If such virus-containing particles are much larger than single virions, they also could be removed by straining and sedimentation. In field studies on virus movement in a rapid infiltration system for wastewater, Hurst et al.[35] noted that viruses accumulated in the

Table 4
FACTORS INFLUENCING VIRUS TRANSPORT OR RETENTION IN SOILS

Factor	Comments	Ref.
Virus aggregation and association with solids	Solids-associated viruses may be retained by entrapment and sedimentation as well as adsorption	36,63
Soil composition	Clay content, metal oxides, and exchangeable ion capacity may increase virus retention	35,41, 64—71
Virus type	Different virus types and strains differ in adsorption and transport patterns	24,35,66, 74—76
pH	Lower pH enhances adsorption	21,25,66
Salt species and conc.	Increased cation valency and ion concentration increase retention; low ionic strength promotes desorption and transport	25,30,78,79, 87,88
Organic matter	Some organics (e.g., humic and fulvic acids) interfere with adsorption and cause desorption	41,61,87, 95,96
Hydraulic conditions and moisture content	Increased flow rates and saturated flow decrease adsorption	22,29, 97—99

wastewater sludge layer that developed on the soil surface as well as in the upper soil levels, possibly because the viruses were already solids-associated and removed by filtration or because they became adsorbed to the sludge solids and soil particles. The relative importance of these different mechanisms of virus removal by soils is largely unknown and has not been adequately investigated in either laboratory or field studies.

B. Soil Composition

Soils vary considerably in their textural, chemical, and mineralogical properties, and hundreds of soil types have been classified and mapped in the U.S. and elsewhere. Furthermore, both vertical and horizontal variability is a normal characteristic of many soils.

A number of studies have shown that the extent of virus transport through soils varied with the type of soil materal. In an early study on poliovirus removals from oxidation pond effluent by columns of three soil materials, Laak and McLean[64] found that removals were greatest with sandy loam, least with sand and intermediate with "garden soil." Hori et al.[65] found that poliovirus removals from distilled water in 6-in. columns of three Hawaiian soils, including two, low-humic latosols (Lahaina and Wahiawa) and a volcanic cinder (Tantalus), averaged >99, >99, and 22%, respectively. With the two, low-humic latosols, there was a trend of decreased retention over the 5-day test period. Goyal and Gerba[66] noted considerable differences in the abilities of nine different soils to adsorb a number of enteric viruses. Statistical analysis indicated that pH was the most important soil characteristic influencing virus retention, with soils having a pH <5 giving consistently high retention. Exchangeable aluminum was another factor that correlated with the adsorption efficiency of many of the viruses. Burge and Enkiri[67] found variations among five different soils in adsorbing bacteriophage ϕX174. With four of the five soils, adsorption rates correlated with cation exchange capacity, specific surface area, and organic matter content.

In studies by Sobsey et al.[25] the retention of poliovirus type 1 by 4-in.-long columns of four soil materials intermittently receiving virus-laden settled sewage varied with soil type. Sandy (Fripp and Lakeland) and organic (Ponzer) soils gave virus removals of only 98 to 99.7%, but a sandy clay loam (Norfolk) gave removals of ≥99.996%. In contrast, fecal coliform bacteria removals were much greater in all four soils, averaging about 99.996% for the two sands, 99.999% for the organic soil, and >99.9998% for the sandy clay loam. These findings suggest that fecal coliform bacteria may not be suitable indicators of enteric virus retention in soils.

Moore et al.[68] recently reported that poliovirus type 2 adsorption to 34 different soil materials suspended in synthetic freshwater was negatively correlated with soil organic matter content and with available negative surface charge as measured by adsorption capacity for a cationic polyelectrolyte. Soil pH, surface area, and elemental composition were not significantly correlated with virus adsorption. Additional studies by the same group[41,68-70] indicated that minerals were generally better adsorbents than were soils. The most effective adsorbents were magnetite sand and hematite, both predominantly oxides of iron. Among the soils tested, the two poorest adsorbents for both poliovirus 2[69] and reovirus 3[68] were muck and silt loam, both of which had high (200 and 31.5 mg/g, respectively) organic matter content.

Funderburg et al.[71] constructed soil columns of various lengths according to the vertical distribution and bulk density of eight different soils as they were found in the field. Columns were challenged with poliovirus 1, reovirus 3, or phage φX174 suspended in wastewater, followed by distilled water flooding to simulate rainfall. A statistical analysis of the data indicated that poliovirus and reovirus recovery in the percolates was correlated most favorably with low soil cation exchange capacity, while the amount of phage φX174 recovered in the percolates did not appear to be related to the soil characteristics, but was related to the retention time of phage in the columns.

The results of these studies indicate that soil type greatly influences the extent of virus transport or retention. It may be possible to distinguish soils into general classes with respect to virus retention, based upon their textural, mineralogical, and chemical properties. However, further studies with a wide range of soil types and viruses are needed to determine if such classifications are possible and to identify the soil characteristics that most influence virus retention.

C. Virus Type and Strain

A number of studies over the last decade have shown that different types and strains of viruses are not equally retained by soils. These virus-specific differences in adsorption to soils are probably related to physicochemical differences in virus capsid surfaces. Although all enteric viruses possess outer capsids comprising polypeptide subunits and generally behave as charged, amphoteric, colloidal particles, the surfaces of the virions differ in the details of their configuration, charge density and distribution, and other features. In fact, even the same virus can display different surface properties that will influence its physicochemical behavior as a result of conformational changes brought about by pH effects and interactions with soluble chemicals and particulate surfaces.[44,50,72,73]

Landry et al.[24] studied the adsorption and subsequent elution with rainwater of several different enterovirus types, including several strains of poliovirus type 1, from columns (123 cm long) or intact cores of coarse, sandy soil dosed with tertiary effluent. There were differences among some enterovirus types and also among strains of poliovirus type 1 in the extent of adsorption to the soil columns and the extent of elution by application of simulated rainwater or sewage effluent. It was concluded that poliovirus type 1, strain LSc, was an inadequate model for predicting the behavior of all enteroviruses in soil columns because it was adsorbed efficiently and not readily eluted.

Recent studies by Hazard and Sobsey[74] and Sobsey et al.[51] compared the adsorption and retention of HAV to that of poliovirus type 1 (strain LSc) and echovirus type 1 (strain V239) in a number of soil suspensions and in unsaturated columns of sandy soils (10 cm deep) dosed twice weekly with 2.5 cm of virus-laden groundwater or primary effluent. Poliovirus was retained most extensively, echovirus was retained the least, and HAV retention was intermediate between these viruses. A similar pattern of differences among these viruses also was observed in studies on short-term adsorption to a number of soils suspended in groundwater or sewage effluents.[51]

Studies by Goyal and Gerba[66] and Gerba et al.[75,76] showed that different enterovirus types and strains varied in their ability to adsorb to soils. For example, adsorption efficiencies of six different strains of echovirus type 1 in suspensions of sandy soil in deionized water ranged from 0 to 99.7%. Type and strain dependence of enterovirus adsorption to a sandy loam soil suspended in distilled water was also reported in another study from the same laboratory.[75] Adsorption efficiencies of ten different virus types and strains ranged from 0% for echovirus type 1, strain V239, and coxsackievirus B4, strain V216, to 99.9% for echovirus type 7, Wallace strain, and poliovirus type 1, strain LSc.[66]

Subsequently, these same authors statistically reexamined their earlier data.[76] This analysis indicated that different types and strains of viruses can be grouped by their ability to be similarly affected by certain soil characteristics. For group I viruses, which included coxsackievirus B4 and echovirus 1, the most important factors affecting adsorption were pH, organic matter, and exchangeable ion content of the soil. For group II viruses, which included poliovirus 1, echovirus 7, and coxsackievirus B3, no studied soil characteristic was found to be significantly associated with virus adsorption. These authors hypothesized that the adsorption-elution profiles of group I viruses would tend to be more sensitive to certain soil characteristics than group II viruses.

In contrast to the findings from batch laboratory studies by the same group, Hurst et al.[35] found that under field conditions at a rapid infiltration site, echovirus type 1, Farouk strain, did not migrate to as deep a point in the soil as did poliovirus type 1, strain LSc. They suggested that the adsorptive behavior of viruses in laboratory batch studies may not be totally reflective of their behavior under field conditions, possibly because of virus adsorption to soil particles prior to infiltration.

Additionally, Lance et al.[77] found that data from batch adsorption experiments could not be correlated to virus migration results from soil columns 250 cm long. In this case, two viruses, echoviruses 1 and 29 (both found to adsorb poorly to soil in batch suspensions[66]), as well as poliovirus 1, were tested for comparative movement in loamy sand columns. When viruses suspended in dechlorinated secondary effluent were applied to the columns, 90% of echovirus 29 and poliovirus type 1 adsorbed to the top 2 cm of the column, but only 77% of echovirus 1 adsorbed to the soil in this region. However, below the 40-cm column depth, the leaching patterns for all three viruses were the same. It was concluded that differences among viruses in their adsorption and movement in batch and short-column studies are not great enough to be demonstrable in long-column studies, which are probably representative of many field conditions.

Because these studies indicate that different enteric viruses are not always transported or retained in soils to the same extent, additional research with a wide variety of viruses is needed to establish the range of differences in virus transport through various soils, especially under field or fieldlike conditions. Such studies are needed particularly for those viruses likely to cause outbreaks of groundwater-borne disease, such as HAV, Norwalk and other Norwalk-like viruses, and possibly rotaviruses.

D. pH

The results of a number of studies indicate that virus retention by soils generally increases at lower pH levels. In an early report Drewry and Eliassen[78] found decreased adsorption of bacteriophages T1, T2, and f2 to Arkansas and California soils at higher pH levels. More recently, Burge and Enkiri[67] found that the rates of bacteriophage ϕX174 adsorption to five soils were significantly correlated with soil pH. In batch adsorption studies with a variety of viruses and nine soils by Goyal and Gerba,[66] pH was found to be the single most important soil factor influencing adsorption. Soils having a saturated pH <5 were the best adsorbers. Studies by Sobsey et al.[25] showed that poliovirus type 1 and reovirus type 3 adsorption to eight different soil materials suspended in settled sewage at pH levels between 3.5 and 7.5 was generally greater at the lower pH levels. In studies by Duboise et al.[21] using cores of sandy forest soil receiving poliovirus in sewage effluent at various pH levels between 5.5

and 9.0, virus retention was best at pH 5.5, and the release and migration of retained viruses by subsequent distilled water applications was lower from the cores that received sewage effluent having lower pH values. Taylor et al.[79] studied the influence of pH on the adsorption of poliovirus 2 to three soils, a sand, and a clay mineral. They found that for each adsorbent, there was a characteristic pH region of transition from strong to weak virus uptake. For all adsorbents, the transition region was above pH 7.5, the isoelectric point of the virus. In electrolytes virus adsorption to soils and minerals was poor above pH 9 and extensive below pH 7.

Although the mechanism by which pH influences virus adsorption to soils has not been investigated in detail, the effects of pH on virus adsorption to soils can be explained on the basis of electrochemical features of virus and soil surfaces. The surface charge of a virus is influenced primarily by ionization of the carboxyl, amino, and other ionizable functional groups on the outer surface of the virion protein capsid, and at neutral pH most viruses are at least weakly negatively charged. Many soils also tend to be generally electronegative at neutral pH, and therefore virus adsorption is not favored due to repulsion of the two negatively charged surfaces. However, if the pH of the surrounding medium is lowered, protonation causes decreased ionization of virion carboxyl and other negatively charged functional groups and increased ionization of amino groups. As a result, viruses become less electronegative or even electropositive at lower pH levels. The isoelectric points of soil particles are generally lower than those of viruses. Thus, at lower pH levels, the viruses may be electropositive, but the soils are still electronegative, thereby resulting in electrostatic attraction and increased adsorption. From studies on adsorption of poliovirus type 1 by hydrous oxides, Murray and Parks[80] concluded that the process can be described by the Dejaguin-Landau-Verwey-Overbeek-Lifshitz theory of coagulation. According to this theory, van der Waals forces are responsible for attraction and adsorption, while electrostatic repulsion due to overlap of the ionic double layers around both virus and adsorbent surfaces inhibits adsorption.

The importance of electrostatic interactions in the promotion of virus adsorption to soils is indicated by the influence of pH, and salt species and concentration (see below). However, recent studies indicate that hydrophobic interactions, as well as electrostatic interactions, may play a role in the association of virus with soils. Dizer et al.[81] found that the presence of surfactants, such as nonyl phenol, decreased the adsorption of poliovirus 1 to sand by over 50%. Studies involving other solids, such as membrane filters,[82,83] wastewater sludges,[84] clays,[85] and estuarine sediments,[86] have indicated that hydrophobic interactions play a role in the association of viruses with these solids. Additional research is needed to establish the relative importance of these hydrophobic interactions in virus adsorption to and retention by soils.

E. Salt Species and Concentrations

The types and concentration of ionizable salts in the soil-water environment greatly influence the extent of virus retention or transport. In general, increasing concentrations of ionic salts and increasing cation valencies enhance virus adsorption.

In an early report, Carlson et al.[87] found that bacteriophage T2 and poliovirus type 1 adsorption to clay suspensions in water was enhanced by low concentrations of $CaCl_2$. The effect was attributed to the role of calcium ions in mediating adsorption between the negatively charged clay and virus surfaces, possibly through clay-cation-virus bridges. Similarly, Schaub et al.[88] found that addition of $CaCl_2$ to water enhanced virus adsorption to clays. They speculated that metal cations served to reduce repulsive forces between the viruses and clay particles, allowing these particles to come close enough to allow the shorter-range attractive forces, such as van der Waals forces, to operate, thus resulting in virus adsorption.

Enhanced adsorption of a number of viruses to a variety of soil materials at increased ionic strength (or concentrations) and in the presence of cations with higher valencies has also been reported by other workers using either soil suspensions or soil columns.[25,30,78,88] Taylor et al.[79] reported that pH and the nature and concentration of several simple electrolytes, including $CaCl_2$, NaCl, and Na_2SO_4, influenced poliovirus type 2 adsorption to five soil materials. Virus adsorption from electrolyte solutions was extensive below pH 7 but not above pH 9. The critical pH region where virus adsorption efficiency changed was characteristic for each soil material. Differences in virus uptake among soils were related to their pH-dependent charge properties as measured by microelectrophoresis. The extent of virus adsorption increased with increasing electrolyte concentrations, but only when the pH of the suspension was near or above the critical region for the soil. Poor virus adsorption obtained at high pH was attributed to electrostatic repulsion between the highly negatively charged virus and soil surfaces.

Recently, Lance and Gerba[89] studied the effect of ionic composition of various suspending solutions on poliovirus adsorption to a soil column 250 cm deep. When viruses were suspended in chloride solutions, adsorption increased as the cation concentration and valence increased. Viruses suspended in sewage were found to penetrate the soil column to below the 80-cm depth. However, when $AlCl_3$ was added to the sewage to provide an Al^{3+} concentration of 0.1 mM, the depth of virus penetration was reduced to 40 cm.

Although solution ions appear to play a role in virus adsorption to soils, the role of exchangeable ions associated with soil particles remains uncertain. In an early report[78] bacteriophage T1, T2, and f2 adsorption to nine soils in suspension was found to increase with increasing ion-exchange capacity and glycerol retention capacity. More recently Burge and Enkiri,[67] using soil suspensions, reported that adsorption rates of bacteriophage ϕX174 correlated with cation exchange capacity for four of the five soils tested. Schiffenbauer and Stotsky[90] reported that the adsorption of bacteriophage T7 to kaolinite and bentonite clays was found to be related to the cation exchange capacity (CEC) of these clays, while adsorption of bacteriophage T1 appeared to be related to the anion exchange capacity (AEC) of these clays. Investigators in the same laboratory[91] further reported that the adsorption of reovirus 3 to these clays was related to CEC. These investigators also studied the adsorption of reovirus 3 to bentonite made homoionic to various cations. It was shown that virus adsorption was increased in the presence of bentonite made homoionic to higher valency ions. Specifically, adsorption of reovirus was greater to bentonite homoionic to aluminum than to bentonite homoionic to calcium or magnesium. The authors suggested that this increase in viral adsorption was due to a reduction of the net negative charge on the clays, which allowed viruses to come close enough to the clays to become protonated and subsequently adsorb by cation exchange.[91] However, Goyal and Gerba[66] found that the adsorption of a number of different viruses to nine different soils was not significantly correlated with cation exchange capacity, total phosphorus, total and exchangeable iron, and exchangeable magnesium.

Decreases in salt concentration or ionic strength can cause adsorbed viruses to be released from soils and then be transported more extensively. In one report it was found that under simulated cycles of intermittent sewage effluent application and rainfall (distilled water application) to cores of sandy forest soil, the extent of poliovirus type 1 migration depended on ionic gradients produced in the waters as they passed through the cores.[21] Applications of distilled water caused decreases in ionic strength of the column effluents and associated increases in virus elution. Virus elution also coincided with increased elution of total organic carbon. Similar findings have also been reported by other investigators.[24,25,92]

The migration of poliovirus in soil columns under changing ionic conditions has also been studied by Lance et al.[22] When poliovirus type 1 in secondary effluent was applied to columns of calcareous sand, most of the viruses were retained in the upper 5 cm of soil, regardless of the infiltration rate. When deionized water was subsequently applied, viruses desorbed from the soil and migrated further down the columns. However, when this was followed by another application of sewage effluent, the viruses were readsorbed by the soils further down the column, and they did not appear in column effluents. These findings suggest that virus movement through soils as a result of heavy rains may be minimized by appropriate management of flooding and drying cycles in rapid infiltration systems.

The results of a number of studies indicate that virus transport through soils is greatly influenced by ionic conditions of the soil-water environment, and they provide a basis for explaining the sudden appearance of natural virus contaminants in groundwaters following heavy rains at field sites where wastewaters are being applied to the land. For example, studies by Wellings et al.[26,93] demonstrated the presence of viruses in wells 10 to 20 feet below soil that had been treated with sewage effluents. These viruses appeared after a period of heavy rainfall, and the investigators hypothesized that the increase in the water-soil ratio resulting from the rainfall led to the desorption of viruses attached to the soil. However, considerably more field work is needed with a variety of soil types, geohydrologic conditions, climatic conditions, and wastewater application systems to better establish the extent of virus movement in soils under changing ionic conditions.

F. Organic Matter

At least certain types of soluble organic matter in water and soil will decrease virus adsorption to soils. In an early report Carlson et al.[87] found that proteinaceous organic matter interfered with bacteriophage T2 adsorption to three types of clay suspended in $CaCl_2$ solutions.

Humic and fulvic acids are highly colored organic compounds that are naturally present in both water and soils.[94] Recent studies indicate that these compounds can cause increased virus transport through soils not only by interfering with virus adsorption but also by causing desorption. Bitton et al.[95] found that poliovirus retention by columns of sandy soil was extensively reduced when applied in highly colored (high concentrations of humic and fulvic acids) cypress dome water compared to its retention from tap water. More recently, Scheuerman et al.[96] reported extensive interference by humic and fulvic acid with poliovirus type 1 retention in columns of organic sediment, muck soil, and brown-red sand. Soils that were capable of retaining all or most of the applied virus in the absence of these organics retained considerably less virus in their presence. The extent of virus transport through the columns correlated with the color of the column effluents.

In batch studies. Bixby and O'Brien[61] found that fulvic acids complexed with bacteriophage MS2 and extensively interfered with its adsorption to a fine loamy sand.

There is conflicting information in the literature concerning the ability of wastewater organics to interfere with virus retention by soils. In an early study by Carlson et al.[87] bacteriophage T2 adsorption to kaolinite was somewhat reduced when the suspensions contained as much as 10 mg/ℓ of sewage. Moore et al.[41] reported that organics in wastewater effluent as well as organics in soils interfered with the adsorption of reovirus type 3 by soil materials. Dizer et al.[81] found that various enteroviruses suspended in secondary treated wastewater were adsorbed to a lesser extent by laboratory columns of sand than were the same viruses suspended in cleaner, tertiary treated water. Recently, Sobsey et al.[51] reported that the extent of adsorption of HAV and echovirus type 1 to kaolinite clay was lower in primary effluent suspensions than in secondary effluent suspensions. However, the extent of adsorption of poliovirus type 1 to kaolinite was the same in both primary and secondary effluent suspensions. In suspensions of other soil types adsorption of these same three viruses also was not lower in primary effluent than in secondary effluent. Gerba and Lance[92] found that removal of poliovirus type 1 from primary and secondary sewage effluent in columns of loamy sand was similar for both types of wastewater.

The results of a number of studies suggest that organic soils and other soils with high concentrations of humic and fulvic acids do not always retain viruses efficiently, and therefore some organic soils may not be suitable for land application of wastewater. However, further studies are needed to establish the validity of this point. The effects of other organics in water, wastewater, and soils on virus retention in soils remain uncertain. Additional studies are needed to further understand and quantify the effects of humic and fulvic acids in water and soil on the retention and inactivation of a variety of viruses in different soils. Such studies are also needed for other classes of water, wastewater, and soil organics.

G. Hydraulic Conditions and Moisture Content

Hydraulic conditions in soils receiving wastewater appear to have a considerable effect on virus transport for at least some soils. Such conditions as flow rate, geohydrologic conditions, hydraulic loading, and application frequency may all influence the extent of virus migration through soils.

In an early report Robeck et al.[97] found that the extent of poliovirus removal from tap water applied to a sand column decreased with increasing flow rates above 1.2 m/day. Recently, Vaughn et al.[98] reported that infiltration rate greatly influenced poliovirus removal in a groundwater recharge system where tertiary effluent was applied to a soil consisting of coarse sand and fine gravel. Recharge at 75 to 100 cm/hr resulted in considerable virus movement into groundwater, while at two lower recharge rates, 6 and 0.5 to 1.0 cm/hr, there was considerably less virus movement. At the lower infiltration rates, the surface mat of sewage solids that formed on the soil surface may have contributed to the greater virus removals observed.

Using sand columns dosed with poliovirus in septic tank effluent, Green and Cliver[29] found virus retention to be much lower under saturated flow conditions created by high hydraulic loadings compared to those under unsaturated flow conditions at lower hydraulic loadings.

Somewhat in contrast to these findings are those of Lance et al.[22] and Lance and Gerba,[99] who studied poliovirus movement through a coarse sand column flooded with sewage effluent at a high rate. In an early study Lance et al.[22] found that poliovirus type 1 removal was not affected by infiltration rates in the range of 15 to 55 cm/day. In a subsequent study, Lance and Gerba[99] found that increasing flow rates from 0.6 to 1.2 m/day resulted in increased movement of polioviruses down the column. However, there was no further increase in virus movement at flow rates up to 12 m/day. Similar results were found for echovirus 1 adsorption at increased flow rates. They concluded that virus retention by soils is increased above some breakpoint flow rate, but flow rate changes above and below this breakpoint do not affect the extent of virus retention.

More recently, Lance and Gerba[100] investigated virus movement in soil columns during saturated (50 to 100 cm/day) and unsaturated (35.2 cm/day) flow. Poliovirus penetration during unsaturated flow was much less than during saturated flow (virus penetration depths of 40 cm and 160 cm, respectively). These investigators suggested that during unsaturated flow, viruses may move through the soil in thinner films of water and be drawn nearer to the soil particles than during saturated flow. This would increase the potential for virus adsorption and thus decrease the virus penetration depth.

Although the results of at least some studies suggest that virus migration increases with increasing hydraulic loads and flow rates and under conditions of saturated flow, further studies are needed with a wide range of soil types and field conditions to quantify the extent of virus movement through soils under different hydraulic conditions.

V. INDICATORS OF ENTERIC VIRUS SURVIVAL AND TRANSPORT IN SOIL SYSTEMS

As indicated by the previous discussion, there is considerable evidence for virus survival and transport in soils under some conditions. Therefore, there is a need to better monitor and predict the behavior of disease-causing enteric viruses in soil systems. Because detection of enteric viruses in such systems is not yet practical, and in the case of some viral pathogens not even possible at this time, it is desirable to have an appropriate indicator organism for viral pollution of soil and groundwater.

The coliform group of bacteria is the established indicator for the sanitary quality of water. In particular, analysis for total or fecal coliforms has become standard practice in determining the extent of fecal contamination of the environment. However, it has become increasingly apparent that coliforms are not always suitable indicators for enteric viruses. The isolation of enteric viruses from groundwater supplies meeting coliform standards[10,13,101] indicates that viruses can survive longer and travel farther in soils than can coliform bacteria.

Coliforms, like other bacteria, are cellular organisms requiring a set of appropriate environmental conditions for physiological activity and viability. In contrast, enteric viruses consist of a nucleic acid genome surrounded by a protein coat or capsid. They are inert but relatively stable in a variety of extracellular environments, and their ability to retain infectivity in soil and groundwater environments is well documented. The relative instability of coliforms to environmental stress makes them a poor candidate for predicting the survival and transport of viruses in soil systems. Another inadequacy of coliforms as models of virus behavior in soils is their relatively large size (1 to 2 μm) compared to enteric viruses (0.025 to 0.1 μm). This size difference is important because of the mechanisms by which bacteria and viruses are removed in soils. Bacteria removal occurs by entrapment or filtration and sedimentation as well as adsorption in the soil matrix. In contrast, virus removal in soils is primarily by adsorption, as they are too small to be removed efficiently by entrapment or sedimentation. Furthermore, physicochemical differences in the outer surfaces of bacteria and viruses may contribute to differences the extent to which they will adsorb to soils.

Given these differences, it is not surprising that several studies have shown fecal coliforms to be unsuitable indicators for enteric virus reductions in soil systems. Using miniature columns of several soils dosed intermittently with wastewater, Sobsey et al.[25] found that reductions of fecal coliforms exceeded reductions of poliovirus by up to 3.3 orders of magnitude. More recently, Vaughn et al.[13] evaluated the movement of naturally occurring enteroviruses and coliform bacteria from septic tank leach fields through a shallow, sandy-soil aquifer. Statistical analyses of the data indicated no significant correlations between virus occurrence and the occurrence of total and fecal coliform bacteria. Indeed, coliforms were rarely detected in wells more than 1.5 m from the effluent distribution source while enteric viruses were detected in 6.2 to 50% of the samples from wells that were 3.0 to 60 m from the effluent source. Although sample sizes for enteric virus isolations were considerably larger than those for coliforms, the fact remains that enteric viruses were detected in samples with relatively low levels of coliform bacteria.

In contrast to the aforementioned studies, other investigators have reported the movement and survival patterns of fecal coliforms in large columns and in the field to be predictive of enterovirus survival and movement in soils. Lance et al.[77] studied the movement of poliovirus type 1 and fecal coliforms in soil columns 250 cm long flooded with wastewater. Fecal coliforms were removed to a greater extent than poliovirus in the upper regions of the columns, but below a column depth of 40 cm, the movement of both fecal coliforms and poliovirus was roughly parallel. Fattal et al.[102] reported on the persistence of enteroviruses and fecal coliforms in surface layers of soil irrigated with wastewater by the drip method. They found that both agents declined at a similar rate, with >4 $\log_{10}$ (>99.99%) reduction in 17 days. These findings suggest that coliforms may be useful indicators for enteric viruses in some soil systems. Unfortunately, the reasons for observed differences in the comparative behavior of coliforms and enteric viruses in different studies are unknown, but given such discrepancies, further studies are certainly needed.

Because some studies have shown that coliform bacteria are inadequate indicators of enteric virus behavior in soil systems, other agents, such as fecal streptococci and bacteriophages, have been suggested as alternative indicators for enteric viruses. However, investigators have found that these agents also are not adequately indicative of enteric virus behavior in soil systems.

The removal of indigenous enteroviruses, bacteriophage f2 and indicator bacteria from wastewater by rapid infiltration through silty sand soil was determined by Schaub and Sorber.[103] Although the bacterial indicators studied — total and fecal coliforms and fecal streptococci — were removed almost entirely at the soil surface, more than 50% of the f2 bacteriophage was not removed at all and appeared in the soil effluent. Removal of indigenous enteric viruses was greater than that of f2 (nearly 90%) but less than that of the bacterial indicators. Subsequently, investigators from the same laboratory evaluated the removal of indigenous enteric viruses and bacteriophage f2 by overland runoff treatment of sewage in a fine, sandy loam soil.[104] Again, removal of f2 was considerably lower than that of enteric viruses. In contrast to these findings, Wang et al.[105] compared the removals of poliovirus type 1, fecal coliforms, and fecal streptococci by four different sandy soils under saturated flow conditions. In all four soils poliovirus removal was substantially greater than the removals of the indicator bacteria studied.

The results of these studies indicate that indicator bacteria and coliphages are not always adequate in predicting the behavior of enteric viruses in soil systems. However, before rejecting indicator bacteria and bacteriophages as enteric virus indicators, additional studies are needed using alternative candidate indicators or a wider range of such candidates as well as a variety of soils, geohydrologic conditions, climatic conditions, and wastewater application systems.

VI. SUMMARY AND CONCLUSIONS

This chapter has shown that a variety of factors influence the survival and transport of viruses in soils. A major factor influencing virus survival is temperature, with generally increased survival at lower temperatures. Soil moisture also influences virus survival, with decreasing survival in dry or drying soils. However, information in the literature is not entirely consistent regarding the moisture levels at which maximum virus inactivation occurs, and more studies are needed in this area using a variety of viruses, soils, and moisture levels. In this regard, consideration should be given to measuring water activity as well as moisture levels, because the former may be more indicative of the effects of moisture on virus survival. Aerobic microbial activity also contributes to virus inactivation in soils at moderate to high temperatures. Further research is needed to precisely identify what is responsible for microbially mediated virus inactivation in soils and to determine the relationships between microbial activity levels and virus inactivation rates.

Although pH and ionic effects may play a role in virus inactivation in soils, these properties are related to the chemical and mineralogical characteristics of soils as well. Information in the literature is not entirely consistent regarding which chemical, mineralogical, and textural properties of soils most influence virus survival or inactivation. However, virus survival often is prolonged in many fine-textured soils to which viruses adsorb readily, presumably because adsorption affords protection against inactivation. It may be possible to predict relative survival of viruses in different soils, based upon their specific properties, but further studies are needed in this area.

Viruses differ in their ability to survive in soils, and more viruses must be evaluated to determine the overall range of virus survival. Such studies are especially needed with those viruses that have been responsible for outbreaks of disease associated with contaminated soil and groundwater, such as HAV, Norwalk virus, other Norwalk-like gastroenteritis viruses, and rotaviruses. Indeed, recent studies on HAV survival in soils, groundwater, and wastewater indicate that it survives longer than model enteroviruses such as poliovirus type 1, at least under certain conditions. In addition, the effects of virus aggregation on survival in soils have not been investigated in any detail.

The role of organic matter in virus survival in soils is uncertain. Although interaction with humic and fulvic acids may cause reversible loss of virus infectivity, there is also some evidence that such virus complexation with organics may actually protect viruses from inactivation by preventing their adsorption to soil particles. The latter effects may result in greater virus survival and mobility in soils. Furthermore, the ability of humic and fulvic acids to mask virus infectivity may result in underestimations of virus levels in soils and groundwater. The effects of organic matter on virus survival in soils clearly need further investigation.

Virus retention by soils is also influenced by a variety of factors, some of which influence virus survival as well. Virus association with particulate matter may enhance retention in soils but may also result in longer survival. The net result of these two effects remains uncertain, especially under field conditions. Soil type also influences virus retention, and adsorption is an important mechanism of virus removal in soils. Although virus retention is generally lower in sand and organic soils than in clay soils, information in the literature is inconsistent on which specific soil properties are most important for virus retention. Further research is needed in this area, especially under field or field-like conditions.

Ionic conditions and pH also influence virus retention by soils, and retention is generally greater at lower pH levels and at higher salt concentrations or with salts having higher cation valences. The effects of these factors are also related to the characteristics of the soil material and other environmental factors, such as hydraulic conditions and rainfall. Therefore, they must be evaluated for a variety of soils under a range of conditions encountered in the field.

Humic and fulvic acids interfere with virus retention by soils, and therefore both water and soils with high concentrations of these organics may be unsuitable for land application of wastewater. The effects of other organics on virus retention by soils remain uncertain. Additional studies are needed to determine the effects of humic and fulvic acids and other organics on both the transport and the survival of viruses in soils, especially under field conditions.

Increased hydraulic loads and flow rates may increase virus transport in soils, especially when the soils become saturated. However, additional laboratory and field studies are needed to quantify virus movement and survival in a variety of soils having different vertical depths of unsaturated flow and subjected to a wide range of hydraulic conditions.

Although a variety of factors that influence virus survival and transport in soils have been identified, there is as yet insufficient information in the scientific literature to accurately predict the extent to which viruses will survive and travel in the soil environment of a particular field site receiving virus-contaminated water, sludge, or solid waste. Such predictive information is a desirable objective for future research efforts in order to develop suitable models that will predict the potential for virus survival and transport under a defined set of soil, geohydrologic, climatic, and waste application conditions.[106]

REFERENCES

1. **Gerba, C. P., Wallis, C., and Melnick, J. L.,** Viruses in water: the problem, some solutions, *Environ. Sci. Technol.*, 9, 1122, 1975.
2. **Melnick, J. L., Wallis, C., and Gerba, C. P.,** Viruses in water, *Bull. WHO*, 56, 499, 1978.
3. **Sabin, A. B.,** Behavior of chimpanzee-avirulent poliomyelitis viruses in experimentally infected human volunteers, *Am. J. Med. Sci.*, 230, 1, 1955.

4a. **Wigand, R. and Sabin, A. B.,** Properties of epidemic strains of ECHO type 9 virus and observations on the nature of human infection, *Arch. Ges. Virusforsch.*, 11, 683, 1962.

4b. **Melnick, J. L. and Rennick, V.,** Infectivity titers of enterovirus as found in human stools, *J. Med. Virol.*, 5, 205, 1980.

5. **Purcell, R. A., Feinstone, S. M., Ticehurst, J. R., Daemer, R. J., and Baroudy, B. M.,** Hepatitis A virus, in *Viral Hepatitis and Liver Disease*, Vyas, G. N., Dienstag, J. L., and Hoofnagle, J. H., Eds., Harcourt Brace Jovanovich, New York, 1984, 9.

6. **Konno, T., Suzuki, H., Imai, A., and Ishida, N.,** Reovirus-like agent in acute epidemic gastroenteritis in Japanese infants: fecal shedding and serologic response, *J. Infect. Dis.*, 135, 259, 1977.

7. **England, B.,** Concentration of reovirus and adenovirus from sewage and effluents by protamine (salmine) treatment, *Appl. Microbial.*, 24, 510, 1972.

8. **Malina, J. F., Jr.,** Viral pathogen inactivation during treatment of municipal wastewater, in *Virus Aspects of Applying Municipal Wastes to Land*, Baldwin, L. B., Ed., University of Florida, Gainesville, 1976, 9.

9. **Craun, G. F.,** A summary of waterborne illness transmitted through contaminated groundwater, *J. Environ. Health*, 48, 122, 1985.

10. **Hejkal, T. W., Keswick, B. H., LaBelle, R. L., Gerba, C. P., Sanchez, Y., Dreesman, G., Hafkin, B., and Melnick, J. L.,** Viruses in a community water supply associated with an outbreak of gastroenteritis and infectious hepatitis, *J. Am. Water Works Assoc.*, 74, 318, 1982.

11. **Sobsey, M. D., Oglesbee, S. E., Wait, D. A., and Cuenca, A. I.,** Detection of hepatitis A virus (HAV) in drinking water, *Water Sci. Tech.*, 17, 23, 1985.

12. **Keswick, B. H. and Gerba, C. P.,** Virus in groundwater, *Environ. Sci. Technol.*, 14, 1290, 1980.

13. **Vaughn, J. M., Landry, E. F., and Thomas, M. Z.,** Entrainment of viruses from septic tank leach fields through a shallow, sandy soil aquifer, *Appl. Environ. Microbiol.*, 45, 1474, 1983.

14. **Goyal, S. M., Keswick, B. H., and Gerba, C. P.,** Viruses in groundwater beneath sewage irrigated cropland, *Water Res.*, 18, 299, 1984.

15. **Lippy, E. C. and Waltrip, S. C.,** Waterborne disease outbreaks 1946—1980: a thirty-five year perspective, *J. Am. Water Works Assoc.*, 76, 60, 1984.

16. **Bitton, G.,** Fate of viruses in land disposal of wastewater effluents, in *Introduction to Environmental Virology*, John Wiley & Sons, New York, 1980, chap. 9.

17. **Duboise, S. M., Moore, B. E., Sorber, C. A., and Sagik, B. P.,** Viruses in soil systems, *Crit. Rev. Microbiol.*, 7, 245, 1979.

18. **Gerba, C. P., Wallis, C., and Melnick, J. L.,** Fate of wastewater bacteria and viruses in soil, *J. Irrig. Drain. Div. Am. Soc. Civ. Eng.*, 101, 157, 1975.

19. **Sobsey, M. D.,** Transport and fate of viruses in soil, in Microbial Health Considerations of Soil Disposal of Domestic Wastewaters, EPA-600/9-83-017, Canter, L. W., Akin, E. W., Kreissl, J. F., and McNabb, J. F., Eds., U.S. Environmental Protection Agency, Cincinnati, 1983, 174.

20. **Vaughn, J. M. and Landry, E. F.,** Viruses in soils and groundwaters, in *Viral Pollution of the Environment*, Berg, G., Ed., CRC Press, Boca Raton, Fla., 1983, 163.

21. **Duboise, S. M., Moore, B. E., and Sagik, B. P.,** Poliovirus survival and movement in a sandy forest soil, *Appl. Environ. Microbiol.*, 31, 536, 1976.

22. **Lance, J. C., Gerba, C. P., and Melnick, J. L.,** Virus movement in soil columns flooded with secondary sewage effluent, *Appl. Environ. Microbiol.*, 32, 520, 1976.

23. **Landry, E. F., Vaughn, J. M., and Penello, W. F.,** Poliovirus retention in 75-cm soil cores after sewage and rainwater application, *Appl. Environ. Microbiol.*, 40, 1032, 1980.

24. **Landry, E. F., Vaughn, J. M., Thomas, M. Z., and Beckwith, C. A.,** Adsorption of enteroviruses to soil cores and their subsequent elution by artificial rainwater, *Appl. Environ. Microbiol.*, 38, 680, 1979.

25. **Sobsey, M. D., Dean, C. H., Knuckles, M. E., and Wagner, R. A.,** Interactions and survival of enteric viruses in soil materials, *Appl. Environ. Microbiol.*, 40, 91, 1980.

26. **Wellings, F. M., Lewis, A. L., and Mountain, C. W.,** Virus survival following spray irrigation of sandy soils, in *Virus Survival in Water and Wastewater Systems*, Malina, J. F., Jr. and Sagik, B. P., Eds., Center for Research in Water Resources, University of Texas, Austin, 1974, 253.

27. **Bagdasaryan, G. A.,** Survival of viruses of the enterovirus group (poliomyelitis, echo, coxsackie) in soil and on vegetables, *J. Hyg. Epidemiol. Microbiol. Immunol.*, 8, 497, 1964.

28. **Hurst, C. J., Gerba, C. P., and Cech, I.,** Effects of environmental variables and soil characteristics on virus survival in soil, *Appl. Environ. Microbiol.*, 40, 1067, 1980.

29. **Green, K. M. and Cliver, D. O.,** Removal of virus from septic tank effluent by sand columns, in *Proc. Natl. Home Sewage Disposal Symposium*, American Society of Agricultural Engineers, St. Joseph, Mich., 1974.

30. **Lefler, E. and Kott, Y.,** Virus retention and survival in sand, in *Virus Survival in Water and Wastewater Systems*, Malina, J. F., Jr. and Sagik, B. P., Eds., Center for Research in Water Resources, University of Texas, Austin, 1974, 84.

31. **Sagik, B. P., Moore, B. E., and Sorber, C. A.**, Infectious disease potential of land application of wastewater, in State of Knowledge in Land Treatment of Wastewater, McKim, H. L., Ed., U.S. Army Corps of Engineers, Cold Regions Research and Engineering Laboratory, Hanover, N.H., 1978, 35.

32. **Hermann, J. E., Kostenbader, K. D., Jr., and Cliver, D. O.**, Persistence of enteroviruses in lakewater, *Appl. Microbiol.*, 28, 895, 1974.

33. **Bitton, G., Farrah, S. R., Pancorbo, O. C., and Davidson, J. M.**, Fate of viruses following land application of sewage sludge. I. Survival and transport patterns in core studies under natural conditions, in *Viruses and Wastewater Treatment*, Goddard, M. and Butler, M., Eds., Pergamon Press, New York, 1981, 133.

34. **Bitton, G., Pancorbo, O. C., and Farrah, S. R.**, Virus transport and survival after land application of sewage sludge, *Appl. Environ. Microbiol.*, 47, 905, 1984.

35. **Hurst, C. J., Gerba, C. P., Lance, J. C., and Rice, R. C.**, Survival of enteroviruses in rapid-infiltration basins during the land application of wastewater, *Appl. Environ. Microbiol.*, 40, 192, 1980.

36. **Jacobs, R. R.**, Evaluation of the "Tentative Standard Method" for Concentrating Adenoviruses from Tap Water and the Development of a Modified Procedure, Ph.D. thesis, University of North Carolina, Chapel Hill, 1980.

37. **Murphy, W. H., Jr., Eylar, O. R., Schmidt, E. L., and Syverton, J. T.**, Adsorption and translocation of mammalian viruses by plants. I. Survival of mouse encephalomyelitis and poliomyelitis viruses in soil and plant root environment, *Virology*, 6, 612, 1958.

38. **Cords, C. E., James, C. G., and McLaren, L. C.**, Alteration of capsid proteins of coxsackievirus A13 by low ionic conditions, *J. Virol.*, 15, 244, 1975.

39. **Salo, R. J. and Cliver, D. O.**, Effect of acid, pH, salts and temperature on the infectivity and physical integrity of enteroviruses, *Arch. Virol.*, 52, 269, 1976.

40. **Gerba, C. P. and Schaiberger, G. E.**, The effects of particulates on virus survival in seawater, *J. Water Pollut. Control Fed.*, 47, 93, 1975.

41. **Moore, R. S., Taylor, D. H., Chen, M., Sturman, L. S., and Reddy, M. M.**, Adsorption of reovirus by minerals and soils, *Appl. Environ. Microbiol.*, 44, 852, 1982.

42. **Sharp, D. G., Floyd, R., and Johnson, J. D.**, Nature of the surviving plaque-forming unit of reovirus in water containing bromine, *Appl. Microbiol.*, 29, 94, 1975.

43. **Young, D. C. and Sharp, D. G.**, Poliovirus aggregates and their survival in water, *Appl. Environ. Microbiol.*, 33, 168, 1977.

44. **Gerba, C. P. and Bitton, G.**, Microbial pollutants: their survival and transport pattern in groundwater, in *Groundwater Pollution Microbiology*, John Wiley & Sons, New York, 1984, 65.

45. **Yeager, J. E. and O'Brien, R. T.**, Enterovirus inactivation in soil, *Appl. Environ. Microbiol.*, 38, 694, 1979.

46. **Larkin, E. P., Tierney, J. T., and Sullivan, R.**, Persistence of virus on sewage-irrigated vegetables, *J. Environ. Eng. Div. Proc. Am. Soc. Civ. Eng.*, 1, 29, 1976.

47. **Tierney, J. T., Sullivan, R., and Larkin, E. P.**, Persistence of poliovirus 1 in soil and on vegetables grown in soil previously flooded with inoculated sewage sludge or effluent, *Appl. Environ. Microbiol.*, 33, 109, 1977.

48. **Dimmock, N. J.**, Differences between the thermal inactivation of picornaviruses at "high" and "low" temperatures, *Virology*, 31, 338, 1967.

49. **Breindl, M.**, The structure of heated poliovirus particles, *J. Gen. Virol.*, 11, 147, 1971.

50. **Yaeger, J. E. and O'Brien, R. T.**, Structural changes associated with poliovirus inactivation in soil, *Appl. Environ. Microbiol.*, 38, 702, 1979.

51. **Sobsey, M. D., Shields, P. A., Hauchman, F. S., Hazard, R. L., and Caton, L. W., III**, Survival and transport of hepatitis A virus in soils, groundwater and wastewater, *Water Sci. Tech.*, in press.

52. **McConnell, L. K., Sims, R. C., and Barnett, B. B.**, Reovirus removal and inactivation by slow rate sand filtration, *Appl. Environ Microbiol.*, 48, 818, 1984.

53. **Cliver, D. O. and Hermann, J. E.**, Proteolytic and microbial inactivation of enteroviruses, *Water Res.* 6, 797, 1972.

54. **Hermann, J. E. and Cliver, D. O.**, Degradation of coxsackievirus type A9 by proteolytic enzymes, *Infect. Immun.*, 7, 513, 1973.

55. **Malina, J. F., Jr., Ranganathan, K. R., Sagik, B. P., and Moore, B. E. D.**, Poliovirus inactivation in activated sludge, *J. Water Pollut. Control Fed.*, 47, 2178, 1975.

56. **Prage, L., Pettersson, V., Hoglund, S., Lonberg-Holm, K., and Philipson, L.**, Structural proteins of adenoviruses. IV. Sequential degradation of the adenovirus type 2 virion, *Virology*, 42, 341, 1970.

57. **Bitton, G. and Mitchell, R.**, Effects of colloids on the survival of bacteriophages in seawater, *Water Res.*, 8, 227, 1974.

58. **LaBelle, R. L. and Gerba, C. P.**, Investigations into the protective effect of estuarine sediment on virus survival, *Water Res.*, 16, 469, 1982.

59. **Rao, V. C., Seidel, K. M., Goyal, S. M., Metcalf, T. G., and Melnick, J. L.,** Isolation of enteroviruses from water, suspended solids and sediments from Galveston Bay: survival of poliovirus and rotavirus adsorbed to sediments, *Appl. Environ. Microbiol.*, 48, 404, 1984.
60. **Murray, J. P. and Laband, S. J.,** Degradation of poliovirus by adsorption on inorganic surfaces, *Appl. Environ Microbiol.*, 37, 480, 1979.
61. **Bixby, R. L. and O'Brien, D. J.,** Influence of fulvic acid on bacteriophage adsorption and complexation in soil, *Appl. Environ. Microbiol.*, 38, 840, 1979.
62. **McGaughy, P. H.,** *Engineering Management of Water Quality,* McGraw-Hill, New York, 1968.
63. **Bitton, G.,** Adsorption of viruses onto surfaces in soil and water, *Water Res.*, 9, 473, 1975.
64. **Laak, R. and McLean, D. M.,** Virus transfer through a sewage disposal unit, *Can. J. Public Health,* 58, 172, 1967.
65. **Hori, D. H., Burbank, N. C., Jr., Young, R. H. F., Lau, L. S., and Klemmer, H. W.,** Migration of poliovirus type 2 in percolating water through selected Oahu soils, Tech Rep. No. 36, Water Resources Research Center, University of Hawaii, Honolulu, 1970.
66. **Goyal, S. M. and Gerba, C. P.,** Comparative adsorption of human enteroviruses, simian rotavirus, and selected bacteriophages to soils, *Appl. Environ. Microbiol.*, 30, 241, 1979.
67. **Burge, W. D. and Enkiri, N. K.,** Virus adsorption by five soils, *J. Environ. Qual.*, 7, 73, 1978.
68. **Moore, R. S., Taylor, D. H., Sturman, L. S., Reddy, M. M., and Fuhs, G. W.,** Poliovirus adsorption by 34 minerals and soils, *Appl. Environ. Microbiol.*, 42, 963, 1981.
69. **Fuhs, G. W., Moore, R. S., Reddy, M. M., and Sturman, L. S.,** A laboratory study of virus uptake by minerals and soils, in *Proc. Water Reuse Symp.*, American Water Works Association, Washington, D.C., 1979.
70. **Taylor, D. H., Fuhs, G. W., Reddy, M. M., and Moore, R. S.,** *The Relationship Between Poliovirus Adsorption and Soil Properties,* Vol. 20, American Chemical Society, Washington, D.C., 1980, 158.
71. **Funderburg, S. W., Moore, B. E., Sagik, B. P., and Sorber, C. A.,** Viral transport through soil columns under conditions of saturated flow, *Water Res.*, 15, 703, 1981.
72. **Mandel, B.,** Characterization of type 1 poliovirus by electrophoretic analysis, *Virology*, 44, 554, 1971.
73. **Fujioka, R. S. and Ackermann, W. W.,** Evidence for conformational states of poliovirus: effects of cations on reactivity of poliovirus to guanidine, *Proc. Soc. Exp. Biol. Med.*, 148, 1071, 1975.
74. **Hazard, R. L. and Sobsey, M. D.,** Reduction of infectious hepatitis virus, poliovirus and echovirus in miniature soil columns, in *Am. Water Works Association, Annu. Conf. Proc.*, American Water Works Association, Denver, 1985, 435.
75. **Gerba, G. P., Goyal, S. M., Hurst, C. J., and LaBelle, R. L.,** Type and strain dependence of enterovirus adsorption to activated sludge, soils and estuarine sediments, *Water Res.*, 14, 1197, 1980.
76. **Gerba, C. P., Goyal, S. M., Cech, I., and Bogdan, G. F.,** Quantitative assessment of the adsorptive behavior of viruses to soils, *Environ. Sci. Technol.*, 15, 940, 1981.
77. **Lance, J. C., Gerba, C. P., and Wang, D.-S.,** Comparative movement of different enteroviruses in soil columns, *J. Environ. Qual.*, 11, 347, 1982.
78. **Drewry, W. A. and Eliassen, R.,** Virus movement in groundwater, *J. Water Pollut. Control Fed.*, 40, R257, 1968.
79. **Taylor, D. H., Moore, R. S., and Sturman, L. S.,** Influence of pH and electrolyte composition on the adsorption of poliovirus by soils and minerals, *Appl. Environ. Microbiol.*, 42, 976, 1981.
80. **Murray, J. P. and Parks, G. A.,** Poliovirus adsorption on oxide surfaces — correspondence with the DVLO-Liftshitz theory of colloid stability, *Amer. Chem. Soc. Adv. Chem. Ser.*, 189, 97, 1980.
81. **Dizer, H., Nasser, A., and Lopez, J. M.,** Penetration of different human pathogenic viruses into sand columns percolated with distilled water, groundwater or wastewater, *Appl. Environ. Microbiol.*, 47, 409, 1984.
82. **Farrah, S. R., Shah, D. O., and Ingram, L. O.,** Effect of chaotropic and antichaotropic agents on elution of poliovirus adsorbed to membrane filters, *Proc. Natl. Acad. Sci. U.S.A.*, 62, 1129, 1981.
83. **Shields, P. A. and Farrah, S. R.,** Influence of salts on electrostatic interactions between poliovirus and membrane filters, *Appl. Environ. Microbiol.*, 45, 526, 1983.
84. **Farrah, S. R., Scheuerman, P. R., and Bitton, G.,** Urea-lysine method for recovery of enteroviruses from sludge, *Appl. Environ. Microbiol.*, 41, 455, 1981.
85. **Lipson, S. M. and Stotsky, G.,** Specificity of virus adsorption to clay minerals, *Can. J. Microbiol.*, 31, 50, 1985.
86. **Wait, D. A. and Sobsey, M. D.,** Method for recovery of enteric viruses from estuarine sediments with chaotropic agents, *Appl. Environ. Microbiol.*, 46, 379, 1983.
87. **Carlson, G. F., Jr., Woodward, F. E., Wentworth, D. F., and Sproul, O. J.,** Virus inactivation on clay particles in natural waters, *J. Water Pollut. Control Fed.*, 47, 93, 1968.
88. **Schaub, S. A., Sorber, C. A., and Taylor, G. W.,** The association of enteric viruses with natural turbidity in the aquatic environment, in *Virus Survival in Water and Wastewater Systems*, Malina, J. F., Jr. and Sagik, B. P., Eds., Center for Research in Water Resources, University of Texas at Austin, 1974, 71.

89. **Lance, J. C. and Gerba, C. P.**, Effect of ionic composition of suspending solution on virus adsorption by a soil column, *Appl. Environ. Microbiol.*, 47, 484, 1984.

90. **Schiffenbauer, M. and Stotsky, G.**, Adsorption of coliphages T1 and T7 to clay minerals, *Appl. Environ. Microbiol.*, 43, 590, 1982.

91. **Lipson, S. M. and Stotsky, G.**, Adsorption of reovirus to clay minerals: effects of cation exchange capacity, cation saturation and surface area, *Appl. Environ. Microbiol.*, 46, 673, 1983.

92. **Gerba, C. P. and Lance, J. C.**, Poliovirus removal from primary and secondary sewage effluent by soil filtration, *Appl. Environ. Microbiol.*, 36, 247, 1978.

93. **Wellings, F. M., Lewis, A. L., Mountain, C. W., and Pierce, L. V.**, Demonstration of virus in groundwater after effluent discharge onto soil, *Appl. Microbiol.*, 29, 751, 1975.

94. **Schnitzer, M. and Khan, S. O.**, *Humic Substances in the Environment*, Marcel Dekker, New York, 1972.

95. **Bitton, G., Masterton, N., and Gifford, G. E.**, Effect of secondary treated effluent on the movement of viruses through a cypress dome soil, *J. Environ. Qual.*, 5, 370, 1976.

96. **Scheuerman, P. R., Bitton, G., Overman, A. R., and Gifford, G. E.**, Transport of viruses through organic soil and sediments, *J. Environ. Eng. Div. Proc. Am. Soc. Civ. Eng.*, 629, 1979.

97. **Robeck, G. C., Clarke, N. A., and Dostall, K. A.**, Effectiveness of water treatment processes in virus removal, *J. Am. Water Works Assoc.*, 54, 1275, 1962.

98. **Vaughn, J. M., Landry, E. F., Beckwith, C. A., and Thomas, M. Z.**, Virus removal during groundwater recharge: effects of infiltration rate on adsorption of poliovirus to soil, *Appl. Environ. Microbiol.*, 41, 139, 1981.

99. **Lance, J. C. and Gerba, C. P.**, Poliovirus movement during high rate land filtration of sewage water, *J. Environ. Qual.*, 9, 31, 1980.

100. **Lance, J. C. and Gerba, C. P.**, Virus movement in soil during saturated and unsaturated flow, *Appl. Environ. Microbiol.*, 47, 335, 1984.

101. **Marzouk, Y., Goyal, S. M., and Gerba, C. P.**, Prevalence of enteroviruses in groundwater of Israel, *Ground Water*, 17, 487, 1979.

102. **Fattal, B., Katzenelson, E., Guttman-Bass, N., and Sadiviski, A.**, Relative survival rates of enteric viruses and bacterial indicators in water, soil, and air, *Monogr. Virol.*, 15, 184, 1984.

103. **Schaub, S. A. and Sorber, C. A.**, Virus and bacterial removal from wastewater by rapid infiltration through soil, *Appl. Environ. Microbiol.*, 33, 609, 1977.

104. **Schaub, S. A., Kenyon, K. F., Bledsoe, B., and Thomas, R. E.**, Evaluation of the overland runoff mode of land wastewater treatment for virus removal, *Appl. Environ. Microbiol.*, 39, 127, 1980.

105. **Wang, D. -S., Gerba, C. P., Lance, J. C., and Goyal, S. M.**, Comparative removal of enteric bacteria and poliovirus by sandy soils, *J. Environ. Sci. Health*, A20, 617, 1985.

106. **Yates, M. V., Yates, S. R., Wagner, J., and Gerba, C. P.**, Modelling virus survival and transport in the subsurface, *J. Contam. Hydrol.*, in press.

Chapter 13

VIRAL ASPECTS OF APPLYING SLUDGES TO LAND

James J. Bertucci, Salvador J. Sedita, and Cecil Lue-Hing

TABLE OF CONTENTS

I. INTRODUCTION

The disposal of sludges from wastewater treatment is one of the major problems facing municipal and private treatment facilities in the U.S. Advancements in technology and the imposition of more stringent regulations for the removal of various physical, chemical, and biological pollutants from wastewater have interacted synergistically to make sludge disposal increasingly difficult. Estimates of municipal sludge production in the U.S. range up to 7.3 $\times$ 10^6 dry tons for 1985, or almost 20,000 dry tons per day.[1]

Several methods of municipal sludge disposal or utilization are currently practiced in the U.S.; these include incineration, landfilling, land application, and ocean disposal.[2,3] About 27% of the sludge is processed by incineration, which is not, strictly speaking, a direct disposal or use method but a treatment method that converts the sludge into ash which must be disposed of.[3]

Landfilling involves the deposition of sludge in or onto a dedicated area, alone or together with other solid wastes, which are then covered with soil. About 15% of the municipal sludge generated in the U.S. is managed or disposed of in this fashion.[3]

Ocean disposal from barges and outfall pipes accounts for only 4% of the sludge produced in the U.S. This percentage soon will be reduced even more because outfall pipe discharges are being eliminated under the Clean Water Act.[3]

Except for incineration, all of these sludge disposal methods may present a viral risk to humans. Incineration, of course, will destroy all infectious agents, including viruses. The viruses which might be present in landfill sludges could enter groundwaters and surface waters. Viruses released by ocean disposal might be transmitted to seawater and shellfish such as oysters, which may be consumed by humans. The aspects of viral transmission to groundwater resulting from the landfilling of sludge and ocean disposal of municipal sludge are considered in detail in other chapters of this book and will not be considered further here.

Land application is the spreading of sludge on or just below the surface of land as a soil conditioner and fertilizer. It accounts for about 24% of the sludge produced in the U.S. Strictly speaking, this is not a method of sludge disposal but rather sludge utilization.

Distribution and marketing of sludge products as a fertilizer and soil conditioner are common practices. Approximately 18% of the sludge produced yearly is sold or distributed for use by commercial and municipal organizations and by the public. This use of sludge is similar to the land application method discussed above. Together these two methods are a means of resource recovery in that the agricultural nutrient value of the sludge used in this manner is ultimately realized.

The value of municipal sludge as a source of soil conditioning material, plant nutrients, and water is considerable. Table 1 lists the plant nutrient content of municipal sludges from 24 cities in Illinois[4] and from 7 other states in the U.S.[5] Most sludges are a good source of nitrogen and phosphorus and are high in calcium and magnesium. The fertilizer value of the plant nutrient in the sludge produced nationally amounts to several hundred million dollars per year.[6] Municipal sludge also contains stable organic matter which when used as a fertilizer or soil conditioner increases the humus content of soil.

Despite the beneficial uses of municipal sludge, several modes of virus or other pathogen transmission come to mind when considering its application to land. Aerosols, which may contain viruses, could be generated if sludge is applied to land by spraying. Runoff and percolation from sludge-treated fields may transmit viruses to surface water or groundwater. People may come into direct contact with sludge applied to fields. Finally, viruses could be transmitted to crops grown on sludge-amended fields.

The U.S. Congress has made a strong commitment to land utilization of municipal sludge. The land application of municipal sludge has received encouragement in the U.S. by the

Table 1
MAJOR PLANT NUTRIENTS PRESENT IN ILLINOIS SLUDGES[4] AND SEVEN STATES OF THE U.S. (% DRY WEIGHT)[5]

	Illinois: 24 cities			Seven states		
	Minimum	Maximum	Mean	Minimum	Maximum	Mean
Total N	2.6	9.8	5.4	0.03	17.6	3.2
NH$_3$-H	0.1	6.1	1.8	0.0005	6.7	0.7
P	0.7	4.9	2.4	0.04	6.1	1.8
K	ND[a]	ND	ND	0.008	1.9	0.3
Ca	ND	ND	ND	0.1	25.0	5.1
Mg	ND	ND	ND	0.03	2.0	0.5

[a] No data.

Table 2
REGULATORY DEFINITION OF PROCESSES TO SIGNIFICANTLY REDUCE PATHOGENS[7]

Aerobic digestion: the process is conducted by agitating sludge with air or oxygen to maintain aerobic conditions at residence times ranging from 60 days at 15°C to 40 days at 20°C, with a volatile solids reduction of at least 38%.

Air drying: liquid sludge is allowed to drain and/or dry on underdrained sand beds, or on paved or unpaved basins in which the sludge depth is a maximum of 9 in.; a minimum of 3 months is needed, for 2 months of which temperatures average on a daily basis above 0°C.

Anaerobic digestion: the process is conducted in the absence of air at residence times ranging from 60 days at 20°C to 15 days at 35°C to 55°C, with a volatile solids reduction of at least 38%.

Composting: using the within-vessel, static aerated pile, or windrow composting methods, the solid waste is maintained at minimum operating conditions of 40°C for 5 days; for 4 hr during this period the temperature exceeds 55°C.

Lime stabilization: sufficient lime is added to produce a pH of 12 after 2 hr of contact.

Other methods: other methods of operating conditions may be acceptable if pathogens and vector attraction of the waste (volatile solids) are reduced to an extent equivalent to the reduction achieved by any of the above methods.

federal Water Pollution Control Act amendments of 1972 and the Clean Water Act of 1977. Pursuant to the Clean Water Act of 1977, the U.S. Environmental Protection Agency (EPA) has established regulations for the application of municipal sludge to land in regulation 40 CFR, Part 257, "Criteria for Classification of Solid Waste Disposal Facilities and Practices".[3,7] These regulations include microbial standards.

The microbial standards require that sewage sludge which is to be applied to land surfaces or which is to be incorporated into soil be treated by a process that significantly reduces pathogens (PSRP) prior to application or incorporation (Table 2).[3,7] Such a process is defined by the standards as one which will effect a one-log reduction in the number of pathogens present or a two-log reduction in indicator bacteria, such as fecal coliforms. Public access to a sludge application site must also be controlled for at least 12 months; grazing by animals whose products are consumed by humans must be prevented for at least 1 month; and crops for direct human consumption cannot be grown for at least 18 months following sludge application. Sludges must be treated by processes that further reduce pathogens (PFRP; Table 3) alone or in addition to PSRP if the above conditions are not met. Such processes (PFRP) are considered by the EPA to "...essentially destroy all bacteria and viruses and greatly reduce the number of parasites...."[7]

While these regulations apply to the land application of sludge, there are no federal regulations which cover the distribution and marketing of sludge products. The EPA advises

Table 3
REGULATORY DEFINITION OF PROCESSES TO FURTHER REDUCE PATHOGENS[7]

Composting: using the within-vessel composting method, the solid waste is maintained at operating conditions of 55°C or greater for 3 days; using the windrow composting method, the solid waste attains a temperature of 55°C or greater for at least 15 days during the composting period; also during the high-temperature period, there will be a minimum of five turnings of the windrow.

Heat drying: dewatered sludge cake is dried by direct or indirect contact with hot gases, and moisture content is reduced to 10% or lower; sludge particles reach temperatures well in excess of 80°C, or the wet bulb temperature of the gas stream in contact with the sludge at the point where it leaves the dryer is in excess of 80°C.

Heat treatment: liquid sludge is heated to a temperature of 180°C for 30 min.

Thermophilic aerobic digestion: liquid sludge is agitated with air or oxygen to maintain aerobic conditions at residence times of 10 days at 55° to 60°C, with a volatile solids reduction of at least 38%.

Other methods: other methods or operating conditions may be acceptable if pathogens and vector attraction of the waste (volatile solids) are reduced to an extent equivalent to the reduction achieved by any of the above methods.

Any of the processes listed below, if added to a PSRP, further reduce pathogens:

β-Ray irradiation: sludge is irradiated with β rays from an accelerator at dosages of at least 1.0 Mrad at room temperature (about 20°C).

γ-Ray irradiation: sludge is irradiated with γ rays from certain isotopes, such as ^{60}Co and ^{137}Ce, at dosages of at least 1.0 Mrad at room temperature (about 20°C).

Pasteurization: sludge is maintained for at least 30 min at a minimum temperature of 70°C.

Other methods: other methods or operating conditions may be acceptable if pathogens are reduced to an extent equivalent to the reduction achieved by any of the above add-on methods.

that the sludges in this category and composted sludges which meet the PFRP criteria are typically used because they have a high solids content and are therefore relatively easy for the user to handle. The pathogen content of these sludges would be very low.[3]

II. ASSESSMENT OF THE VIRUS LEVELS IN SLUDGE APPLIED TO LAND AND THE POTENTIAL HUMAN RISK

Despite the extensive processing of sludge that is required or recommended prior to its application to land (Tables 2 and 3), the question of the actual transmission of harmful viruses resulting from the use of sludge has generated much discussion in the scientific and lay communities. A numerical characterization of the probability of risk along with confidence limits would be the desired answer to this question. This type of risk assessment is often developed for toxic and carcinogenic substances utilized in an industrial environment.[8-10] The numerical assessment of risk from environmental viruses, however, is much more difficult than such an assessment for specific toxic or carcinogenic chemicals. Methodologies for the quantitative measurement of viruses in various environmental substrates are much less well developed than quantitative chemical measurements. There are more than 100 viral entities with distinct characteristics that must be considered in a viral assessment, while chemical assessments can be considered individually. Finally, viruses may become inactivated in various environments, while some chemicals tend to be much more conservative in nature. The difficulties inherent to viral, as compared to chemical, risk assessments necessarily reduce the viral risk assessment to a qualitative or at best semiquantitative exercise.

There are, however, certain generally quantitative aspects of assessing a potential hazard to humans from viruses in sludge applied to land. The study of risk may take either of two approaches. The first approach consists of the study of the fate of viruses from their environmental origin in feces through the sewage treatment and environmental processes that culminate in the application of sludges to land. The fate of viruses in sludge might then be

Table 4
VIRUS LEVELS REPORTED IN SEWAGE[a]

Year	Virus level in raw sewage or primary effluent (pfu/ℓ)[b]	Conc method	Cell system[c]	Ref.
1962	17,200—62,800	Direct inoculation	MK	14
1969	5—11,184	Phase separation	BSC$_1$ PMK	15
1972	1,075—11,575	Filter adsorption-elution	PMK	16
1974	0—400,000	Direct inoculation	Vero MK$_2$	17
1974	300—570	Membrane filter	RMK	18
1975	27—19,000	PE-60, protamine sulfate	PMK WI-38	19
1976	6,000—1,060,000	Direct inoculation	MK$_2$ Vero BGM	20
1977	250—2,738	Filter adsorption, direct inoculation	PMK	21
1977	7,364	Phase separation	PMK	22
1978	109—427	Filter adsorption-elution	BGM	23
1983	0—80	Direct inoculation, aluminum hydroxide	BGM Hep-2 WI-38 PMK	24

[a] Minimum, maximum, or average.
[b] pfu, plaque-forming unit.
[c] MK$_2$, BSC$_1$, Vero, and BGM, monkey kidney cell lines; PMK, primary monkey kidney; Hep-2, human cell line; WI-38, human cell strain.

followed from land deposition until humans somehow become exposed. Finally, an estimate of the probability of human exposure together with virus minimal infective dose information might be combined with the above in order to model the fate of viruses from their introduction into the environment to their production of infection and disease in humans. Information concerning the various stages of the chain of infection might be combined to bridge gaps in our current knowledge of the fate of environmental viruses throughout the chain of infection.

The second approach to assessing the human health risk associated with viruses in land-applied sludges involves the study of exposed human populations. This approach utilizes epidemiological techniques to study a possible link between human infection and disease and exposure to land-applied sludge.

Considering the first approach, the human viruses that find their way into sludge applied to land originate, for the most part, in the feces of infected persons.[11] The viruses in feces are the major source of viruses in sewage. Theoretical and empirical estimates of virus levels in sewage have varied widely, as shown by an examination of the literature over the past 25 years.[12-24] Table 4 lists some reported virus levels in raw sewage from several workers around the world. Reported virus concentrations have ranged from undetectable to more than 10^6 pfu/ℓ of sewage. Several factors are undoubtedly responsible for this wide variation. Though the more recently reported values have a greater reliability and accuracy because of technological and methodological refinements, one cannot postulate a typical virus level in sewage, as can be done for other sewage constituents.

The sewage treatment processes that result in the sludges produced in the U.S. typically

Table 5
ENTEROVIRUS REMOVAL DURING
MUNICIPAL WASTEWATER
TREATMENT[25]

Treatment	Expected virus removal
Primary settling	50%
Activated sludge	90%
Trickling filter	50%
Stabilization ponds and aerated lagoons	90%

include primary and secondary treatment consisting of sedimentation followed by activated sludge treatment. The final product of activated sludge treatment is termed the secondary effluent. During primary and secondary treatment, virus removal depends to a large extent on virus association with solids and the subsequent removal of the solids to produce the effluent. Approximately 0 to 50% of viruses are removed from sewage by primary treatment, and 90 to 99% are removed from primary effluent by secondary treatment.[25] Some virus inactivation may also occur during the activated sludge process, depending on the retention time.[26] Table 5 lists some expected enterovirus removals from sewage based on normal operating conditions. The viruses removed from the sewage during primary and secondary treatment are concentrated onto the solids, which are in turn removed from the processes as sludge. The primary sludge and the waste-activated sludge which constitutes the secondary sludge are usually treated by processes which meet the PSRP and/or PFRP criteria before being applied to land. In an extensive literature survey Pederson, in 1981, reviewed the density levels of organisms, including viruses, in municipal sludges and their reduction by various treatment processes.[27] The author concluded that there was a "dearth of reliable data available on the density levels of . . . enteroviruses . . . in raw sludge and septage". From the available data the author was able to demonstrate that anaerobic digestion reliably produced consistent reductions in the virus content of sludges. Pederson also concluded that mesophilic composting may be an effective means of inactivating viruses under conditions specified in regulation 40 CFR Part 257. Other chapters in this book examine in more detail the recent literature on the removal of viruses during sludge processing and the mechanisms of virus inactivation in wastewater sludges.

Kowal[28] has reviewed the possible routes of human exposure and infection by viruses in sludge applied to land. Other than a listing of the oral infective doses of enteric viruses to humans, no quantitative data which might be useful in a numerical risk assessment model were presented. The minimal infective doses ranged from 1 $TCID_{50}$ (50% tissue culture infective dose) per gram to $10^{7.5}$ $TCID_{50}$.

We are faced with the conclusion that insufficient data are available to develop a realistic numerical risk assessment model describing the fate of environmentally significant viruses that might present a human hazard in land application situations. We will for the present, therefore, examine the human health implications of viruses in sludge applied to land using the data from field studies in which viruses were sought in the actual sludge products applied to land, and from epidemiological information relating health and exposure to land application of sludge operations.

Several recent field and epidemiological studies have been conducted to address the concern of a possible human health threat related to the application of municipal sludge to land. The various large-scale projects selected for discussion here have all utilized some combination of environmental virus monitoring, health questionnaires, and serological surveys to detect viruses in the environment or human virus infections related to the land application of municipal sludges.

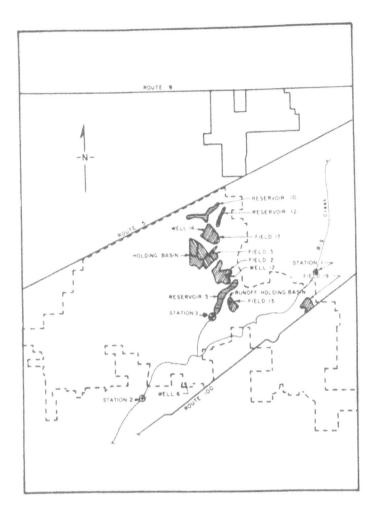

FIGURE 1.   Fulton County, Ill. land reclamation site.

III. FIELD STUDIES: VIRAL AND HUMAN HEALTH MONITORING

A. Case Study 1

In the fall of 1970, the Metropolitan Sanitary District of Greater Chicago initiated a program whereby processed sludge was used for the reclamation of strip-mined fields and the fertilization of row crops in Fulton County, Ill.[29] Waste-activated sludge and some primary sludge were thickened and anaerobically digested with an average retention time of 15 days at 35°C. The liquid digested sludge containing approximately 4% solids was barged 200 mi down the Illinois River from Chicago and pumped through a 10.8-mi pipeline to holding basins at the utilization site. The site consisted of about 15,000 acres, of which about 5000 acres received sludge as fertilizer and soil conditioner. During the application season, principally April through October, liquid sludge from the holding basins was applied to land at first by spray irrigation and later by soil incorporation. Rates of up to 30 dry ton/acre/year were applied during the growing season to fields which were planted principally in corn.

Virus concentrations in the surface water which drained the sludge application sites were monitored from April 1972 to August 1979.[24,29,30] Samples for virus analysis were taken monthly until September 1977; after this time samples were taken quarterly. The three sampling stations are shown in Figure 1. Stations 1 and 2 were at points where Big Creek receives, directly and indirectly, the flow from several runoff retention basins that drain

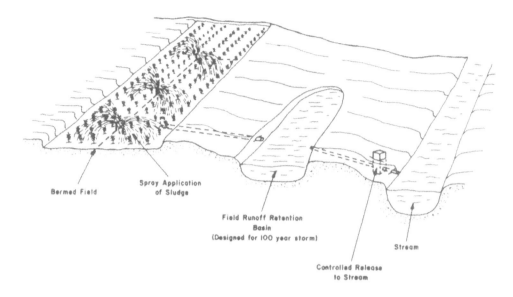

FIGURE 2.   Typical field design with runoff water capture system, Fulton County, Ill.

sludge-treated fields. Station 3, situated at the outlet of reservoir 3, also receives direct and indirect discharges from field runoff retention basins (Figure 2). Samples of up to 20 ℓ were concentrated by the PE 60 method[29] or the aluminum hydroxide continuous flow centrifuge method.[29] The concentrates were plaque assayed in primary monkey kidney cultures or BGM cell cultures. During the period from April 1972 to September 1975, all plaques were confirmed by a second passage in the same type of cell culture used for primary isolation to ensure that they were virus induced.[24,29,30] After this change in procedure, only confirmed plaques were scored as viruses. These results are plotted in Figure 3. The early results before plaque confirmation suggested that relatively large concentrations of viruses were present at each station sampled. After the initiation of the confirmation procedures, viruses were detected only once in the surface water of stations 1, 2, and 3.

In 1976 and 1977, under an EPA contract, a more comprehensive study of the virus and bacterial levels resulting from sludge application was conducted.[29] In addition to the virus monitoring described above, other virus monitoring included:

1.   Virus survival in a sludge lagoon: virus levels, in fresh digested sludge, were monitored over a 6-month period to determine the rate of virus die-off.
2.   Virus levels in groundwater: three wells were monitored; two test wells (W14 and W12) were adjacent to a sludge application site, and the third (W6) was a control well not influenced by sludge application (Figure 1 and Table 6).
3.   Aerosol studies: aerosols were captured during application of digested, lagooned sludge to several fields (2, 3, 17, and 19). These were examined for animal virus, coliphage, and bacterial levels (Figure 1).

No estimate of animal virus die-off rate could be made because no confirmed animal viruses were detected in any of the samples from the sludge lagoon at the site over a 2-month period. After 2 months no further animal virus tests were performed on the lagooned sludge.

During the 1 year period from October 1975 through September 1976, 16 test well samples and two control well samples were analyzed for animal viruses. No confirmed viruses were detected in any of these samples (Table 6).

The animal virus aerosol data were obtained by analyzing samples collected by employing

187

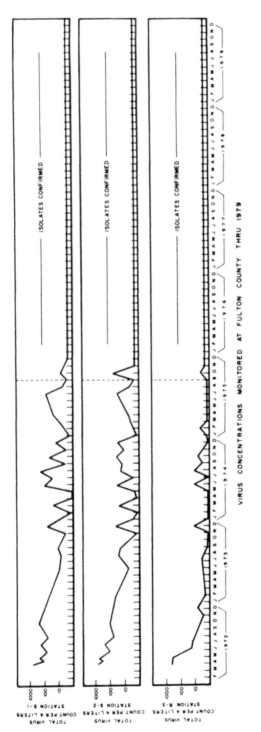

FIGURE 3.   Virus concentrations monitored at Fulton County through 1979.

Table 6
VIRUS DATA FROM FULTON COUNTY WELL WATER
ANALYSIS, 1975 AND 1976[29]

Sample source	Date	Sample volume assayed (ℓ)[a]	No. of plaques observed	Confirmed (%)	Confirmed viruses/ℓ
W-14	10-30-75	36.7	696	0 (0/9)[b]	0
W-14	11-20-75	17.5	2	0 (0/2)	0
W-14	12-18-75	18.4	0	—	0
W-6	12-18-76	28.0	0	—	0
W-14	1-06-76	16.3	0	—	0
W-12	1-22-76	17.9	0	—	0
W-12	3-09-76	14.2	0	—	0
W-12	3-16-76	16.9	4	0 (0/2)	0
W-14	3-18-76	18.2	2	0 (0/2)	0
W-12	3-30-76	35.1	0	—	0
W-12	4-13-76	36.3	0	—	0
W-6	4-27-76	37.1	6	0 (0/6)	0
W-12	5-20-76	37.6	0	—	0
W-12	7-07-76	35.0	0	—	0
W-12	7-22-76	37.2	0	—	0
W-12	9-02-76	34.6	0	—	0
W-12	9-08-76	31.1	0	—	0
W-14	9-22-76	34.0	0	—	0

[a] Samples were processed to concentrate viruses, and a volume of concentrate equivalent to the sample volume assayed was tested for viruses.
[b] Plaques confirmed/plaques picked.

Litton high-volume samplers. These devices were placed at various distances up- and down-wind of the sludge sprayer during application. This study protocol was expected to provide data on conditions prior to spraying (upwind) and on the rate at which the animal virus levels decreased as a function of distance (downwind) from the source of spray.

Forty-six samples were analyzed for animal viruses, 5 from the sludge-spray source, 5 upwind control samples, and 36 downwind samples. No confirmed animal viruses were detected in any of the sludge-spray source samples. Of the remaining 41 ambient air samples analyzed, four had detectable virus levels. One of those samples was an upwind control. In each case the animal virus was identified as poliovirus type 1.

A rigorous statistical analysis of these data was conducted which suggested that upwind or prior conditions affected the downwind conditions. This was conservatively interpreted to mean that contamination of the air samples which were positive for animal virus could not be ruled out. In any event, analysis of the data available did not show the presence of significant concentrations of indigenous viruses in aerosols from the spray applications of digested, lagooned sludge.[29,31]

B. Case Study 2

In another field study conducted in Florida, aerobically digested sludge (15-day detention time) from the Montclair sewage treatment plant and anaerobically digested sludge (30-day detention time) from the Main Street sewage treatment plant in Pensacola, Fla. were transported by tank truck to a sludge-holding lagoon located at the Agricultural Research Center of the University of Florida in Jay. Sludge was removed from the lagoon and spread on a field which was subdivided into 72 plots of 4800 ft² each. Each plot received from 0 to 15 acre-in. of sludge per year. Addition of digested sludge to the lagoons as well as to the

Table 7
ENTEROVIRUSES IN LAGOONED SLUDGE FROM 2/78 TO 1/79, UNIVERSITY OF FLORIDA, AGRICULTURAL RESEARCH CENTER AT JAY[32]

Sampling date	Virus conc.[a] ($TCID_{50}$/g)	Viruses detected (no.)
2-17-78[b]	15[c]	Polio-1 (4)
3-31-78[b]	80	Polio-1 (12), polio-2 (7), echo-7 (1)
5-09-78[b]	64	Not done
6-02-78	2	Polio-2 (2)
7-10-78	0.1	Not done
8-08-78	<0.1	No isolates
9-13-78	0.3	Polio-1 (1)
10-03-78	<0.1	No isolates
11-06-78	<0.1	No isolates
12-06-78[b]	42	Polio-1 (1)
12-14-78[b]	33	Polio-1 (1), polio-2 (2), echo-7 (3), echo-15 (1), coxsackie B4 (3)
1-14-79	36	Polio-1 (2), polio-2 (1), echo-7 (1), coxsackie B4 (1)
1-24-79	100	Polio-1 (1), polio-2 (1), coxsackie B4 (3)

[a] Additions to the sludge lagoon were aerobically digested sludge (30%) with a range of virus concentrations of from 14-260 $TCID_{50}$/g and anaerobically digested sludge (70%) with a range of virus concentrations of from 2 to 7 $TCID_{50}$/g.
[b] Dates when sludge was added to lagoon.
[c] Viruses could not be detected in overlying lagoon water when sludge addition was suspended.

plots was suspended at certain stages of plant growth. The survival of viruses associated with sludge during lagooning and land disposal was determined throughout the study.[32,33]

The results of these studies (Tables 7 and 8) showed that the aerobically digested sludge, which made up nearly 30% of the total sludge mixture, had a concentration of viruses ranging from 14 to 260 $TCID_{50}$/g. The anaerobically digested sludge component had virus concentrations ranging from 2 to 7 $TCID_{50}$/g. Viruses were readily detected when digested sludge was added to the on-site holding lagoon, but decreased to low levels or were not detectable following sludge deposition on land. In the absence of sludge additions, virus levels decreased in the sludge lagoon from an average of 47.5 $TCID_{50}$/g in February and March 1978 to 0.1 $TCID_{50}$/g by November 1978. This indicates that lagooning provides an acceptable means of further virus reduction, at least under the conditions which prevailed in north-central Florida for the duration of these studies. No viruses were detected in test wells located on the site near the sludge land application area and the sludge-holding lagoon. The authors did not indicate whether they confirmed the viral etiology of the cytopathic effects observed in their assays.

C. Case Study 3
A project consisting of two related studies was carried out by the Kananaskis Center for Environmental Research, University of Calgary, Alberta, Canada.[34,35] These investigators took a unique approach to the assessment of health risk and sludge application to land. Realizing that proving the existence of an actual pathogen hazard was virtually impossible, this group decided to determine how long and under what conditions known pathogens

Table 8
VIRUS SURVIVAL IN SLUDGE
AFTER APPLICATION TO LAND,
10/79, UNIVERSITY OF FLORIDA
AGRICULTURAL RESEARCH
CENTER AT JAY[32,33]

Days after spreading on land	Water (%)	Average virus conc (TCID$_{50}$/g)[a]
0	91	3.0
1	39	0.008
2	40	0.41
5	Not done	<0.01
7	15	0.05
9	19	0.01

[a]   Averages derived from data from two different virus assay procedures used by Farrah and Bitton.[32] The two procedures gave essentially the same number for viruses, within experimental limits.

persisted on or in the soil to which sludge had been applied. A total of 56 samples were analyzed for viruses by the Cadham Provincial Laboratories, Winnipeg, Manitoba, Canada. The sludged fields were considered safe when they could not be distinguished from control fields with respect to selected indicator organisms and pathogens, including enteroviruses. The types of samples included 8 control field samples, 7 pasture samples, 33 samples from fields to which sludge had been added previously for various times ranging from 7 days to 18 months, 5 samples of lagoon sludge of indeterminate age, and 3 samples of the mesophilic digester sludge used as the sludge source. All samples were negative for viruses by the technique used. Acknowledging that their methodology for detecting viruses in sludge and soil was less than perfect, these investigators nevertheless concluded that no potential hazard from viruses existed on sludge-applied fields.

D. Case Study 4

The Cadham Provincial Laboratory carried out two studies for the disposal of sewage sludge and effluent on land during 1981 and 1982, a sludge injection and an effluent irrigation study. Only the sludge injection study will be discussed here.[36]

The sludge injection study consisted of the subsurface injection of raw and anaerobically digested sludge into soil at an application rate of 44.7 ℓ per linear meter using an injector developed by the Agricultural Engineering Department of McGill University. Comparisons were made between raw and anaerobically digested sludges obtained from two of the sewage treatment plants (South End and North End) of a large city (population 604,269) in Manitoba. The following types and numbers of samples were analyzed for enteric viruses, bacteria, and intestinal parasites.

1. Raw and digested sludges: 29 raw and 6 digested samples
2. Postinjection soil samples (at 7.5-, 15-, 30-, and 45-cm depths): 1 week, 56 samples; 1 month, 56 samples; 3 months, 56 samples; and 12 months, 42 samples
3. Background soil samples: 9 samples
4. Wheat grown on injected and background plots: 8 samples
5. Ditch water: 10 samples (bacterial analysis only)
6. Well water: 15 samples

Table 9
**SAMPLE SOURCE AND TYPE, AND VIRUS CONCENTRATIONS OF
SAMPLES FROM THE SLUDGE INJECTION STUDY DESCRIBED IN
CASE STUDY 4, WINNIPEG, MANITOBA, CANADA, IN 1981 AND 1982[36]**

Sample source type	No. of samples tested	No. of samples positive (%)	Virus conc[a] (MPN/100 g or MPN/10 ℓ)
South End plant raw sludge[b]	22	5 (22.7)	2.6
North End plant raw sludge[b]	7	1 (14.2)	1.5
Combined South End and North End plants digested sludge[b]	6	0 (0)	2.0
Control soils[b]	19	0 (0)	0.5
Soils injected with raw sludge[b,c]	4	1 (25)	3.0
Soils injected with digested sludge[d] after			
1 week	56	0 (0)	<0.2[e]
1 month	56	0 (0)	<0.2
3 months	56	0 (0)	<0.2
12 months	42	0 (0)	<0.2
Wheat grown on injected soil and from background plots[b]	8	9 (0)	<1.25
Well water[f]	15	0 (0)	<0.02

a   Virus concentrations calculated by authors using data from Sekla and Stackiw.[36]
b   Ten-gram sample size.
c   Soil injected with raw sludge from South End plant and sampled 1 week postinjection at a depth of 7.5 cm.
d   100-g sample size.
e   Virus MPN values with < are the detection limit for the number of samples analyzed.
f   Well water samples 450 ℓ each.

The results of the sludge injection study (Table 9) indicated that the confirmed enteric virus content of raw sludge ranged from 2.6 MPN/100 g at the South End plant to 1.5 MPN/100 g at the North End plant. None of the digested sludge samples, however, was found to contain enteric viruses by the method used. None of the background soil samples was found to contain enteric viruses. One soil sample, collected 1 week after the injection of raw sludge from one of the two treatment plants (the South End plant), was positive for enteric virus, indicating that the virus had survived one week. None of the soil samples injected with digested sludge was shown to contain enteric viruses. No enteric viruses were detected in any of the wheat samples or the well water samples (Table 9).

E. Case Study 5

More recently, Sorber and his colleagues[37] have reported the results of their studies of sludge application sites in the U.S. Anaerobically digested sludges from four sewage treatment plants were held in lagoons and then applied to land using either tank trucks or by spray application. The confirmed enteric virus concentrations of the sludges used in these studies ranged from undetectable (<0.03 pfu/g TSS) to 2.1 pfu/g TSS (Table 10). During a special aerosol run at one of the sites, nine high-volume samples were operated simultaneously for six consecutive 30-min periods during sludge spraying. A total of 1470 m³ of air was sampled for enteric viruses. No human enteric viruses were detected in the aerosol samples. This represents a theoretical aerosol concentration of less than 0.0016 pfu/m³ of air sampled (Table 10). Aerosolization of microorganisms other than enteric viruses at the

Table 10
GEOGRAPHICAL LOCATION, TYPE OF SITE, AND
ENTEROVIRUS CONCENTRATIONS IN SLUDGE
AND AEROSOLS AT FOUR STUDY SITES: CASE
STUDY 5[37]

Site designation and geographical location	Site type	Av. enteroviruses[a] (pfu/g TSS)		Enteroviruses in aerosol (pfu/m³)
		Hela	**Hep-2**	
Rocky Mountain	Tank truck	0.87	ND[b]	ND
Southwest	Tank truck	1.91	0.79	ND
Northwest	Spray gun (intermittent)	0.96	0.22	ND
Southeast	Spray gun	0.96	ND	ND
Special enterovirus aerosol study	Spray gun, site D	0.70	ND	<0.0016[c]

[a] Data from Sorber et al.,[37] values at detection limit (i.e., < values not included in averages.

[b] Not done.

[c] A total of 1470 m³ of air was sampled.

spray application sites, notably fecal coliforms, fecal streptococci, and mycobacteria, how-ever, was detected. These findings are consistent with previous studies of this nature.[31]

The conclusions from these studies were that microbiological aerosols generated in the application of sludge to land did not appear to represent a serious threat to human health for individuals located more than 100 m downwind of the site.

F. Case Study 6

The final case study describes a comprehensive multiyear demonstration project involving the controlled disposal of sewage sludge on farmlands in the Columbus, Ohio area.[38-40] Digested municipal sludge was applied to the land on family-operated farms at rates of from 2 to 10 dry ton/ha/year. Forty-seven farms received sludge, and 46 other similar farms received no sludge and acted as controls. The project included a prospective epidemiological survey comparing, by disease symptoms and serum neutralization procedures, the occurrence of viral infections in individuals living on control farms and on farms where sludge had been applied. Samples of sludge were also analyzed for viruses, and isolates were confirmed and identified. The sludge sources were four sewage treatment plants located in Columbus (one plant), Medina County (two plants), and Springfield (one plant), Ohio. These plants treat sewage from communities with populations of 372,000, 3690, 40,900 and 70,000 people, respectively.

Over a period of 3 years, 293 enterovirus isolates were recovered from 307 sludge samples (Table 11).[39] However, no significant differences between individuals on farms receiving sludge and those on control farms were found with respect to symptoms of respiratory, digestive, or other disease. No significant differences in the frequency of viral infections, as measured by serological examinations, were found between the two groups.[34,40]

IV. SUMMARY AND CONCLUSIONS

The construction of a quantitative concept or model to assess the risk from human viruses in municipal sludge applied to land is very difficult. The results from field studies, several

Table 11
SUMMARY OF VIRAL ISOLATIONS FROM ALL SLUDGE SAMPLES ACCORDING TO SEROTYPE AND LOCATION: CASE STUDY 6[39]

Virus		Medina 300 plant (n = 63)	Medina 500 plant (n = 63)	Columbus plant (n = 123)	Springfield plant (n = 58)
			No. of isolations		
Echovirus type	3	0	0	4	2
	5	3	1	3	0
	6	1	1	3	2
	7	28	22	13	2
	11	7	3	1	0
	13	0	1	0	3
	15	0	0	1	0
	17	0	0	1	0
	19	1	0	1	0
	20	0	0	5	0
	21	0	0	4	1
	22	0	0	1	0
	24	1	1	13	3
	25	0	1	1	1
	27	1	1	2	1
	30	1	1	4	1
Coxsackievirus	A9	0	0	3	0
	A16	0	0	1	1
	B2	4	0	4	1
	B3	15	10	15	7
	B4	3	3	2	1
	B5	0	0	4	3
Reovirus		3	2	ND[a]	ND
Poliovirus type	1	2	2	3	1
	2	17	6	15	7
	3	ND	4	5	1
Total		87	59	109	38

[a] No data.

of which we have considered here, are negative with respect to a virus-related hazard to humans. A recent comprehensive literature review and analysis of the overall health effects associated with the land application of sludge points out some possible routes of human virus transmission in such operations, but give no evidence of actual transmission.[28]

Although other workers have argued that the current analytical and epidemiological methods are not adequate to establish a hazard from viruses in the environment,[41-44] a hazard to humans from viruses in sludge has not been demonstrated, despite several studies designed to reveal such a hazard.

The preponderance of negative findings from field and epidemiological studies prevents, at this time, the construction of a realistic model for estimating the possible risk to humans from viruses applied to land. The best information currently available does not, in fact, indicate the existence of an assessable risk.

Despite the current lack of evidence of a virus problem associated with land application of sludge, recent and upcoming advancements in virology will give environmental virologists additional matters to study.[45] The importance of rotaviruses as the etiological agents of diarrhea and gastroenteritis in children and adults has recently been demonstrated. The Norwalk virus and related agents have also been found to be a cause of water-borne gastroenteritis. Following the development of techniques for the laboratory cultivations of these viruses, their fate during sewage and sludge treatment should be elucidated. Environmental studies of these viruses along with further study of the better-known environmental viruses will shed further light on the viral aspects of sludge application to land. Toward this possibility, continued research in the field of environmental virology should be supported.

REFERENCES

1. **Zenz, D. R. and Lue-Hing, C.,** Municipal sludge management, in *Environment and Solid Wastes: Characterization, Treatment and Disposal*, Francis, C. W., Auerbach, S. I,. and Jacobs, V. A., Eds., Butterworths, Woburn, Mass., 1983, chap. 3.
2. Booz-Allen and Hamilton, Inc., Description and Comparison of Municipal Sewage Sludge Generation and Disposal Data Basis, Office of Solid Waste, U.S. Environmental Protection Agency, Washington, D.C., 1982.
3. EPA, Environmental Regulations and Technology, Use and Disposal of Municipal Wastewater Sludge, EPA 625/10-84-003, Environmental Protection Agency, Cincinnati, 1984.
4. **Zenz, D. R., Lynam, B. T., Lue-Hing, C., Rimkus, R. R., and Hinesly, T. D.,** U.S. EPA guidelines on sludge utilization and treatment agencies, paper presented at the 48th Annu. Water Pollution Control Federation Conf., Miami Beach, October 1975, Department of Research and Development Rep. No. 75-20, Metropolitan Sanitary District of Greater Chicago, 1975.
5. **Sommers, L. E., Nelson, O. W., and Yost, K. J.,** Variable nature of the chemical composition of sewage sludges, *J. Environ. Qual.*, 5, 303, 1976.
6. **Pahren, H. R.,** Overview of the problem, in *Sludge-Health Risks of Land Application*, Bitton, G., Damron, B. L., Edds, G. T., and Davidson, J. M., Eds., Ann Arbor Science, Ann Arbor, Mich., 1980, 1.
7. Environmental Protection Agency, 40 CFR Part 257, criteria for classification of solid waste disposal facilities and practices; final, interim final and proposed regulations, *Fed. Reg.*, 44, 53438, 1979.
8. **Stokinger, H. E.,** Maintaining professional objectivity in managing risk, *Dangerous Prop. Ind. Mater. Rep.*, 3, 2, 1983.
9. National Research Council, *Risk Assessment in the Federal Government: Managing the Process*, National Academy Press, Washington, D.C., 1983.
10. **Lave, L. B.,** The role of quantitative risk assessment in environmental regulations, in *Risk Quantitation and Regulatory Policy*, Hoel, D. G., Merrill, R. A., and Perera, R. P., Eds., Cold Spring Harbor Laboratory, Cold Spring Harbor, N.Y., 1985, 3.
11. **Rodgers, F. G.,** Concentration of viruses in fecal samples from patients with gastroenteritis, in *Viruses and Wastewater Treatment*, Goddard, M. and Butler, M., Eds., Pergamon Press, Oxford, 1981, 15.
12. **Slade, J. S.,** Viruses in sewage, in *Viruses and Wastewater Treatment*, Goddard, M. and Butler, M., Eds., Pergamon Press, Oxford, 1981, 19.
13. **Rao, V. C.,** Introduction to environmental virology, in *Methods in Environmental Virology*, Gerba, C. P. and Goyal, S. M., Eds., Marcel Dekker, New York, 1982, 1.
14. **Mack, W. N., Frey, J. R., Riegle, B. J., and Mallmann, W. L.,** Enterovirus removal by activated sludge treatment, *J. Water Pollut. Control Fed.*, 34, 1133, 1962.
15. **Shuval, H. I., Fattal, B., Cymbalista, S., and Goldblum, N.,** The phase separation method for the concentration and detection of viruses in water, *Water Res.*, 3, 225, 1969.
16. **Rao, V. C., Chandorkar, U., Rao, N. U., Kunaran, P., and Lakhe, S. B.,** A simple method for concentrating and detecting viruses in wastewater, *Water Res.*, 6, 1565, 1972.
17. **Buras, N.,** Recovery of viruses from wastewater and effluent by the direct inoculation method, *Water Res.*, 8, 19, 1974.
18. **Kott, Y., Rose, N., Sperber, S., and Betzer, W.,** Bacteriophages as viral pollution indicators, *Water Res.*, 8, 165, 1974.
19. **Lau, L. S.,** Water Recycling of Sewage Effluent by Irrigation: A Field Study on Oahu, Technol. Rep. 94, Water Resources Research Center, University of Hawaii, Honolulu, 1974.

20. **Buras, N.** Concentration of enteric viruses in wastewater and effluent: a two-year survey, *Water Res.*, 10, 295, 1976.
21. **Rao, V. C., Lakhe, S. B., Waghmare, S. U., and Dube, P.,** Virus removal in activated sludge treatment, *Prog. Water Technol.*, 9, 113, 1977.
22. **Fattal, B. and Nishimi, M.,** Enterovirus types in Israel sewage, *Water Res.*, 11, 393, 1977.
23. **Berg, G., Dahling, D. R., Brown, G. A., and Berman, D.,** Validity of fecal coliforms, and fecal streptococci and indicators of viruses in chlorinated primary sewage effluents, *Appl. Environ. Microbiol.*, 36, 880, 1978.
24. **Bertucci, J. J., Abid, S. H., Lue-Hing, C., Clark, C. S., Fenters, J. D., and Fannin, K. F.,** Confirmed viruses versus unconfirmed plaques in sewage, *J. Environ. Eng.*, 109, 351, 1983.
25. **Gerba, C. P.,** Virus survival in wastewater treatment, in *Viruses and Wastewater Treatment*, Goddard, M. and Butler, M., Eds., Pergamon Press, Oxford, 1981, 39.
26. **Glass, J. S. and O'Brien, R. T.,** Enterovirus and coliphage inactivation during activated sludge treatment, *Water Res.*, 14, 877, 1980.
27. **Pederson, D. C.,** Density Levels of Pathogenic Organisms in Municipal Wastewater Sludge, a Literature Review, EPA 600/2-81-015, U.S. Environmental Protection Agency, Washington, D.C., 1985.
28. **Kowal, N. E.,** Health Effects of Land Application of Municipal Sludge, EPA 600/1-85-015, U.S. Environmental Protection Agency, Washington, D.C., 1985.
29. EPA, Viral and Bacterial Levels Resulting from the Land Application of Digested Sludge, EPA 600/1-79-015, U.S. Environmental Protection Agency, Cincinnati, 1979.
30. **Bertucci, J. J.,** Virology Report, 1972—1979, for Fulton County, Illinois, Department of Research and Development Rep. No. 81-24, Metropolitan Sanitary District of Greater Chicago, 1981.
31. **Lue-Hing, C., Zenz, D. R., Sedita, S. J., O'Brien, P., Bertucci, J. J., and Abid, S. H.,** Microbial content of sludge soil and water at a municipal sludge application site, paper presented at the Meeting of the Working Group on Sewage Sludge to Land: Health Implication of Microbial Content, Stevenage, England, Department of Research and Development Rep. No. 80-27, Metropolitan Sanitary District of Greater Chicago, 1980.
32. **Farrah, S. R. and Bitton, G.,** Fate of viruses following land application of sewage sludge. II. Field experiments (monitoring of lagooned sludge, soil and groundwater), in *Viruses and Wastewater Treatment*, Goddard, M. and Butler, M., Eds., Pergamon Press, Oxford, 1981, 137.
33. **Edds, G. T. and Davidson, J. M.,** Sewage Sludge Viral and Pathogenic Agents in Soil-Plant-Animal Systems, EPA 600/1-81-026, U.S. Environmental Protection Agency, Cincinnati, 1981.
34. **Lehman, D. L., Wallis, P. M., MacMillan, D. A., and Buchanan-Mappin, J. M.,** When may sludged fields be considered free of potential health hazards?, in *Biological Health Risks of Sludge Disposal to Land in Cold Climates*, Wallis, P. M. and Lehmann, D. L., Eds., University of Calgary Press, Alberta, Canada, 1983, 299.
35. **Wallis, P. M., Lehmann, D. L., MacMillan, D. A., and Buchanan-Mappin, J. M.,** Sludge application to land compared with a pasture and hayfield: reduction of biological health hazard over time, *J. Environ. Qual.*, 13, 645, 1984.
36. **Sekla, L. and Stackiw, W.,** Land disposal of municipal sludge and effluent: two microbiological studies, in *Biological Health Risks of Sludge Disposal to Land in Cold Climates*, Wallis, P. M. and Lehmann, D. L., Eds., University of Calgary Press, Alberta, Canada, 1983, 337.
37. **Sorber, C. A., Moore, B. E., Johnson, D. E., Harding, H. J., and Thomas, R. E.,** Microbiological aerosols from the application of liquid sludge to land, *J. Water Pollut. Control Fed.*, 56, 830, 1984.
38. **Hamparian, V. V., Ottolenghi, A. C., and Hughes, J. H.,** Viral infections in farmers exposed to sewage sludge, paper presented at the Annual Meeting of the American Society for Microbiology, Atlanta, 1982.
39. **Hamparian, V. V., Ottolenghi, A. C., and Hughes, J. H.,** Enteroviruses in sludge: multiyear experience with four wastewater treatment plants, *Appl. Environ. Microbiol.*, 50, 280, 1985.
40. **Brown, R. E.,** A Demonstration of Acceptable Systems of Land Disposal of Sewage Sludge, EPA 600/1-85-015, U.S. Environmental Protection Agency, Washington, D.C., 1985.
41. **Gerba, C. P. and Goyal, S. M., Eds.,** *Methods in Environmental Virology*, Marcel Dekker, New York, 1982.
42. **Melnick, J. L.,** Are conventional methods of epidemiology appropriate for risk assessment of virus contamination of water?, in *Proc. Conf. Risk Assessment and Health Effects of Land Application of Municipal Wastewater and Sludges*, Sagik, B. P. and Sorber, C. A., Eds., University of Texas, San Antonio, 1978, 61.
43. WHO, *Human Viruses in Water, Wastewater and Soil*, WHO Tech. Rep. Ser. 639, World Health Organization, Geneva, 1979.
44. IAWPRC Study Group on Water Virology, The health significance of viruses in water, *Water Res.*, 17, 121, 1983.
45. **Gerba, C. P., Rose, J. B., and Singh, S. N.,** Waterborne gastroenteritis and viral hepatitis, in *Crit. Rev. Environ. Control*, 15(3), 213, 1985.

Chapter 14

INTERACTIONS BETWEEN VIRUSES AND CLAY MINERALS*

Steven M. Lipson and G. Stotzky

TABLE OF CONTENTS

* The literature review for this chapter was completed in April 1986.

I. INTRODUCTION

The existence of a disease in a population is contingent on the persistent infection of susceptible individuals in the same or different population that might transmit, either directly or indirectly, the infectious agent (e.g., a virus) to noninfectious individuals.[1] This concept of virus persistence in perpetuity is characterized, in part, by the persistence of a low level of infection in a population, which periodically or during specific seasons may increase to epidemic proportion. In temperate climates, for example, enterovirus infections prevail during the summer and early autumn, whereas rotavirus infections are prevalent in the winter months.

The mode of virus transmission to humans is varied and is generally related to the portal of entry, sites of multiplication, and stability of the viral particle. Viruses may be broadly characterized by their ability to be transmitted directly through human-to-human contact and fomites (e.g., Herpesviridae),[2] or indirectly through vectors (e.g., Togaviridae), aerosols (e.g., Myxoviridae), and vehicles such as drinking water and shellfish (e.g., viral gastroenteritis, Enteroviridae).[3]

The persistence and transmission of human enteric viruses through the fecal-water-oral route have been reviewed.[4-8] Several investigators have suggested that clay minerals and other naturally occurring particles present in terrestrial and aquatic systems are responsible, in part, for the prolonged infectivity of viruses in these environments.[9-14]

The purpose of this chapter is to review briefly those variables that influence the interactions between viruses and clay minerals, with special emphasis on the mechanisms of adsorption and binding. Interactions between other surfaces and viruses will also be briefly discussed to illustrate the types of surface interactions that may occur at clay-virus interfaces.

II. BACKGROUND

Outbreaks of viral disease were attributed to contaminated waters during the 1930's and 1940's and were undoubtedly responsible for earlier episodes. However, it was not until the mid-1950's that the transmission of many viruses (albeit only hepatitis A was epidemiologically confirmed) to humans through water and shellfish was generally recognized.[15,16] The application in the late 1970s of radioimmunoassays, enzyme-linked immunosorbent assays, immune electron microscopy, and other techniques to epidemiologic studies established the transmission of two additional types of enteric viruses — rotavirus and Norwalk virus (etiologic agents responsible for acute viral gastroenteritis) — to humans through the water route (Table 1).[17-19] Several outbreaks of viral gastroenteritis, in which these viruses have been seroepidemiologically linked to human illness, have occurred as a result of drinking and recreational waters wherein the source of the contamination was not identified and the numbers of coliforms were within acceptable levels.[19-21] The limitations of coliform levels as an indicator of viral contamination of waters have been discussed.[22]

A causal relation between human disease and the persistence of enteroviruses (genus *Enterovirus* of the group Picornaviridae) in potable and recreational waters and in shellfish has not been established, probably as a result of the self-limiting nature of the syndrome produced by most of the enteric viruses, the limited serologic and virologic capabilities of most public health laboratories,[23] and the difficulties involved in obtaining epidemiologically valid data (e.g., through telephone interviews and questionnaires) directly linking transmission of enteroviruses to humans through fecally contaminated waters and shellfish that are certified "acceptable" by current bacteriologic standards. However, seroepidemiologic data implicate the transmission of rotavirus, Norwalk virus, hepatitis A (enterovirus 72), and possibly other difficult-to-cultivate and noncultivatable viruses to humans through drinking and recreational waters and shellfish, and therefore suggest the transmission of enteroviruses through the water route.

Table 1
INCIDENCE OF HUMAN ENTERIC VIRUSES IN DRINKING AND
RECREATIONAL WATERS: ETIOLOGIC AGENTS OF GASTROENTERITIS

Location	Virus detection or isolation[a]	Serologic identification[b]	Comments	Ref.
Sweden	Rotavirus Norwalk virus	—	30- and 75-nm particles were found in stool specimens; authors suggested sewage-contaminated water supply.	182a
Colorado	Norwalk virus	—	Camp water supply was inadequately chlorinated and contaminated by a leaking septic tank; 27-nm particles were identified in stool specimens; first report of a water-borne outbreak of gastroenteritis attributed to a transmissible virus.	17
New South Wales and Victoria, Australia	Norwalk virus	Norwalk virus	At least 2000 persons were affected by the consumption of oysters harvested from fecally contaminated Georges River.	183, 184
Northeast Georgia	—	Norwalk virus	Radioimmunoassays (RIA) showed a 4-fold rise in 12 of 19 serum pairs from ill persons infected by a community water supply contaminated by a fecally contaminated industrial water supply.	185
Tacoma, Wash.	—	Norwalk virus	Infections were probably caused by sewage-contaminated drinking water supply.	176
Monroe County, Pa.	Rotavirus	—	Virus was identified in stool by electron microscopy; illness was associated with drinking water-based beverage called "bug juice"; source of contaminated water supply was not identified.	21
Recreation park in Macomb County, Mich.	Norwalk virus	Norwalk virus	Confirmed water-borne mode of transmission suggested that individuals are susceptible to low doses of the virus; source of contamination was not identified.	20
Georgia	—	Norwalk virus	Primary outbreak was caused by contaminated swimming pool water; pool chlorinator was inadvertently disconnected.	186
Recreation park in Macomb County, Mich.	—	Norwalk virus	Infection was traced to recreational lake water; coliform count was acceptable; source of contamination was not identified.	19

Table 1 (continued)
INCIDENCE OF HUMAN ENTERIC VIRUSES IN DRINKING AND
RECREATIONAL WATERS: ETIOLOGIC AGENTS OF GASTROENTERITIS

Location	Virus detection or isolation[a]	Serologic identification[b]	Comments	Ref.
Eagle-Vail, Colo.	Rotavirus	Rotavirus	Six of 7 persons tested were infected with rotavirus; infections were caused by the consumption of sewage-contaminated water.	18
Georgetown, Tex.	Coxsackievirus Hepatitis A virus	—	Outbreak was associated with fecal contamination of groundwater; coxsackievirus B3 may have been responsible for the outbreak.	187
Tate, Ga.	—	Norwalk virus	Sewage contaminated drinking water.	188, 189
Various counties, New York	Norwalk virus (?) (shellfish)	Hepatitis A virus	Possible contamination of shellfish beds by heavy rains and subsequent runoff was suspected; harvesting from closed waters could not be ruled out.	190
Jinzhou, China	Rotavirus	Rotavirus	Fecally contaminated drinking water affected more than 12,000 adults; a new adult diarrhea rotavirus (ADRV) was identified by RNA analysis.	191
Cincinnati, Ohio, Memphis, Tenn., and Chicago	—	Norwalk virus	Newly hired wastewater workers (less than 2 years of experience) had higher levels of antibody to Norwalk virus than did experienced workers and control groups.	192
Southeast England	Hepatitis A (shellfish)		Hepatitis virus was linked to cockles harvested from the Thames River Estuary; steaming for 1 to 2 min (to remove the shell), followed by boiling for 4 min, did not inactivate the virus.	193
Meade County, Ken.	—	Hepatitis	Fecal contamination of spring water occurred, possibly by septic tanks and/or animal wastes.	194

[a]   Electron microscopy, enzyme-linked immunosorbent assay, reverse passive hemagglutination.
[b]   Immune electron microscopy, radioimmunoassay, complement fixation, enzyme-linked immunosorbent assay.

Despite the efforts of sanitation engineers and public health agencies, both treated and untreated virus-laden sewage is discharged into aquatic and terestrial ecosystems.[24-26] Primary treatment (i.e., the removal of large solids by screening or sedimentation) removes 0 to 50% of the viruses initally present. Secondary treatment (i.e., a biological process involving mixed cultures of microorganisms in activated sludge, trickling filters, or oxidation ponds) may remove 85 to 90% of the input viruses.[24] Advanced or tertiary treatment, which usually involves some type of physicochemical processing, such as coagulation with lime, alum, iron salts, or polyelectrolytes and/or passage through activated carbon or resin to remove residual organics, removes large numbers of viruses. However, only a small percentage of the population in the U.S. is served by tertiary treatment, and only 43% of the population is served by secondary treatment.[23,27-29] In many other countries, these percentages are considerably lower. Consequently, the isolation of human enteric viruses from waters (including marine, estuarine, lake, river, and well), edible shellfish, and sewage sludge disposal sites on the continental shelf is not unexpected.[4,26]

The presumed sequence of events leading to the viral contamination of waters, soils, and shellfish have been reviewed.[30] These events are related, in part, to the type and efficiency of sewage treatment; the sites of sludge disposal and sewage outfalls; and, perhaps most important, the environmental factors (e.g., temperature, pH, and particulates) of the recipient ecosystem, each of which, under appropriate conditions, will affect the persistence of viruses.[7,11,31] Particulates in terrestial and aquatic environments may not only decrease the rate of virus inactivation,[11] but they may also serve to transport viruses from polluted to nonpolluted and shellfish-rearing marine waters.[32] Clearly, the role of particulates in the persistence of enteric viruses in environmental systems and in their transmission to humans requires further investigation.[14,26,32-34]

The mechanisms by which enteric viruses persist in aquatic and terrestrial environments are not known. However, laboratory and field observations indicate that persistence in these ecosystems is enhanced when the viruses are associated with naturally occurring particulates, such as clay minerals. It has been suggested that viruses released from sewage treatment plants adsorb on particulates (e.g., sludge particles and clays) in sediments and in the water column, and thereby are protected against biological and abiological inactivation.[13,26,35-37] Subsequent fluctuations in water chemistry (e.g., in pH and ionic concentration) and disturbance of the sediments by currents, swimmers, and power boats may cause the release of infectious virus to the water column, thereby subjecting individuals to infection through either direct contact with the water or indirectly through the consumption of contaminated shellfish.[38]

The rate of loss in infectivity of bacteriophages and enteric viruses is decreased when these viruses are associated with clay minerals. For example, bacteriophage ϕM15 of *Staphylococcus aureus* RN450 was completely inactivated after 6 days in natural, autoclaved, or filtered (0.45 μm, Millipore) lake water, whereas inactivation was reduced by 1 to 2 log in the presence of clay minerals, with the sequence of reduction being attapulgite > vermiculite > montmorillonite = kaolinite.[35] The addition of kaolinite to synthetic estuarine and natural seawater decreased inactivation of coliphage T2 after 2 days by 1 to 2 log as compared to seawater without kaolinite.[9] A more dramatic effect was observed with coliphage T7, in that the virus titer after 60 days (input, 5×10^7 plaque-forming units [pfu] per milliliter) was 3×10^4 pfu/mℓ in seawater amended with montmorillonite, whereas it was < 10 pfu/mℓ in seawater alone.[39] The rate of inactivation of reovirus in synthetic estuarine and distilled water was decreased in the presence of kaolinite or montmorillonite: reovirus was detected after 18 and 22 weeks in distilled and synthetic estuarine waters amended with kaolinite and montmorillonite, respectively, whereas the virus was not detected after 7.5 weeks in either water in the absence of clay.[11] Clay minerals may also affect the bioaccumulation of viruses by shellfish; e.g., there was a greater uptake by clams of kaolinite-associated poliovirus than of the virus alone.[40]

III. CLAY MINERALS

A. Distribution

Clay minerals are major components of soils, and because of their unique physicochemical characteristics, they have a profound effect on the ecology and activity of microorganisms[37,41-43] and on the behavior of organic compounds (e.g., enzymes and other proteins)[37,44-47] in this habitat. In many aquatic systems, especially those characterized by tidal currents, clay minerals are a common constituent of the suspended solids[48] and sediments of both fresh and marine systems.[49-52] The behavior of viruses in many terrestrial and aquatic systems is probably the result, in part, of the type and quantity of clay minerals present.

Clay minerals have been detected in the Gulf of Mexico several kilometers from the coast of Florida and more than 100 km from Louisiana and Texas. These clay minerals are probably derived from soils in the parent river basins:[53] the Mississippi and other rivers to the west supply large amounts of montmorillonite; kaolinite gradually becomes abundant in the soils and rivers of the Mobile (Alabama) River basin; and kaolinite is the dominant clay in the Apalachicola (Georgia and Florida) River basin. Water currents then redistribute these clay particles into clear regional patterns.[49] The southern facies of the U.S. characteristically contain montmorillonite and kaolinite, whereas the northeastern coast is characterized by illite and chlorite. Illite is the predominant clay off the California coast, although montmorillonite, chlorite, and, to a lesser extent, kaolinite have also been identified in these sediments.[54]

B. Physicochemical Characteristics

The surfaces of most naturally occurring inorganic particles (e.g., sand and silt) are relatively inert, and consequently they do not display the same physicochemical surface characteristics as clay minerals. A detailed review of the properties of clay minerals, however, is beyond the scope of this chapter, and only those characteristics that apparently affect the behavior (e.g., adsorption, binding and infectivity) of viruses will be presented. The reader is referred elsewhere for more comprehensive discussions of clay mineralogy.[43,54-57]

Clay minerals are primarily crystalline hydrous aluminosilicates composed of a tetrahedron of oxygen atoms surrounding a central cation, usually Si^{4+}, and an octahedron of oxygen and hydroxyl ions surrounding a central cation, usually Al^{3+}, Mg^{2+}, Fe^{3+}, or Fe^{2+}. The geometry of these units is dominated mainly by oxygen (radius, 0.140 nm) or hydroxyl (radius, 0.146 nm) ions, as Si^{4+} (radius, 0.041 nm), Al^{3+} (radius, 0.050 nm), Mg^{2+} (radius, 0.065 nm), and Fe^{3+} (radius, 0.064 nm) are considerably smaller. The sharing of oxygen or hydroxyl ions between horizontal neighbor tetrahedra or octahedra, respectively, results in the formation of sheets called tetrahedral (silica) or octahedral (alumina, magnesium, or iron) sheets, respectively. A near identical symmetry between the tetrahedral and octahedral sheets permits the sharing of oxygen atoms between the sheets in a vertical direction. In two layer clays (e.g., kaolinite), the sheets are associated in a 1.1 (silica-alumina) ratio, and in three-layer clays (e.g., montmorillonite), the sheets are associated in a 2:1 (silica-alumina-silica) ratio. Some clay minerals do not form such platelike structures, but form fibrous needlelike crystals, such as attapulgite (palygorskite).

Montmorillonite, referred to as a 2:1 expanding clay, is composed of a unit layer of two silica tetrahedral sheets with a central alumina octahedral sheet. In the stacking of the silica-alumina units, the oxygen layers of each unit face each other, and the individual units are held together by weak physical bonds (e.g., van der Waals forces and the electrostatic interaction of interlayer cations). Polar molecules can enter between the unit layers, causing them to expand (it is actually the hydration energy of the interlayer charge-compensating cations that forces the layers apart). Water molecules adsorb on the hydration layers of the charge-compensating cations and, possibly, on the oxygen surfaces, by forming hydrogen bonds, and a stable configuration is obtained with two or four monolayers of water between the unit layers. Depending on the associated cations, the basal or c spacing (i.e., the distance between the bottom oxygens of one silica-alumina unit and the bottom of an adjacent unit) may increase on the order of 0.96 nm, with a few layers of a polar molecule between the unit layers, to essentially complete separation of the individual layers to 13 nm or more in the presence of some polar molecules.[58,59]

Montmorillonite maintains a net negative charge as a result of isomorphous substitution (e.g., substitution of trivalent cations for Si^{4+} in the tetrahedral sheet and divalent cations for Al^{3+} in the octahedral sheet without any change in the gross structure of the clay). The negative charge is balanced by exchangeable (charge-compensating) cations adsorbed on and between the expandable unit layers and, to a lesser extent, around the edges. The cation exchange capacity (CEC) of montmorillonite may vary from 80 to 150 meq/100 g clay. The term exchange capacity is used as a quantitative measure of the amount of exchangeable ions that can be retained and exchanged by a clay mineral and is expressed in terms of milliequivalents per gram or, more frequently, per 100 g clay. The specific surface area of montmorillonite on dispersion is approximately 750 to 800 m^2/g clay. The external surface area of the clay is approximately 82 m^2/g.

In some 2:1 clays, the surface charge deficiency, which is located primarily between the unit layers, is neutralized by a positively charged sheet of $Al_2(OH)_6$ (gibbsite) or $Mg_3(OH)_6$ (brucite) or by smaller interlayers of Al or Fe oxyhydroxides, which reduce not only the CEC but also the ability of the clays to swell. In general, the greater the amount of iso-morphous substitution, the more tenaciously are charge-compensating cations retained by the clays. The differences in the physicochemical characteristics (e.g., CEC and specific surface area) of the various 2:1 layer clays (i.e., the phyllosilicates) are the result primarily of differences in the amount of isomorphous substitution.

Positive charges may be present on the edges of clay minerals as a result of the replacement of hydroxyl ions, the adsorption of anions that have about the same size and geometry as the silica tetrahedron (e.g. phosphate, arsenate and borate) on the edges of the tetrahedral sheet,[54] and exposed aluminum ions resulting from broken bonds at the edges.[60] This results in the anion exchange capacity (AEC), which for montmorillonite is approximately 23 meq/100 g clay.[54] There is essentially no AEC on the basal planar surfaces of clay minerals.[66]

The unit layer of kaolinite, referred to as a 1:1 nonexpanding clay mineral, is composed of a single silica tetrahedral sheet and a single alumina octahedral sheet combined so that the tips of the silica tetrahedron and one of the layers of the alumina octahedron form a common layer. Each unit layer is connected to the layers below and above it by hydrogen bonds between the hydroxyl groups of the alumina sheet and the oxygen groups of the silica sheet, which prevents swelling or shrinking, as water or other polar molecules cannot normally enter (intercalate) between the unit layers. The CEC ranges from approximately 3 to 15 meq/100 g; the AEC ranges from approximately 6.6 to 20 meq/100g clay;[54] and the specific surface area is approximately 40 m^2/g clay.[61] The CEC of kaolinite results primarily from the pH dependent ionization of hydroxyl groups at the edges of the clay, although a small amount of isomorphous substitution (a pH-independent reaction) gives kaolinite a net negative charge at pH values from approximately 1.0 to 13.0.[57,63]

The CEC-AEC ratio determines the relative net negativity of different clays and, therefore, the ability of net negatively charged particles (e.g., viruses) to approach the clays. The CEC-AEC ratio of montmorillonite is approximately 6.7 and that of kaolinite is approximately 0.5. In natural environments, however, the CEC-AEC ratios may be considerably different from those of pure clays, as some positively charged edge sites may be blocked by inorganic and organic molecules or as a result of face-to-edge aggregation of the clay particles. The association of metal oxyhydroxides with clays may also change the CEC-AEC ratios, es-pecially when the ambient pH fluctuates.[43]

Most other crystalline clay minerals commonly found in natural habitats have structures and physicochemical characteristics similar to either the 2:1 or 1:1 type of clay minerals. In addition to crystalline hydrous aluminosilicates, some soils and sediments contain amor-phous clays (e.g., allophane and imogolite) and many contain polymeric hydrous oxides of Fe^{3+}, Al^{3+}, Mn^{4+}, and Si^{4+}. Allophane and imogolite are present primarily in young soils and sediments derived from recent volcanic activity, whereas oxyhydroxides are usually present in highly weathered and old soils, characteristically in the tropics. The amorphous clay minerals can be either net positively, negatively, or neutrally charged, and the origin of these charges are not as well characterized as they are for crystalline hydrous aluminosilicates.

IV. SURFACE INTERACTIONS: CONCEPTS AND TERMINOLOGY

Surface interactions between biological entities (e.g., viruses) and clay minerals are usually viewed differently by physical scientists and biologists, including virologists. For example, the physical scientist is interested primarily in the mechanisms of adhesion, whereas the virologist is more concerned with the differences between free and adsorbed viruses in infectivity, susceptibility to inactivation by biological and abiological factors, and transport of the infectious agents through natural systems (e.g., surface and groundwaters, sediments, and soils). These differences in scientific approach and purpose often result in the use of different terminology for the same phenomena. Consequently, to enhance communication between investigators with different backgrounds but similar interests, some of the more relevant terms are defined for the purpose of this chapter.

The term adsorption is commonly used to define the concentration of an adsorbate (e.g., a virus) at an interface with an adsorbent (e.g., a clay mineral). However, equilibrium adsorption and binding will be used in this chapter to define more definitively the interactions at the adsorbent-adsorbate interface. Equilibrium adsorption is characteristically a reversible reaction, exhibits low specificity between adsorbent and adsorbate, usually results primarily from physical forces (e.g., hydrogen bonding and London-van der Waals forces of attraction), and may involve the formation of multilayers of adsorbate.[64] As coverage of an adsorbent surface by the adsorbate increases, it usually becomes progressively more difficult for adsorption to occur. Equilibrium is attained when the rate of adsorption equals the rate of desorption. Binding describes the condition in which the adsorbate is not removed from the adsorbent by repeated washings with appropriate solutions.[11,65,66] Binding involves primarily chemical bonds, but it may also result from the attachment of an adsorbent by numerous physical forces, the sum of which may exceed the tendency of the adsorbate to detach from the adsorbent. Physical forces, especially in conjunction with hydrophobic interactions, may result in binding that is stronger than that resulting solely from chemical bonds.

The term sorption is usually used when the mechanisms involved in a surface interaction between an adsorbate and an adsorbent are poorly defined. Adhesion, a similarly vague term, is often used to imply the sticking of a cell or virus particle to a surface, but it does not indicate whether equilibrium adsorption or binding is involved.[43]

V. FACTORS THAT MEDIATE ADSORPTION OF VIRUSES ON CLAY MINERALS

Several factors appear to mediate the adsorption of viruses on clay minerals and other particulates and surfaces. Some of these factors and the mechanisms involved have been demonstrated experimentally in studies on surface interactions between clays and viruses, and others have been extrapolated from studies on interactions between clays and other particulates and microbes or proteinaceous materials, as well as from basic concepts of colloid chemistry.[43] The adsorption of viruses on clay minerals can not be attributed to a single mechanism. Adsorption appears to be a function of the physicochemical characteristics of both the clay and the virus. For descriptive purposes only, the factors that mediate the adsorption of viruses on clay minerals and other surfaces are categorized and discussed as being either abiologic or biologic. Only a few examples of the influence of each factor are discussed in detail below. A more complete summary of the literature is presented in Table 2.

Table 2
SUMMARY OF THE LITERATURE ON ADSORPTION OF VIRUSES BY CLAY MINERALS

Clay mineral	Virus	Observations	Ref.
Magnetite, kaolinite	Bacteriophage MS2	0.15 mg kaolinite in 1 ℓ distilled water containing 100 mg NaCl (pH 4) decreased 6-fold the amount of phage adsorbed on magnetite. The addition of approximately 0.04 mg/ℓ cationic polyelectrolyte (Catoleum A8101) counteracted the effect of kaolinite. The polyelectrolyte reduced the charge on the kaolinite, thereby eliminating competition between the net negatively charged MS2 and kaolinite for adsorption sites on the net positively charged magnetite.	154
Bentonite[a]	AS-1 cyanophages	0.9 mg bentonite added to 100 mℓ lysate preferentially adsorbed nonviral contaminants from lysate.	155
Attapulgite	Poliovirus type	Attapulgite was a better adsorbent than kaolin. Adsorption by attapulgite was independent of pH and temperature.	156
Bentonite	Coxsackieviruses A9 and B5, adenovirus 5, echovirus 11, influenza virus A2, parainfluenza virus A2, herpes simplex virus, rubella virus	Transport medium containing bentonite maintained virus titer from 3—21 days, depending on the virus, and was superior to a charcoal transport medium. HS-, influenza-, and rubellaviruses were protected against inactivation by serum-coated bentonite. Viruses were eluted with 0.3 M citrate buffer + 20% rabbit serum (pH 5.3—6.2). Coxsackievirus A9 adsorbed onto the face of the clay and appeared monodispersed in the electron microscope.	110
Montmorillonite	Bacteriophage T7	Concentrations of montmorillonite ranged from 75—750 µg/mℓ, but virus survival did not increase significantly above 200 µg/mℓ. Virus survival in clay-sea water mixture was 2×10^6 pfu/mℓ after 16 days and 3×10^4 pfu/mℓ after 60 days; virus inactivation after each period without clays was greater than 99.9%. Input titer: 5×10^7 pfu/mℓ.	39
Kaolinite	Bacteriophage T7	102 mg clay added to 1 ℓ distilled water containing CaCl$_2$ plus 2.0 and 4.0×10^5 pfu/mℓ. 99.9% adsorption was achieved with 0.005 M and 0.010 M CaCl$_2$. Kaolinite did not appreciably protect the virus from chlorine at pH 7.	130
Bentonite	Rubellavirus	During the preparation of ^{125}I- labeled rubellavirus RNA, "all buffers and reagents contained 0.2% sodium dodecyl sulfate (SDS) and bentonite." The bentonite probably adsorbed the RNase.	157

Table 2 (continued)
SUMMARY OF THE LITERATURE ON ADSORPTION OF VIRUSES BY CLAY MINERALS

Clay mineral	Virus	Observations	Ref.
Bentonite	Poliovirus (Sabin type 1), coxsackievirus A9, bacteriophage f2	Adsorption of polio-, coxsackie-, and f2 viruses by bentonite (in 10^{-2} M NaCl) was 10, 13, and 1% of input titer ranging from 8.3×10^5—1.6 $\times 10^6$, 1.5—4.7×10^6, and 8.8×10^8—2.3×10^9 pfu/mℓ, respectively. The viruses were not protected by either 0.007 or 0.035 mg/mℓ bentonite against 0.007 to 0.21 mg/ℓ ozone.	158
Bentonite	Brome mosaic virus, tobacco mosaic virus	Brome mosaic, but not tobacco mosaic, virus coat protein was adsorbed by bentonite. The affinity of bentonite for brome mosaic virus coat protein was higher than the affinity for the viral RNA. Bentonite may increase degradation of tobacco mosaic virus by adsorbing a virus stabilizing protective component from the system.	159
Bentonite	Tobacco mosaic virus	Low concentrations of virus degraded more rapidly in the presence than in the absence of bentonite in buffers of pH 7—9. Degradation of virus in some RNA and protein preparations was complete at pH 8.5 in the presence of bentonite and EDTA. Bentonite may enhance degradation of the virus by adsorbing protein or RNA.	160
Montmorillonite, kaolinite	Bacteriophage T1	Adsorption of bacteriophage T1 to clay was observed by electron microscopy using critical point drying. Adsorption occurred by a tail-to-clay (edge) orientation.	126
Kaolin[b]	Bacteriophage of *Staphylococcus aureus* (not identified)	0.2 g kaolin was added to 2.5 mℓ containing 10^5 phage, of which 38% was adsorbed following shaking. Adsorption was better at pH 8.4 than at 6.8. Virus eluted by 0.002 M K$_2$HPO$_4$, Na$_2$SO$_4$, or Na$_2$CO$_3$. Rate of elution varied.	152
Kaolinite, montmorillonite, illite	Bacteriophage T2, poliovirus type 1	Maximum virus adsorption (50 mg/ℓ kaolinite, 10^6 virus per milliliter) was achieved after 20-min contact time in 0.005 M CaCl$_2$ at pH 7. Virus adsorption increased linearly to a concentration of 50 mg/ℓ kaolinite; at each clay concentration, more adsorption of the virus occurred with 0.01 M CaCl$_2$ than 0.01 M NaCl. Equivalent data not reported for montmorillonite or illite. The effect of varying salt concentrations on the adsorption of coliphage T2 to 50 mg/ℓ clays was investigated. The maximum adsorption obtained with Na was 99% for all three clays with salt concentrations of 0.03 kaolinite, 0.05 montmorillonite and 0.10% illite. Al	76

was superior to Ca or Na. Egg and bovine albumin preferentially adsorbed to clays. The CEC of kaolinite, montmorillonite, and illite was 10.2, 96.2, and 11.0 meq/100 g clay, respectively, as determined by the BaCl$_2$ method.

Adsorbent	Virus	Remarks	Ref.
Bentonite	Coxsackievirus type A9	25-g sample of cottage cheese was artificially contaminated with virus. A slurry was prepared with glycine-NaOH buffer plus 1 M MgCl$_2$. The slurry was treated with Freon TF and bentonite to facilitate centrifugation. Virus concentrated from supernatant by treatment with polyethylene glycol followed by ultracentrifugation. 50% probability existed for detecting viruses at levels less than 5 pfu/25-g sample.	196
Bentonite	Poliovirus 1	Light and turbidity (turbidity produced by 3% beef-coated bentonite) increased the loss of virus infectivity. Loss of infectivity was explained by either adsorption of virus on bentonite or virion aggregation, thus decreasing virus titers.	161
Clay earth, garden soil, Oxfordshire clay (undefined)	Nuclear polyhedrosis virus of *Mamestra brassicae*	Greatest variation in virus elution efficiency occurred with Oxfordshire clay. ''This suggests that the nuclear polyhedrosis virus is adsorbed primarily onto clay particles.''	162
Kaolin	Bacteriophage O$_x$6	Adsorption of the tryptophan-dependent bacteriophage by kaolin required 0.05 M NaCl and a temperature of 37°C. Elution of the phage occurred by altering the NaCl or the temperature.	163
Kaolinite 4 (Ward's)	Coliphage T2	The adsorption of 10^9 virus particles on varying concentrations of kaolinite (0—50 mg) in synthetic sea water was determined. Approximately 85 and 90% of the input virus was adsorbed by kaolinite, 3 and 50 mg/100 mℓ synthetic seawater (pH, 8.1), respectively. Maximum adsorption of virus (approximately 90% of the input population) occurred at 35 g/kg (parts per thousand) synthetic sea salt; adsorption decreased with decreasing salinity. Optimum adsorption occurred at pH 8.1. The virus was eluted from kaolinite by distilled water and nutrient broth.	9
Bentonite, kaolinite, attapulgite, vermiculite	Coxsackievirus B3	Clay minerals may be used in the removal of viruses from sewage.	164
Cherkassian montmorillonite	Bacteriophage T2	Adsorption of virus maximal (99.3—99.6%) at 0.06—0.1 M NaCl, Na$_2$SO$_4$, or CaCl$_2$. Adsorption increased as the valency of the cation increased and was not dependent on the anion composition of the medium.	165
Kaolinite, montmorillonite	Poliovirus type 1	Initial virus and adsorbent concentrations were approximately 5×10^4 pfu/mℓ and 50 mg/ℓ, respectively. Adsorption was directly related to salinity. Approximately 97% of the virus adsorbed on kaolinite in 10^{-2} M NaCl, 10^{-3} M MgCl$_2$, or 10^{-5} M AlCl$_3$. Fetal bovine serum prevented and reversed the adsorption. Adsorption was similar at pH 5 to 9.	166

Table 2 (continued)

SUMMARY OF THE LITERATURE ON ADSORPTION OF VIRUSES BY CLAY MINERALS

Clay mineral	Virus	Observations	Ref.
Bentonite	Poliovirus LSc2ab	Adsorption on membrane filters (0.3—0.45 μm) or an ion exchange resin (AB-17) was greater (74—100%) the lower the virus concentration; with bentonite, adsorption was enhanced (60%) at higher concentrations of virus. The poliovirus titer in this study ranged from 0.05 to 12 pfu/mℓ tap water.	167
Bentonite	Respiratory syncytial virus	Stability of the virus in bentonite transport medium was equivalent to chick and CDC media.	168
Kaolin	"Bacteriophage" (unidentified), fowlpox	Adsorption and subsequent elution yielded virus free of protein. 20% kaolin adsorbed both viruses from suspension. Fowlpox virus and bacteriophage remained infectious in the adsorbed state.	169
Soils containing kaolinite, montmorillonite, quartz, Fe and Al oxides	Bacteriophage MS2	Divalent cations enhanced adsorption by soils; sorptive capacities explained in terms of clay mineral content. Soils containing kaolinite or illite adsorbed more virus than soils containing montmorillonite. Virus input ranged from 1.1×10^5—1.1×10^6 pfu/mℓ.	170
Kaolin and other adsorbents	Fowl "vaccine", fowlpox, chicken sarcoma virus	Kaolin added to sterile Locke solution containing fowl vaccine or fowlpox and sarcoma viruses. Fowlpox adsorbed on 10% (w/v) kaolin "inactivated" fowl vaccine and was inactivated by 20% kaolin. 25 and 50% kaolin adsorbed sarcoma virus, which was not inactivated.	153
Kaolinite, montmorillonite	Reovirus type 3	More virus was adsorbed, in both distilled and estuarine water, by lower concentrations of montmorillonite than of kaolinite containing a mixed complement of cations on the exchange complex. Adsorption on the clays was immediate and was related to the cation-exchange capacity of the clays. The addition of cations to distilled water enhanced adsorption. More virus was adsorbed on montmorillonite made homoionic to various mono-, di-, and trivalent cations (except montmorillonite homoionic to potassium) than on comparable concentrations of kaolinite homoionic to the same cations. The sequence of the amount of virus adsorbed on homoionic M was Al > Ca > Mg > K and that on K was Na > Al > Ca > Mg > K.	70
Kaolinite, montmorillonite	Reovirus type 3	Chymotrypsin and ovalbumin reduced the adsorption of virus on kaolinite and montmorillonite homoionic to Na; lysozyme reduced adsorption on montmorillonite but not on kaolinite. The proteins (except lysozyme	111

Clay	Virus		Ref.
Kaolinite, montmorillonite	Reovirus type 3	bound on kaolinite) apparently competed with reovirus for adsorption sites on the clay. The reduced adsorption of reovirus on kaolinite and montmorillonite was related to the molecular weight of the proteins and/or the adsorption of each protein to cation-exchange sites on the clays. Chymotrypsin and lysozyme markedly decreased reovirus infectivity in distilled water, but this effect was less pronounced when the proteins were bound on montmorillonite, suggesting that the biophysical characteristics of the proteins were altered at the clay-virus interface.	11
Kaolinite, montmorillonite	Reovirus type 3	Desorption studies indicated that more virus was bound on montmorillonite than on kaolinite. Approximately 100% of the input virus population was recovered from montmorillonite and kaolinite homoionic to Na and from kaolinite either homoionic to Mg or with a natural mixed complement of cations. 150% of the input virus population was recovered from montmorillonite homoionic to Mg. The infectivity titer of kaolinite- but not montmorillonite-virus complexes was greater than that of free virus. More virus was adsorbed on the clays below pH 4 than at pH 7. Loss of virus infectivity in estuarine and distilled water containing 10 mg kaolinite or montmorillonite (18 and 22 weeks, respectively) was less than that in either water alone (7.5 weeks).	
Kaolinite, montmorillonite	Reovirus type 3, coliphage T1	Comparative adsorption studies indicated that these viruses do not share common adsorption sites on kaolinite or montmorillonite. The blockage of positive sites on the clays with polyphosphate anions or with supernatants from montmorillonite or kaolinite had no effect on adsorption of the reovirus. These data indicate that there was a specificity in adsorption on clay minerals from mixed populations of reovirus type 3 and coliphage T1.	88
Kaolinite, montmorillonite	Reovirus type 3	Culture filtrates from *Bacillus subtilis* displayed antiviral activity, but those from *Escherichia coli* or *Serratia marcescens* did not. Studies with both live and killed *B. subtilis* indicated that the antiviral component was metabolic in origin. *B. subtilis* and *S. marcescens*, but not *E. coli*, grew on reovirus concentrations $> 3.1 \times 10^6$ TCID$_{50}$/mℓ as the sole source of carbon and energy. Adsorption of reovirus on kaolinite was enhanced by culture supernatants from *S. marcescens* and on montmorillonite by those from *E. coli*. Reovirus was not adsorbed on the bacteria.	171
Clay-size particles in soil	Wheat mosaic viruses	Clay-size particles in soil may protect viruses from degradation.	145, 146

Table 2 (continued)
SUMMARY OF THE LITERATURE ON ADSORPTION OF VIRUSES BY CLAY MINERALS

Clay mineral	Virus	Observations	Ref.
Kaolinite	Tobacco mosaic virus	Electron micrographs of tobacco mosaic virus (TMV) on kaolinite showed that no particles adhered to the sharp edges of the clay. "The TMV was clearly discernible on the surface in more or less irregular array (i.e., without much two-dimensional order)."	132
Kaolinite	Poliovirus type 2 (strain W-2)	The accumulation of virus associated with kaolinite was enhanced in hepatopancreas and siphon tissues of clams. Viruses associated with human feces were accumulated to a greater extent than "free" viruses or viruses associated with kaolinite.	40
Kaolinite, montmorillonite	Wheat yellow and barley yellow mosaic viruses	Viruses were stable in the adsorbed state to 8 weeks.	12, 144
Bentonite, kaolin (USP grade)	Bacteriophage T7, T2, f2, poliovirus type 1 (Mahony)	Percentage of virus adsorbed on 95 mg bentonite or 16 mg kaolin/ℓ distilled water was determined by assay of supernatants. In general, pH extremes (4 and 10) and divalent cations enhanced adsorption, but none of the viruses tested can be used to describe a general pattern of adsorption.	73
Kaolinite, montmorillonite, plus 32 soils and minerals	Poliovirus type 2	Poliovirus type 2 was added to 500 mg each of 34 soils and minerals (including kaolinite and montmorillonite) in 1.9 mℓ synthetic fresh water (containing Na, K, Mg, and Ca as the salt of Cl, SO_4, or HCO_3) to yield a final titer of 10^8—10^9 pfu/2 mℓ. Arizona and Wyoming montmorillonite adsorbed 98.7% of the virus. The weakest adsorbents among the soils contained larger quantities of organic matter. Among the materials with little organic matter, montmorillonite (with a large negative surface charge density) was a less effective adsorbent. Interpretation of data is unclear, due to the large amount of clay and soil added to the system.	172
Attapulgite, sepiolite, zeolite	Poliovirus type 1 (attenuated), herpes simplex virus type 1	10% attapulgite-distilled water suspensions removed all herpes simplex virus from the system (6×10^4 infectious particles/mℓ). 5% attapulgite removed more than 99.9% poliovirus from distilled water suspensions originally containing 4×10^6 particles /mℓ.	173
Kaolinite, montmorillonite	Poliovirus type 1	1 g kaolinite suspended in 9 mℓ of poliovirus containing 5×10^6 $TCID_{50}$/mℓ adsorbed approximately 99.9% of the input virus. Approximately 99.9% of the virus was removed by montmorillonite under identical conditions. The percentage of virus adsorbed was greatest for dilute virus suspensions, but the total amount of virus adsorbed was greatest for concentrated suspensions.	174

Sorbent	Virus	Comments	Ref.
Bentonite, soils, cation exchange resin	Bacteriophages of *Arthrobacter*	Phages appear to adsorb to cation exchange sites by their tails, and inactivation of the phages may be due to damage to their tails. Phages added to particles with high cation exchange capacity could not be eluted in the infectious state.	84
Montmorillonite	Bacteriophage of *E. coli* strain M13	*E. coli* was protected from phage attack at low electrolyte concentrations by an envelope of sorbed clay. High electrolyte concentrations protected the cells, presumably as the result of both the colloid envelope and the sorption of the phages on the clay.	127
Cherkassian montmorillonite	Bacteriophage T2	Under conditions of constant mixing, 50 mg/ℓ clay and 5×10^8 pfu/mℓ achieved equilibrium adsorption in 18 hr. Adsorption followed the Langmuir equation. Maximum adsorption was calculated at 16.98×10^{13} pfu/g clay.	175
Talc-celite complex	Poliovirus type 1 (Sabin)	Samples containing 45 to 60 pfu/mℓ in 10 to 100 mℓ sewage samples (pH adjusted to 6) were passed through 400 mg of polyelectrolyte 60 (a cross-linked copolymer of isobutylene and maleic anhydride; PE 60) and layers of 300 mg talc and 100 mg celite. PE 60 recovered 76% and talc-celite recovered 68% of input virus.	176a
Bentonite (USP grade)	Poliovirus type 1 (vaccine), coxsackievirus B2, encephalomyocarditis (Columbia SK and Mengo) virus	Adsorption of each virus to bentonite was maximal in the presence of 0.005 M CaCl$_2$. Infectivity of Columbia SK was retained after 4 days when adsorbed on bentonite. Infectivity studies were not performed with the other viruses.	177
Montmorillonite	Encephalomyocarditis (Columbia SK and Mengo) virus	10^5 pfu/mℓ Columbia SK virus was added to 36 mg/ℓ clay in the presence of 10^{-3} M CaCl$_2$, and maximum adsorption (i.e., 98.4%) was achieved after 30 min. The virus remained infectious in the tissue culture. Serum inhibited adsorption of Columbia SK virus on the clay. Mengovirus adsorbed to the clay retained infectivity after oral inoculation of suckling mice.	74
Montmorillonite, kaolinite	Coliphages T1 and T7	Positively charged sites (i.e., anion exchange sites) on the clays appeared to be primarily responsible for the adsorption of coliphage T1 on kaolinite but only partially responsible for the adsorption of coliphage T7 on montmorillonite. Reduction in adsorption after pretreatment of the clay with sodium metaphosphate [Na(PO$_3$)$_{13}$] was more pronounced with kaolinite than with montmorillonite. Equilibrium adsorption isotherms of coliphage T7 on kaolinite and montmorillonite suggested a correlation between adsorption and the cation exchange capacity of the clays; this relation was not observed with coliphage T1.	85

Table 2 (continued)
SUMMARY OF THE LITERATURE ON ADSORPTION OF VIRUSES BY CLAY MINERALS

Clay mineral	Virus	Observations	Ref.
Montmorillonite, kaolinite	Coliphages T1 and T7	Kaolinite and montmorillonite were better adsorbents for coliphages T1 and T7 than were nonhost Gram negative and -positive bacteria (early log, late log, stationary phase cultures), actinomycetes, and yeasts. The coliphages were not inactivated by the microorganisms studied.	86
Bentonite	Coxsackievirus, poliovirus types 1, 2, 3	Incorporation of bentonite into the agar overlay enhanced plaque-forming activity. Coxsackievirus produced plaques more readily than poliovirus in a bentonite-agar overlay system.	140
Bentonite	Coxsackieviruses B1 and B6	Maximal adsorption of viruses at acid pH (3.5); maximal elution achieved at alkaline pH (7.5—8.0) in low ionic strength medium (i.e., distilled water).	178
Bentonite	Coxsackieviruses A7, A8, A10, A18, B2, B3, B4, B5, and B6	Adsorption of B1, B2, and A10 occurred at acid pH values (pH 6.0). Other strains adsorbed at all pH values (4.0 to 7.2) tested. B1, B2, and A10 eluted easily at alkaline pH, whereas the other strains were more efficiently eluted with 1 M sucrose.	98
Bentonite, kaolinite	Reovirus type 3, poliovirus type 1	1% bentonite adsorbed reovirus from the liquid removed from the upper half of settled domestic sewage with >99% efficiency at pH levels ranging from 3.5—7.5. However, the efficiency of adsorption of poliovirus on bentonite ranged from 70—91% of the control, with the greatest adsorption occurring at pH 7.5 and 4.5. 1% kaolinite adsorbed both viruses with an efficiency of >92% at pH 3.5 to 7.5. Distilled water, 3.5% NaCl, or nutrient broth did not appreciably elute the viruses from the clays.	179
Bentonite	Poliovirus type 3	Bentonite enhanced virus retention by electropositive (Virosorb 1MDS) and electronegative (Filterite) filters. Enhanced retention by Filterite filters of the virus from waters containing 50 mM MgCl$_2$ and bentonite (10 nephelometric turbidity units) was attributed to the adsorption of the poliovirus on bentonite and the retention of the clay-virus complex on the filters. Enhanced retention by Virosorb filters was attributed to reduced porosity of the filter medium by bentonite, which decreased flow rates and, thereby, increased virus contact times with the filter, or which provided additional adsorption sites for the virus. Bentonite reduced elution of the virus from each filter.	96

Adsorbent	Virus		Ref.
Bentonite	Bacteriophage MS2	Bacteriophages adsorbed on clay were more resistant to HOCl than were free phages. 10^{-2} M MgCl$_2$ caused 80% adsorption in a system containing 35 mg clay and 1.9×10^5 pfu/mℓ; an identical system without Mg had 25% adsorption. Adsorption showed first order kinetics, and an increase in agitation increased the rate of adsorption. Fetal bovine serum was the most effective eluent.	129
Kaolin, montmorillonite	Bacteriophages f6 and f13 of *Streptomyces*	Viruses were added to clays suspended in nutrient broth at concentrations from 0.001—100 mg/mℓ. No significant adsorption occurred at clay concentrations <1 mg/mℓ. Adsorption on kaolin was greater when the clay contained a normal mixed complement of cations or was homoionic to Na than when homoionic to Ca or Al. One wash with nutrient broth removed 1—2% of the adsorbed virus. A fixed proportion of bacteriophage f6 adsorbed on 20 mg/mℓ kaolin, regardless of the virus concentration added to the system, indicating a heterogeneity in the virus population. More virus adsorbed on kaolin than on montmorillonite, possibly as a result of the greater ratio of positively to negatively charged sites on kaolin than on montmorillonite.	125
Kaolin, Al(OH)$_3$, infusorial earth, charcoal, glass, sand	Vaccinia virus	Virus adsorbed on 1% clay and other adsorbents in 0.85% NaCl and in distilled water but not in "hormone" broth (minced beef heart, peptone, 0.85% NaCl). The broth also inhibited the adsorption of vaccinia virus on Berkefeld V filters.	180
Allophane, montmorillonite	Coliphage R17, reovirus type 3	The principal factors influencing adsorption of the viruses were mixing time, pH, and the concentrations and pI of both the virus and the adsorbent. Reovirus adsorption on allophane decreased as the pH$_h$ increased above the pI of the virus. 1 mg/mℓ allophane adsorbed approximately 97% of a 1.1×10^{10} pfu reovirus input at pH 5.5. Montmorillonite adsorbed approximately 50% of the reovirus under identical conditions.[199] Adsorption of coliphage R17 by allophane was approximately 10% less efficient under identical conditions.	93
Montmorillonite, soil particulates, sand	Poliovirus type 2 (vaccine strain p712-Ch-2ab)	More than 99% of the virus adsorbed on montmorillonite below pH 9.0 when it was added to 250 mg montmorillonite suspended in 0.3, 1, 3, 10, or 30 mM NaCl, CaCl$_2$, or Na$_2$SO$_4$ to yield a final viral titer of 10^1—10^2 pfu/mℓ. Adsorption was enhanced by increased electrolyte concentration but was markedly reduced on all particles tested above pH 9. Differences in adsorption paralleled pH-dependent charge properties of the particles.	181

Table 2 (continued)
SUMMARY OF THE LITERATURE ON ADSORPTION OF VIRUSES BY CLAY MINERALS

Clay mineral	Virus	Observations	Ref.
Montmorillonite, kaolinite	Poliovirus type 1 (Sabin), coliphage T2	Clays enhanced the flocculation of the cationic polymer, Nalcolyte 605 (Nalco Chemical Company, Chicago). Positively charged amino groups on the polymer probably linked with either clay or virus to form settleable agglomerates (i.e., by a mechanism similar to the "clay-cation-virus bridge" theory). Increasing concentrations (0.001—0.1 M) of divalent cations (i.e., Mg, Ca) in the system enhanced adsorption.	182
Montmorillonite, illite, kaolinite, attapulgite	Stem-mottle virus, tobacco mosaic virus	0.5 g of each clay was added individually to 10 mℓ of clarified sap from diseased plants. Montmorillonite and illite were more effective in adsorbing stem-mottle virus than were attapulgite or kaolinite. Illite, attapulgite, and kaolinite did not significantly adsorb tobacco mosaic virus from sap, whereas montmorillonite showed a greater tendency to adsorb tobacco mosaic virus.	135
Montmorillonite	Poliovirus type 1 (Sabin)	0.04 mg/mℓ montmorillonite homoionic to Na remained monodispersed at 10^6 pfu/mℓ virus in phosphate buffered saline. Clay aggregates formed at 10^7 pfu/mℓ, which resulted in less virus adsorption on the clay. Saturation-limited adsorption (determined by linear regression analysis) was measured at 3.2×10^8 pfu/mg clay. Adsorption of poliovirus on montmorillonite-Na fit the Freundlich isotherm, and the amount of virus adsorbed corresponded to approximately 0.2 virions/particle of montmorillonite.	197
Montmorillonite	Poliovirus type 1 (Sabin)	SEM indicated that the virus adsorbed to the edges of the clay. Adsorption of the net negatively charged virus may have resulted from electrostatic interaction with the positively charged edge sites of the clay.	133
Kaolinite, montmorillonite	Herpes simplex virus type 1	The adsorptive capacity of kaolinite for HSV-1 was approximately 2 times greater than that of montmorillonite, but montmorillonite was more stabilizing for the virus than was kaolinite. The adsorbed virus remained infectious. HSV-1 was more firmly bound on montmorillonite than on kaolinite. Adsorption of HSV-1 was not related to the CEC of the clays. A heterogeneity within the virus population was observed.	87

[a] Bentonite is the geologist's name for a mineral that contains predominantly montmorillonite.
[b] Kaolin is the geologist's name for a mineral that contains predominantly kaolinite.

A. Abiologic Factors

1. Electric Double Layer

The concept of the double layer and its implication in interactions between clay minerals and viruses had its origin some 60 years ago, when Stern[67] described the existence of an electric double layer around charged surfaces, the components of which consisted of earlier proposed diffuse (Gouy-Chapman) and fixed (Helmholtz) layers. Most biocolloids (e.g., viruses and bacteria) have a net negative charge at pH values above their isoelectric point (pI) that results primarily from the dissociation of carboxyl groups on the surface. Clay minerals also have a net negative charge, but in contrast to biocolloids, this charge results primarily from isomorphous substitution within the crystal (pH independent) or from dissociation of exposed hydroxyl groups (pH dependent). To attain a net neutral charge, counterions are attracted to the surface of these net negatively charged particles to produce a compact layer of cations and then a diffuse layer of cations and anions termed the Stern and Guoy-Chapman layers, respectively. This arrangement of counterions around a net negatively charged particle, termed the Stern-Guoy-Chapman diffuse double layer,[68,69] is used to explain the electrokinetic or zeta potential (defined as the potential at the plane of shear) of a colloid. Increases in the valance or concentration of the counterions compresses the thickness of the diffuse double layer, in accordance with the Schulze-Hardy rule, and allows net negatively charged particles (e.g., clay minerals, viruses, bacteria, and eukaryotic cells) to come close enough together for hydrogen bonding, van der Waals forces, and other mechanisms of adsorption and binding to occur at the particle-particle interface.[37,43,55,70-72]

These concepts of colloid chemistry have been used to explain the adsorption of encephalomyocarditis and polioviruses on clay minerals,[13,73,74] of avian[75] and bacterial viruses (coliphages T1, T2, and T7 and bacteriophage MVL-1 of Gourlay *Acholeplasma*) on their host cells,[73,76-78] of coliphage T2 and influenza virus PR8 on a cation exchange resin,[79] of microbial cells on clay minerals,[71,72] the greater adsorption of reovirus type 3 on clay minerals in synthetic estuarine than in distilled water,[70] and similar surface interactions.[43]

2. Cation Exchange Capacity

The CEC of clay minerals and other net negatively charged colloids appears to be an important factor in the adsorption of animal and bacterial viruses. For example, studies on the mechanisms of adsorption of reovirus type 3 on clay minerals indicated that adsorption occurred primarily to negatively charged (i.e., cation exchange) sites on montmorillonite and kaolinite, although both the virus and the clays were net negatively charged. This apparent paradox may be explained by the fact that the surface pH (pH_s) of colloidal clays may be 3 to 4 units lower than the bulk pH (pH_b) of the suspension,[64,80] and the acidity near the clay surface probably resulted in protonation of the virus (e.g., proton [H^+] transfer) followed by cation exchange at the clay surface. Adsorption of bacteriophages ϕX174,[81] T1, T2, and f2[82,83] by soils and of a bacteriophage of *Arthrobacter* sp.[84] on montmorillonite was also related, in part, to the CEC of the adsorbents.

In contrast, coliphage T1 adsorbed primarily to positively charged (i.e., anion exchange) sites on montmorillonite and kaolinite. However, the adsorption of coliphage T7 on montmorillonite and kaolinite appeared to be primarily to negatively charged sites on both clays, as an increase in the CEC of the experimental systems resulted in an increase in the adsorption of coliphage T7.[85,86] The adsorption of herpesvirus hominis type 1 (herpes simplex virus type 1 [HSV-1]) on montmorillonite and kaolinite was not related to the CEC of the systems, as two distinct adsorption isotherms, one for each day, were obtained when the quantity of sorbed virus was plotted against CEC.[14,87]

The apparent specificity of these viruses for different adsorption sites on montmorillonite and kaolinite may have resulted from differences in the amino acid composition of the capsid proteins of reovirus type 3 and coliphages T1 and T7 and, possibly, from the phospholipid envelope surrounding HSV-1.[88] The differential adsorption of several enteric viruses and bacteriophages on soils, sludges, and sediments has been shown to be dependent on the type of virus and even on the strain within a virus type,[89,90] further indicating the importance of the polypeptide structure of the virus capsid in the adsorption of virus particles on charged surfaces.

3. Electrostatic (Coulombic) Interactions

Electrostatic attraction between viruses and oppositely charged surfaces (e.g., positively charged filters and clay minerals with a high pI, such as allophane) has been exploited in the development of methodologies to concentrate viruses from tap water, lake water, and sewage at ambient pH and in the absence of cationic salts (e.g., poliovirus;[91,92] reovirus;[93] and coliphages MS-2, ϕX174, T2, T4 and R17[90,93,94]). The use of positively charged filters (e.g., chrysotile asbestos, 30S Zeta-plus, and Virosorb 1MDS) in the concentration of viruses from sewage and waters appears to be relatively efficient and simple. However, the efficacy of these and other positively charged filters for the concentration of viruses under various environmental conditions requires more evaluation,[94-96] as the numerous diverse ionic and particulate inorganic components present in many natural systems can interfere with coulombic interactions.

The amorphous clay, allophane (pI = pH 6.3), which is the predominant clay mineral species in the Pacific Basin and northern New Zealand, is net positively charged at the pH_b of many nonsaline waters and soils. Consequently, the adsorption of viruses with a low pI (e.g., reovirus [pI = pH 3.9] and coliphage R17 [pI = pH 3.0]) on this clay species would be expected to increase in the absence of excess amounts of cationic salts as the pH_b approaches the pI of the clay.[93] Although allophane adsorbed approximately twice the number of reovirus particles (1 $\times$ 10[10]/mℓ) as did montmorillonite,[199] adsorption on montmorillonite might have been comparable, or perhaps even greater, had the montmorillonite been made homoionic to a di- or trivalent cation. The increased adsorption of viruses on clays in the presence of di- or trivalent cations, either in solution[76] or on the clays,[70] has been shown. For example, twice the quantity of reovirus was removed from suspension by 0.1 mg montmorillonite homoionic to Al than by the same concentration of montmorillonite homoionic to Ca.[70]

Colloidal silica, modified to have either a net negative or net positive surface charge, was used to evaluate the effect of electrostatic interactions on the adsorption of viruses with a high (poliovirus 1 strains LSc and Brunhilde) or low (reovirus types 1 and 3, and coliphage MS-2) pI.[97] Each virus tested adsorbed extensively on either negatively or positively charged silica particles at pH values below or above their pI, respectively, showing, at least in this system, the importance in adsorption of electrostatic interactions at the particle-virus interface.

Adsorption of reovirus on kaolinite, with a net negative charge that is predominantly pH dependent, and to a much lesser extent on montmorillonite, with a pH-independent net negative charge, was inversely related to the pH_b of the system.[11] The probable protonation of exposed dissociated hydroxyl groups on kaolinite during the decrease in pH_b from 7 to 3 apparently reduced the net negative charge on the clay (i.e., the kaolinite became more positively charged), and this in turn enhanced the adsorption of the net negatively charged reovirus primarily by electrostatic attraction. Inasmuch as the hydroxyl groups of montmorillonite are located in the middle octahedral sheet of the 2:1-unit layer and are not exposed at the surface of the intact clay particle, variations in pH would have little effect on this mechanism of adsorption of reovirus on montmorillonite.

The relative adsorption of various types of coxsackievirus on the mineral, bentonite (which contains primarily montmorillonite), was differentially affected by variations in pH. This variation in adsorption apparently reflected differences in the pI among the coxsackievirus types and again indicated the importance of electrostatic interactions as a mechanism in the adsorption of viruses on charged surfaces.[98]

Artificially modified charged surfaces may mimic some natural surfaces, and therefore they can serve as models to estimate the behavior (e.g, adsorption and persistence) of viruses in soils, sediments, and waters containing different types and concentrations of clay minerals. However, the unique surface characteristics of different clay mineral species *in situ* clearly differentiate these particles from relatively homogeneous membrane filters, cation and anion exchange resins, purified clay minerals, or modified silica. These difference must be considered in all extrapolations from model to ''real-world'' systems.

The CEC and AEC of clay minerals affect their adsorption interactions with net positively and net negatively charged organic compounds, respectively.[43,64] However, the CEC-AEC ratio of clays is probably more important in surface interactions between clays and charged biocolloids (e.g., viruses), as this ratio affects the net negativity of clays, and therefore, the ability of net negatively charged biocolloids to come close enough to the clays for adsorption to occur. Alteration of the CEC-AEC ratio of clay minerals has been shown to affect the adsorption of viruses. For example, pretreatment of kaolinite and montmorillonite with nutrient broth (average pI = pH 4.2) or egg albumin (pI = pH 4.6) reduced the adsorption of both coliphages T1 and T7 on each clay (coliphages T1 and, to a lesser extent, T7 adsorb primarily to positively charged sites on clay minerals[85]), with the reduction in adsorption on kaolinite being greater than on montmorillonite. These net negatively charges proteins (the pH of the clay-phage systems was 6.9) apparently blocked positively charged sites on the clays, and the effect was more pronounced with kaolinite with a CEC-AEC ratio of approximately 0.5 than with montmorillonite with a CEC-AEC ration of approximately 6.7. Conversely, lysozyme (pI = pH ll) enhanced the adsorption of both phages on both clays, probably because this net positively charged protein reduced the overall negativity of the clays, which in turn facilitated the approach of the net negatively charged coliphages (pI = pH 4.0) to the postively charged sites on the clays.[37,43,200]

The importance of the CEC-AEC ratio in the adsorption of viruses on clays was further demonstrated by the use of polyphosphates, which block postively charged sites (primarily edge sites) on clay minerals.[99,100,101] The adsorption of coliphages T1 and T7 on kaolinite was markedly reduced by pretreating the clay with sodium metaphosphate [$Na(PO_3)_{13}$], whereas the reduction in adsorption on similarly pretreated montmorillonite was significantly less. Furthermore, a nondialyzable but filterable (<0.2 μm) component in the supernatants from suspensions of montmorillonite (probably negatively charged colloidal particles of montmorillonite) reduced the adsorption of the phages on kaolinite but not on montmorillonite. The supernatants from suspensions of kaolinite had no such effect.[200] Inasmuch as the ratio of positively to negatively charged sites is larger with kaolinite than with montmorillonite, the greater reduction in the adsorption of the coliphages on kaolinite pretreated with $Na(PO_3)_{13}$ or with the montmorillonite supernatants was not unexpected. These data confirmed not only that these viruses adsorbed primarily to postively charged sites on the clays,[85] but also the importance of the CEC-AEC ratio of the clays in surface interactions. Conversely, the blockage of positively charged sites on kaolinite and montmorillonite by $Na(PO_3)_{13}$ or the montmorillonite supernatants had no effect on the adsorption of reovirus, which confirmed that adsorption of this virus was primarily to negatively charged sites on these clays.[70]

4. Hydrophobic Interactions

Hydrophobic interactions appear to influence the adsorption on and the elution from membrane filters, estuarine sediments, and probably also clay minerals of some viruses.[88,102,103] For example, detergents (e.g., cetyltrimethylammonium bromide) and some ions (e.g., tricholoroacetate, thiocyanate) enhanced the efficiency of elution of poliovirus from membrane filter.[104,105] These and some other low-molecular-weight ionic compounds, termed chaotropic agents, presumably decrease the structure of water around hydrophobic groups, thereby decreasing the entropy that restricts the solubilization of hydrophobic substances in water. In contrast, antichaotropic ions (e.g., acetate, sulfate, calcium, and magnesium) apparently increase the structure of water, which reduces the ability of aqueous solutions to accommodate nonpolar groups and favors hydrophobic interactions.[106,107]

Hydrophobic interactions may have been involved in the adsorption of HSV-1 which has a phospholipid envelope, on clay minerals. The adsorption of this virus on either kaolinite or montmorillonite showed no correlation with the CEC, AEC, CEC-AEC ratio, or specific surface area of these clays.[201] Furthermore, pretreatment of the clay minerals with bovine serum albumin had no effect on the subsequent adsorption of HSV-1.

In the monitoring of water quality, bound viruses must be effectively eluted, especially as the infectivity of some virus suspensions (e.g., of poliovius, reovirus, and coliphage T4) has been shown to increase after the desorption of the viruses from particulate adsorbents.[11,73,108] Although hydrophobic interactions do not appear to be primary mechanisms in the adsorption of enteric viruses on estuarine sediments[103] and clay minerals,[70] the use of chaotropic agents in conjunction with other soluble organics for the elution of viruses from naturally occurring particulates requires further investigation.

5. Cation Bridging

The possibility of cation bridges mediating the adsorption of organic anions on cation exchange sites of clay minerals was suggested in the early 1930's. This theory was subsequently extended to interactions between clay minerals and viruses by some investigators.[64,76] For example, the adsorption of coliphage T2 on kaolinite, montmorillonite, and illite and of poliovirus type I (Sabin) on kaolinite was suggested to occur through a cation-mediated reduction of the net electonegative potential of the viruses and the clays, which permitted the particles to approach each other sufficiently to effect the formation of a clay-cation-virus bridge.[76] However, evidence for direct cation bridging as a mechanism in the adsorption of viruses on clay minerals is not convincing: not only are clay-associated cations usually hydrated under natural and even most laboratory conditions, but the neutralization of the charges of two separate negatively charged particles by a single mutivalent cation is unlikely.[43] Adsorption probably occurred (1) indirectly through protonation of the viruses at the clay-virus interface or (2) directly through electrostatic interactions between the net negatively charged viruses and exposed aluminum ions at broken edges of the clay or between the positively charged tail fibers of the coliphage and cation exchange sites on the clay.

B. Biological Factors

1. Proteins

Clay minerals in soils and aquatic systems may be associated with monomeric and polymeric organics and inorganics.[43] Consequently, viruses compete with proteins and other organics in the environment for adsorption sites on clays and other particulates, which could affect their surface interactions and persistence. For example, egg and bovine albumin and fetal bovine serum decreased the adsorption of coliphages and murine enteroviruses on clay minerals.[74,76] In contrast, the coating of bentonite (montmorillonite) with rabbit serum enhanced the adsorption of lipid-containing viruses (i.e., herpes simplex virus[HSV], influenza virus, and rubella virus) which in turn decreased the rate of loss of infectivity of these viruses.[110] Other studies showed that preadsorption of bovine serum albumin on montmorillonite or kaolinite had no effect on the adsorption of HSV-1, but increased the amount of virus bound (i.e., the quantity of virus remaining on the clays after repeated washing of the clay-virus complexes with distilled water).[14,201]

These and other studies that have attempted to simulate environmental conditions have employed either heterogeneous mixtures of proteins or only a few individual proteins with similar biophysical characteristics. Hence, the mechanisms involved in the adsorption of viruses at protein-clay interfaces have been difficult to interpret. Consequently, defined proteins (chymotrypsin [pI = pH 8.1 to 8.6], lysozyme [pI = pH 11], and ovalbumin [pI = pH 4.6]) were used to examine the effects of these organics on the adsorption of a human enteric virus on clay minerals. The binding of chymotrypsin, lysozyme or ovalbumin on clay minerals before exposure of the clay-protein complex to reovirus type 3 reduced the adsorption of the virus on montmorillonite and, with the exception of lysozyme, on kaolinite. This may be due to blocking by the proteins of negatively charged sites on the clays to which the protonated virus apparently adsorbed.[111]   As the adsorption of lysozyme on kaolinite does not appear to be the result of a cation exchange reaction,[112] lysozyme did not block the adsorption of the reovirus to the negatively charged sites on kaolinite. Although the pI of ovalbumin was below the pH of the experimental systems (pH, 5.6 to 6.3), this relatively large globular molecule (molecular weight, 43,500 to 54,000) was probably adsorbed by van der Waal forces and hydrogen bonding, rather than by cation exchange,[112] and it effectively blocked the adsorption of the reovirus on the clays by steric interference.

Chymotrypsin and lysozyme markedly decreased the infectivity of reovirus in distilled water. However, this effect was less pronounced when the proteins were bound on montmorillonite.[111] Adsorption of the net positively charged proteins on the net negatively charged reovirus in the absence of clay probably reduced the adsorption of the virus on the test cells, thereby decreasing the infectivity titer. The lower inhibition of the infectivity titer of the virus when the proteins were complexed with montmorillonite may have been the result of conformational changes and the consequent alteration of the biophysical characteristics of the proteins in the bound state. Conformational changes in the proteins bound on clay mineral have been reported.[42,43,113]

These studies suggest that the adsorption and persistence of viruses in natural environments depend on the type of clay mineral and on the type of organic matter associated with the clays with which viruses come in contact.

2. Virus Heterogeneity

More than 20 years ago, some molecular virologists proposed the existence of heterogeneous particles within an animal virus population to explain the occurrence of nonadsorbable viruses in cell culture.[114,115] Heterogeneity within a viral population has now been recognized in many virus groups, as shown, for example, by the differences in the degree of neuropathogenicity caused in mice by clones of vaccinia virus and the qualitative differences in the capsid polypeptides of the progeny of Venezuelan encephalitis and measles viruses from single host cell systems.[116-118] Noninfectious (or potentially infectious)[119] or defective interfering (DI)[120,121] particles are also recognized in populations of reovirus.

A heterogeneity in the net charge of individual virus particles within a population of poliovirus was indicated by the results of migration studies of the virus in soil.[122] The percentage of virus adsorbed on soil at different depths was unchanged regardless of the initial virus concentration added, which was attributed to differences in the net negative charge of individual virus particles within the population. Hence, dilution of the virus populations did not change the relative numbers of differently charged particles and, consequently, the percentage of virus adsorbed at each soil depth. Burge and Enkiri,[123] using adsorption kinetics, showed that at least two populations of coliphage ϕX174 were distinguishable by the rate at which they were adsorbed on soil.

Experiments designed to investigate the effect of population heterogeneity on the adsorption of reovirus type 3 on clay minerals showed that more than 90% of infectious reovirus particles that were not adsorbed initially on kaolinite and, to a lesser extent, on montmorillonite were adsorbed when fresh quantities of each clay were added to the virus suspension.[70] The removal by fresh clay, but not by the initial clay, of large quantities of infectious virus may have been the result of the blockage of adsorption sites on the initial clay by preferentially adsorbed noninfectious or DI reovirus particles. Some DI reovirus particles contain a mutation in the S4 RNA segment, which encodes for the σ3 outer capsid polypeptide.[120] The σ3 virion polypeptide is the most abundant virion protein and is a major constituent of the reovirus outer capsid.[124] A difference in the distribution of charge on the reovirus particles, as the result of quantitative differences in capsid polypeptides, may explain, in part, the difference in behavior (e.g., adsorption to fresh but not to the initial clays) of individual virions with the population.

A heterogeneity among viral particles was also observed in a population of HSV-1. Approximately 10% of the input viral population was not adsorbed on either kaolinite or montmorillonite, even when the nonadsorbed viruses were exposed to fresh clays. The nonadsorbed HSV-1 particles were capable of infecting their host cells, indicating that this subpopulation was not composed of DI particles but of infective virions that had surface characteristics different from those of the other 90% of the population.[37,201] Sykes and Williams[125] reported that 50 to 60% of the input population of bacteriophage f6 of *Streptomyces* was not adsorbed on kaolinite and that this nonadsorbed population is infectious.

3. Differential Adsorption of Mixed Virus Populations

Sewage and contaminated receiving waters contain multiple virus populations, as many enteric virus groups and types within each group are routinely isolated from patients (Table 3). Consequently, the occurrence of numerous types of enteric viruses in a single environmental sample is not unexpected.[25,32]

Although the adsorption of populations of individual enteric viruses and other virus groups on clay minerals has been studied fairly extensively (Table 2), the relative adsorption of different viruses from samples containing multiple viral populations has been addressed only recently and then only with dual populations.[88] Studies on the competitive adsorption of coliphage T1 and reovirus type 3 from a mixture of these viruses on kaolinite and montmorillonite indicated a specificity for adsorption sites on the clays for each virus, confirming the results of adsorption studies with individual populations that showed that the coliphage adsorbed primarily to positively and the reovirus primarily to negatively charged sites on the clays.[70,85] Competitive adsorption between related (e.g., coxsackieviruses B4 and B5) or unrelated viruses (e.g., coliphage T7 and reovirus type 3) that presumably adsorb to the same sites on clays has not been studied. More studies on the adsorption and persistence of individual component viral populations within multiple populations in soils and sediments should be conducted, as the extent of adsorption of such multiple virus populations on clays, and their subsequent persistence, is important in the interpretation of results obtained from environmental monitoring and impact studies.

4. Location of Viruses on Clays

The location of adsorbed viruses on clay minerals and other net negatively charged particles apparently depends on the type of adsorbent and the class or type of virus (Table 4). Coliphage T1, for example, appears to adsorb by an edge- (tail of virus) to-edge (kaolinite and montmorillonite)[126] as well as by a face- (head of virus) to-face (montmorillonite)[127] arrangement. Electrostatic forces were probably involved in the adsorption of the positively charged tail fibers of coliphage T1 to negatively charged sites on the clays (tail fibers of most tailed bacteriophages, including T7, T3, P22, and φII, are positively charged[128]). However, as data from equilibrium adsorption isotherms indicated that the adsorption of coliphage T1 on kaolinite and montmorillonite occurs primarily to positively charged sites on the clay,[85] several mechanisms of adsorption are apparently involved. Face-to-face attachment of coliphage T1 (a net negatively charged virus) to negatively charged sites on kaolinite probably occurs by cation exchange after protonation, a mechanism similar to that described for the adsorption of reovirus on clays.[70]

Table 3
ISOLATION OF ENTERIC VIRUSES FROM PATIENT POPULATIONS IN EASTERN QUEENS, NASSAU, AND SUFFOLK COUNTIES, N.Y.

Virus	No. of isolates/month							
	May	June	July	Aug.	Sept.	Oct.	Nov.	Dec.
Coxsackievirus								
A[a]	1	2	5	8	1	—[b]	—	—
A-9	—	—	1	—	—	1	—	—
A-16	—	—	1	—	—	1	—	—
B-2	—	—	—	—	11	1	3	—
B-4	—	—	—	—	1	—	—	—
B-5	—	—	14	18	10	2	—	—
Echovirus								
5	—	—	—	—	1	2	1	—
6	—	—	—	—	—	—	4	—
7	—	—	—	—	2	—	3	—
9	1	—	2	4	3	2	2	—
11	—	—	4	—	15	—	6	—
12	—	—	—	—	—	12	—	—
17	—	—	—	—	—	—	—	1
20	—	—	—	—	—	—	4	—
22	—	—	—	—	—	—	6	2
30	—	—	—	—	1	1	1	1
Poliovirus (Lsc)								
1	—	—	1	—	—	—	1	1
2	—	—	1	—	1	4	1	—
3	—	—	1	—	—	5	1	—
Untyped enterovirus	4	—	8	40	15	10	—	3
Adenovirus								
1	1	1	4	5	1	—	1	—
2	4	2	—	3	—	1	5	8
5	—	1	—	1	—	—	—	—
7	1	—	3	—	1	—	—	—
16	—	1	3	—	—	—	—	—
19	—	—	—	1	2	—	—	2
Untyped adenovirus	—	1	5	2	—	—	—	4

[a]   Newborn mice.
[b]   Not isolated.

Adapted from Lipson, S.M. and Szabo, K., Virus Update, May to December 1984, Virology Laboratory, Department of Pathology and Laboratories, Nassau County Medical Center, East Meadow, N.Y.[203]

The protection by clay minerals of viruses against inactivation may be related to the location of the adsorbed viruses on the clays. For example, Stagg et al.[129] reported that coliphage MS-2 (a tailless phage) was protected from the virucidal action of chlorine when adsorbed on montmorillonite, but Boardman and Sproul[130] did not observe this protective effect on coliphage T7 associated with kaolinite. If coliphage T7 was adsorbed on kaolinite by a tail-to-face mode of attachment, its head may have been exposed to the inactivating effect of the chlorine.

However, adsorption itself may adversely affect the survival of some tailed bacteriophages. For example, inactivation of the bacteriophages of *Arthrobacter* occurred following an edge-to-face attachment on montmorillonite, and tail damage was observed by electron micros-

Table 4
LOCATION OF VIRUSES ADSORBED TO CLAY
MINERALS AND OTHER PARTICULATES

Virus	Adsorbent	Mode of attachment (virus to clay)	Ref.
Cyanophage AS-1	Montmorillonite	Tail[a] to edge[b]	155
Coxsackievirus A9	Montmorillonite	Virion to face[c]	110
Coliphage T1, T7	Kaolinite	Tail to edge	126
	Montmorillonite	Tail to edge	
Coliphage T1	Montmorillonite	Tail to face	127
		Head to face	
Coliphage T4	Celkate T-21	Head to face	195
Tobacco mosaic virus	Kaolinite	Virion to face	132
Bacteriophage AN31s-1 of *Arthrobacter*	Montmorillonite soil particulates	Tail to face	84
Reovirus type 3	Kaolinite	Virion to face	131

[a]   Bacteriophage attached by its tail.
[b]   Edge of clay mineral.
[c]   Planar surface of clay mineral.
[d]   Bacteriophage attached by its head.

copy.[84] However, coliphages T1 and T7 were protected against thermal inactivation in distilled water by montmorillonite and kaolinite,[200] indicating that tail damage resulting from adsorption on clays may occur with some bacteriophages but not with others.

Nonlipid and lipid-containing viruses (e.g., coxsackie-, echo-, and adenoviruses; and influenza, parainfluenza, HSV, and rubella viruses, respectively) were protected against inactivation during transport by the use of a medium containing bentonite. Electron microscopy indicated that coxsackievirus A9 was adsorbed to the bentonite by a face-to-face attachment of monodispersed viral particles.[110] This mode of attachment was similar to that of reovirus[131,202] and tobacco mosaic virus on clays,[132] suggesting that tailless viruses having a low pI adsorb to crystalline clay minerals by similar mechanisms (i.e., protonation followed by cation exchange).[31,70]

Electron microscopy also indicated that the adsorption of poliovirus type 1 (Sabin) on montmorillonite occurs at the edge surfaces of the clay, again suggesting that adsorption resulted from electrostatic attraction between the net negatively charged virus and the net positively charged edges of the clay.[133] However, this proposed mechanism was not confirmed by competitive adsorption experiments (e.g., pretreatment of the clay with polyphosphate ions).

5. Modification of Virus Infectivity

The association of some viruses with clay minerals may not only prolong their infectivity, but it may also enhance their infectivity as well as aid in their isolation and purification. For example, clay minerals have been effectively used in the isolation of plant viruses from sap;[134,135] in the isolation and purification of RNA from plant viruses;[136,137] in the enhancement of infectivity of some plant and animal viruses;[11,73,74,76,110,139-141] and in the enhancement of transfection by poliovirus RNA.[137,142] The implications of the results of these in vitro studies for the promotion of disease are obvious, as the transmission of water- (e.g., enterovirus 72 and rotavirus) and soil- (e.g., wheat and barley mosaic viruses) borne viruses to humans and plants, respectively, may not only be enhanced but also prolonged.[12,31,32,143-146]

Some mechanisms to explain how clay minerals enhance the specific infectivity (i.e., the measurable plaque-forming units [pfu] or tissue culture infective dose-50 [TCID$_{50}$] relative to the number of virions in the inoculum) of viruses have been proposed, although there is a paucity of definitive studies. For example, it has been suggested that clay minerals and other particulates increase the infectivity of some viruses by adsorbing and inactivating RNase or DNase,[133,134,136,198] by abrading the surface of the host cell and thereby facilitating penetration,[137] and by enhancing the transport of clay-bound nucleic acids[142] or viruses[11,135] to host cells, Scanning electron microscopy (SEM) indicated that kaolinite enhanced the efficiency of plating of reovirus type 3 by increasing the rate and number of particles reaching the test cell surface, primarily as the result of the more rapid settling of the kaolinite-virus complex on the cell monolayer.[202] However, extrapolation of this possible mechanism to explain the enhanced specific infectivity of other particulate-virus-cell systems should be made cautiously, as the adsorption of viruses to particulates differs with both the type of virus and the type of adsorbent.[31,70,85,147]

The influence of clay minerals on transduction in natural environments has not been demonstrated. However, the influence of clay minerals in soils and sediments on other forms of gene transfer has been indicated.[148-151] Consequently, the possibility of clay minerals mediating the transfer of novel genes, via transduction, into autochthonous microbes must be considered and investigated thoroughly.[149]

The apparent persistence of viruses of plants, animals, and microbes in soil and other natural ecosystems implies the possibility of long-term transmission of these infectious agents to their hosts. Furthermore, levels of viruses that are initially too low to cause an infection may develop enhanced infectivity as a result of their sorption on clay minerals and subsequent transport to susceptible host cells.

VI. CONCLUSION

Clay minerals were suggested as adsorbents for viruses as early as the 1920s. However, it has only been within the last two decades that the mechanisms that mediate the adsorption of viruses on clays have been studied. It is now apparent that no single mechanism is responsible for the adsorption of all viruses on all clay minerals, although some mechanisms (e.g., hydrogen bonding, hydrophobic interactions, and anion or cation exchange) may predominate in many clay-virus systems. Adsorption of viruses on clay minerals appears to be the result of several interrelated mechanisms, each of which is dependent on the physicochemical characteristics of the adsorbent (i.e., the clay mineral) and the biophysical characteristics of the adsorbate (i.e, the virus).

The persistence in and transmission of viruses through soils, sediments, and waters is related, in part, to their adsorption on clays and other particulates in these environments. An understanding of the mechanisms involved in these sorptive interactions is necessary to explain the behavior of pathogenic viruses in these environments and to enhance the efficacy of strategies to eliminate or decrease virus transmission to the human host.

ACKNOWLEDGMENTS

This work was supported, in part, by grants R809067 and CR812484 from the U.S.E.P.A. The views expressed are not necessarily those of the Agency.

REFERENCES

1. **Shope, R. E.**, Transmission of viruses and epidemiology of viral infections, in *Viral and Rickettsial Infections of Man*, Horsfall, F. L., Jr. and Tamm, I., Eds., Lippincott, Philadelphia, 1965, 385.

2. **Pass, R. F., Cecelia, S. C., Reynolds, D. W., and Pulhill, R. B.**, Increased frequency of cytomegalovirus infection in children in group day care, *Pediatrics*, 74, 121, 1984.

3. **Ray, C. G.**, Viruses, rickettsia, and chlamydiae, in *Manual of Clinical Microbiology*, 4th ed., Lennette, E. H., Balows, A., Hansley, W. J., Jr., and Shadomy, H. J., Eds., American Society for Microbiology, Washington, D.C., 1985, 705, 743, 755, 785, 705.

4. **Berg, G.**, *Viral Pollution of the Environment*, CRC Press, Boca Raton, Fla., 1983.

5. **Berg, G., Bodily, H. L., Lennette, E. H., Melnick, J. L., and Metcalf, T. G., Eds.**, *Viruses in Water*, American Public Health Association, Washington, D.C., 1976.

6. **Bitton, G.**, *Introduction to Environmental Virology*, John Wiley & Sons, New York, 1980.

7. **Melnick, J. L. and Gerba, C. P.**, The ecology of enteroviruses in natural waters, *Crit. Rev. Environ. Control*, 10, 65, 1980.

8. **Melnick, J. L., Ed.**, Enteric viruses in water, *Monogr. Virol.*, 15, 1, 1984.

9. **Gerba, C. P. and Schaiberger, G. E.**, Effect of particulates on virus survival in seawater, *J. Water Pollut. Control Fed.*, 47, 93, 1975.

10. **Kapusinski, R. B. and Mitchell, R.**, Processes controlling virus inactivation in coastal waters, *Water Res.*, 14, 363, 1980.

11. **Lipson, S. M. and Stotzky, G.**, Infectivity of reovirus adsorbed to homoionic and mixed-cation clays, *Water Res.*, 19, 227, 1985.

12. **Miyamoto, Y.**, Further evidence for the longevity of soil-borne plant viruses adsorbed by soil particulates, *Virology*, 9, 290, 1959.

13. **Schaub, S. A. and Sorber, C. A.**, Viruses on solids in water, in *Viruses in Water*, Berg, G., Bodily, H. L., Lennette, E. H., Melnick, J. L., and Metcalf, T. G., Eds., American Public Health Association, Washington, D.C., 1976, 128.

14. **Stotzky, G., Schiffenbauer, M., Lipson, S. M., and Yu, B. H.**, Surface interactions between viruses and clay minerals and microbial cells: mechanisms and implications, in *Viruses and Wastewater Treatment*, Goddard, M. and Butler, M. Eds., Pergamon Press, Oxford, 1981, 199.

15. **Goldfield, M.**, Epidemiological indicators for transmission of viruses by water, in *Viruses in Water*, Berg, G., Bodily, H. L., Lennette, E. H., Melnick, J. L., and Metcalf, T. G., Eds., American Public Health Association, Washington, D.C., 1970, 70.

16. **Mosley, J. W.** Transmission of viral diseases by drinking water, in *Transmission of Viruses by the Water Route*, Berg, G., Ed., John Wiley & Sons, New York, 1967.

17. **Morens, D. M., Zweighaft, R. M., Vernon, T. M., Gary, G. W., Eslien, J. J., Wood, B. T., Holman, R. C., and Dolin, R.**, A waterborne outbreak of gastroenteritis with secondary person-to-person spread, *Lancet*, 1, 964, 1979.

18. **Hopkins, R. S., Gaspard, G. B., Williams, F. P., Jr., Karlin, R. J., Cukor, G., and Blacklow, N. R. A.**, Community waterborne gastroenteritis outbreak: evidence for rotavirus as the agent, *Amer. J. Public Health*, 74, 263, 1984.

19. **Baron, R. C., Murphy, F. D., Greenberg, H. B., Davis, C. E., Bregman, D. J., Gary, G. W., Hughes, J. M., and Schonberger, L. B.**, Norwalk gastroenteritis illness. An outbreak associated with swimming in recreational lake and secondary person-to-person transmission, *Amer. J. Epidemiol.*, 115, 163, 1982.

20. **Koopman, J. S., Eckert, E. A., Greenberg, H. B., Strohm, B. C., Isaccson, R. E., and Morita, A. S.**, Norwalk virus enteric illness acquired by swimming exposure, *Amer. J. Epidemiol.*, 115, 173, 1982.

21. **Wilson, R., Anderson, L. J., Holman, R. C., Gary, G. W., and Greenberg, H. B.**, Water-borne gastroenteritis due to the Norwalk agent. Clinical and epidemiological investigations, *Am. J. Public Health*, 72, 72, 1982.

22. **Gerba, C. P., Keswick, B. H., Dupont, H. L., and Fields, H. A.**, Isolation of rotavirus and hepatitis A virus from drinking water, *Monogr. Virol.*, 15, 119, 1984.

23. **Scarpino, V. P.**, *Human Enteric Viruses and Bacteriophages as Indicators of Sewage Pollution*, Pergamon Press, New York, 1975.

24. **Irving, L. G. and Smith, F. A.**, One-year survey of enteroviruses, adenoviruses, and reoviruses isolated from effluent at an activated-sludge purification plant, *Appl. Environ. Microbiol.*, 41, 51, 1981.

25. **Lund, E., Lydholm, B., and Nielsen, A. L.**, Demonstration of virus in treated wastewater, sludges, and liquid manures, *Monogr. Virol.*, 15, 87, 1984.

26. **Goyal, S. M., Adams, W. N., O'Malley, M. L., and Learl, D. W.**, Human pathogenic viruses at sewage sludge disposal sites in the middle Atlantic region, *Appl. Environ. Microbiol.*, 48, 758, 1984.

27. **Eisenhardt, A., Lund, E., and Nissen, B.**, The effect of sludge digestion on virus infectivity, *Water Res.*, 11, 579, 1977.

28. **Gerba, C. P.,** Virus survival in wastewater treatment, in *Viruses and Wastewater Treatment*, Goddard, M. and Butler, M., Eds., Pergamon Press, New York, 1981.

29. **Turk, C. A., Moore, B. E., Sagik, B. P., and Sorber, C. A.,** Recovery of indigenous viruses from wastewater sludges, using a bentonite concentration procedure, *Appl. Environ. Microbiol.*, 40, 423, 1980.

30. **Melnick, J. L., Gerba, C. P., and Wallis, C.,** Viruses in water, *Bull. WHO*, 56, 499, 1978.

31. **Lipson, S. M. and Stotzky, G.,** Adsorption of viruses to particulates: possible effects on virus persistence, in *Virus Ecology*, Misra, A. and Polasa, H., Eds., South Asian Publ., New Delhi, 1984, 165.

32. **Rao, V. C., Seidel, K. M., Goyal, S. M., Metcalf, T. G., and Melnick, J. L.,** Isolation of enteroviruses from waters, suspended solids, and sediments from Galveston Bay: survival of poliovirus and rotavirus adsorbed to sediments, *Appl. Environ. Microbiol.*, 48, 404, 1984.

33. **Bitton, G., Pancorbo, O. C., and Farrah, S. R.,** Virus transport and survival after land application of sewage sludge, *Appl. Environ. Microbiol.*, 47, 905, 1984.

34. **Metcalf, T. G., Rao, V. C., and Melnick, J. L.,** Solid-associated viruses in a polluted estuary, *Monogr. Virol.*, 15, 97, 1984.

35. **Babich, H. and Stotzky, G.,** Reductions in inactivation rates of bacteriophages by clay minerals in lake water, *Water Res.*, 14, 185, 1980.

36. **Denis, F.,** Epidemiological consequences of virus contamination of waters, *Rev. Epidemiol. Med. Soc. Sante Publ.*, 21, 273, 1973.

37. **Stotzky, G.,** Surface interactions between clay minerals and microbes, viruses, and soluble organics, and the possible importance of these interactions to the ecology of microbes in soil, in *Microbial Adhesion to Surfaces*, Berkeley, R. C. W., Lynch, J. M., Melling, J., Rutter, P. R., and Vincent, B., Eds., Ellis Harwood, Chichester, England, 1980, 231.

38. **Sobsey, M. D.,** Human viruses in the marine environment, in *Natural Toxins and Human Pathogens in the Marine Environment: Problems and Research Needs*, Colwell, R. R., Ed., University of Maryland, College Park, 1982, 12.

39. **Bitton, G. and Mitchell, R.,** Effect of colloids on the survival of bacteriophages in seawater, *Water Res.*, 8, 227, 1974.

40. **Metcalf, T. G., Mullin, B., Eckerman, D., Moulton, E., and Larkin, E. P.,** Bioaccumulation and depuration of enteroviruses by the soft-shelled clam, *Mya arenaria*, *Appl. Environ. Microbiol.*, 38, 275, 1979.

41. **Filip, Z.,** Clay minerals as a factor influencing the biochemical activity of soil microorganisms, *Folia Microbiol.*, 18, 56, 1973.

42. **Stotzky, G.,** Activity, ecology, and population dynamics of microorganisms in soil, *Crit. Rev. Microbiol.*, 2, 59, 1972.

43. **Stotzky, G.,** Influence of soil mineral colloids on metabolic processes, growth, adhesion, and ecology of microbes and viruses, in *Interactions of Soil Minerals with Natural Organics and Microbes*, Huang, P. M., et al., Eds., Soil Science Society of America, Madison, Wis., 1986, 305.

44. **Burns, R. G.,** Extracellular enzyme-substrate interactions in soil, in *Microbes in Their Natural Environments*, Slater, J. H., Whittenburg, R., and Wimpenny, J. W. I., Eds., Cambridge University Press, Cambridge, England, 1983, 149.

45. **Haska, G.,** Influence of clay minerals on sorption of bacteriolytic enzymes, *Microb. Ecol.*, 1, 234, 1975.

46. **Morgan, H. W. and Corke, C. T.,** Adsorption, desorption, and activity of glucose oxidase on selected clay species, *Can. J. Microbiol.*, 22, 684, 1976.

47. **Novakova, J. and Sisa, R.,** Effect of clays on the cellulolytic activity of soil, *Zentralbl. Mikrobiol.*, 139, 505, 1984.

48. **Edzwald, J. K., Upchurch, J. B., and O'Melis, C. R. O.,** Coagulation in estuaries, *Environ. Sci. Technol.*, 8, 58, 1974.

49. **Griffin, G. M.,** Regional clay-mineral facies. Products of weathering intensity and current distribution in the northeastern Gulf of Mexico, *Bull. Geol. Soc. Am.*, 73, 737, 1962.

50. **Hathaway, J. C.,** Regional clay mineral facies in estuaries and continental margins of the United States, in *Environmental Framework of Coastal Plain Estuaries*, Memoir 113, Nelson, B. W., Ed., Geological Society of America, 1972, 293.

51. **O'Brien, N. R. and Ali, B. A.,** Clay mineral composition of bottom sediments: Western Great South Bay and South Oyster Bay, Long Island, New York, *Mar. Sediment.*, 10, 107, 1974.

52. **Weaver, C. E.,** The distribution and identification of mixed-layer clays in sedimentary rocks, *Amer. Mineral.*, 41, 202, 1965.

53. **Manheim, F. T. and Hathaway, J. C.,** Suspended matter in surface water of the northern Gulf of Mexico, *Limnol. Oceanogr.*, 17, 17, 1972.

54. **Grim, R.,** *Clay Mineralogy*, McGraw-Hill, New York, 1968.

55. **Mysels, K. J.,** *Introduction to Colloid Chemistry*, John Wiley & Sons, New York, 1967.

56. **Swartzen-Allen, S. and Matijevic, E.,** Surface and colloid chemistry of clays, *Chem. Rev.*, 74, 385, 1974.

226 *Human Viruses in Sediments, Sludges, and Soils*

---

57. **Van Olphen, H.,** *An Introduction to Clay Colloid Chemistry*, Wiley Interscience, New York, 1977.
58. **Falconner, J. and Mattson, S.,** The law of soil colloid behavior. XIII. Osmotic inhibition, *Soil Sci.*, 36, 317, 1933.
59. **Norrish, K.,** The swelling montmorillonite, *Disc. Faraday Soc.*, 118, 120, 1954.
60. **Parks, G. A.,** The isoelectric points of solid oxides, solid hydroxides, and aqueous hydroxo complex systems, *Chem. Rev.*, 65, 177, 1965.
61. **Nelson, R. A. and Hendricks, S. B.,** Specific surface of some clay minerals, soils, and soil colloids, *Soil Sci.*, 56, 285, 1944.
62. **Bolland, M. D. A., Posner, A. M., and Quirk, J. P.,** pH-independent and pH-dependent surface charges on kaolinite, *Clays Clay Min.*, 28, 412, 1980.
63. **Street, N. and Buchanan, A. S.,** The ζ-potential of kaolinite particles, *Austr. J. Chem.*, 9, 450, 1956.
64. **Theng, B. K. G.,** *Formation and Properties of Clay-Polymer Complexes*, Elsevier, New York, 1979.
65. **Chassin, P.,** Adsorption du glycolle par la montmorillonites, *Bull. Groupe Fr. Argiles*, 21, 71, 1969.
66. **Harter, R. D. and Stotzky, G.,** Formation of clay-protein complexes, *Soil Sci. Soc. Am. Proc.*, 35, 385, 1971.
67. **Stern, O.,** Zur Theorie der elecktrolytischen Doppelschicht, *Z. Electrochem.*, 30, 508, 1924.
68. **Riddick, T. M.,** *Control of Colloid Stability Through Zeta Potential*, Livingston, Wynnewood, Pa., 1968.
69. **Singley, J. E., Birkner, F., Chen, C.-L. Cohen, J. M., Ockershausen, K. E., Shull, K. E., Weber, W. J., Metijevic, E., and Packham, R. F.,** State of the art coagulation. Mechanisms and stoichiometry, committee report, *J. Am. Water Works Assoc.*, 63, 99, 1971.
70. **Lipson, S. M. and Stotzky, G.,** Adsorption of reovirus to clay minerals: effects of cation-exchange capacity, cation saturation, and surface area, *Appl. Environ. Microbiol.*, 46, 673, 1983.
71. **Santoro, T. and Stotzky, G.,** Effect of electrolyte composition and pH on the particle size distribution of microorganisms and clay minerals as determined by the electrical sensing zone method, *Arch. Biochem. Biophys.*, 122, 664, 1967.
72. **Santoro, T. and Stotzky, G.,** Sorption between microorganisms and clay minerals as determined by the electrical sensing zone particle analyzer, *Can. J. Microbiol.*, 14, 299, 1968.
73. **Moore, B. E., Sagik, B. P., and Malina, J. F., Jr.,** Viral association with suspended solids, *Water Res.*, 9, 197, 1975.
74. **Schaub, S. A. and Sagik, B. P.,** Association of enteroviruses with natural and artificially introduced colloidal solids in water and infectivity of solids-associated virions, *Appl. Environ. Microbiol.*, 30, 212, 1975.
75. **Allison, A. C. and Valentine, R. C.,** Virus particle adsorption. III. Adsorption of viruses by cell monolayers and effects of some variables on adsorption, *Biochim. Biophys. Acta*, 40, 400, 1960.
76. **Carlson, G. E., Jr., Woodard, F. E., Wentworth, D. E., and Sproul, O. J.,** Virus inactivation on clay minerals in natural waters, *J. Water Pollut. Control Fed.*, 40, R98, 1968.
77. **Frazer, D. and Fleischman, C.,** Interaction of mycoplasma with viruses. I. Primary adsorption of virus is ionic in mechanism, *J. Virol.*, 14, 1069, 1974.
78. **Puck, T. T., Green, A., and Cline, J.,** The mechanism of virus attachment to host cells, *J. Exp. Med.*, 93, 65, 1951.
79. **Puck, T. T. and Sagik, B. P.,** Virus and cell interaction with ion exchangers, *J. Exp. Med.*, 97, 807, 1953.
80. **McLaren, A. D. and Esterman, E. F.,** Influences of pH on the activity of chymotrypsin at a solid-liquid interface, *Arch. Biochem. Biophys.*, 68, 157, 1957.
81. **Burge, W. D. and Enkiri, N. K.,** Virus adsorption by five soils, *J. Environ. Qual.*, 7, 73, 1978.
82. **Drewry, W. A. and Eliassen, R. E.,** Virus movement in groundwater, *J. Water Pollut. Control Fed.*, 40, 257, 1968.
83. **Funderburg, S. W., Moore, B. E., Sagik, B. P., and Sorber, C. A.,** Viral transport through soil columns under conditions of saturated flow, *Water Res.*, 15, 703, 1981.
84. **Ostle, A. G. and Holt, J. G.,** Elution and inactivation of bacteriophages on soil and cation exchange resin, *Appl. Environ. Microbiol.*, 38, 59, 1979.
85. **Schiffenbauer, M. and Stotzky, G.,** Adsorption of coliphages T1 and T7 to clay minerals, *Appl. Environ. Microbiol.*, 43, 590, 1982.
86. **Schiffenbauer, M. and Stotzky, G.,** Adsorption of coliphages T1 and T7 to host and non-host microbes and to clay minerals, *Curr. Microbiol.*, 8, 245, 1983.
87. **Yu, B. H. and Stotzky, G.,** Adsorption and binding of herpesvirus hominis type 1 (HSV-1) by clay minerals, *Abstr. Annu. Meet. Am. Soc. Microb.*, 188, 1979.
88. **Lipson, S. M. and Stotzky, G.,** Specificity of virus adsorption to clay minerals, *Can. J. Microbiol.*, 31, 50, 1985.
89. **Gerba, C. P., Goyal, S. M., Hurst, C. J., and LaBelle, R. L.,** Type and strain dependence of enterovirus adsorption to activated sludge, soils, and estuarine sediments, *Water Res.*, 14, 1197, 1980.

90. **Goyal, S. M., and Gerba, C. P.**, Comparative adsorption of human enteroviruses, simian rotavirus, and selected bacteriophages to soils, *Appl. Environ. Microbiol.*, 38, 241, 1979.
91. **Garelick, H. and Scutt, J. E.**, Concentration of virus from tap water at ambient salt and pH levels using positively charged filter media, *Water Res.*, 15, 815, 1981.
92. **Sobsey, M. D. and Jones, B. L.**, Concentration of poliovirus from tap water using positively charged microporous filters, *Appl. Environ. Microbiol.*, 37, 588, 1979.
93. **Taylor, D. H., Bellamy, A. R., and Wilson, A. T.**, Interactions of bacteriophage R17 and reovirus type III with the clay allophane, *Water Res.*, 14, 399, 1978.
94. **Rose, J. B., Singh, S. N., Gerba, C. P., and Kelley, L. M.**, Comparison of microporous filters for concentration of viruses from wastewater, *Appl. Environ. Microbiol.*, 47, 989, 1985.
95. **Sobsey, M. D. and Hickey, A. R.**, Effects of humic and fulvic acids on poliovirus concentration from water by microporous filtration, *Appl. Environ. Microbiol.*, 49, 259, 1985.
96. **Sobsey, M. D. and Cromeans, T.**, Effect of bentonite clay solids on poliovirus concentration from water by microporous filter methods, *Appl. Environ. Microbiol.*, 49, 795, 1985.
97. **Zerda, K. S., Gerba, C. P., Hou, K. C., and Goyal, S. M.**, Adsorption of viruses to charge-modified silica, *Appl. Environ. Microbiol.*, 49, 91, 1985.
98. **Shirobokov, V. P.**, Differentiation of coxsackieviruses based on the character of adsorption onto bentonite, *Acta Virol.*, 12, 185, 1968.
99. **Lahav, H.**, Adsorption of sodium bentonite particles on *Bacillus subtilis*, *Plant Soil*, 17, 191, 1982.
100. **Marshall, K.**, Sorptive interactions between soil particles and microorganisms, in *Soil Biochemistry*, Vol. 2, McLaren, A. D. and Skujins, J., Eds., Marcel Dekker, New York, 1971, 429.
101. **Michaels, A. S.**, Deflocculation of kaolinite by the alkali polyphosphates, *Ind. Eng. Chem.*, 50, 951, 1958.
102. **Shields, P. A., Berenfeld, S. A., and Farrah, S. R.**, Modified membrane-filter procedure for concentration of enteroviruses from tap water, *Appl. Environ. Microbiol.*, 49, 453, 1985.
103. **Wait, D. A. and Sobsey, M. D.**, Methods for recovery of enteric viruses from estuarine sediments with chaotropic agents, *Appl. Environ. Microbiol.*, 46, 379, 1983.
104. **Farrah, S. R., Shah, S. O., and Ingram, L. O.**, Effect of chaotropic and antichaotropic agents on elution of poliovirus adsorbed on membrane filters, *Proc. Natl. Acad. Sci. U.S.A.*, 78, 1229, 1981.
105. **Shields, P. A. and Farrah, S. R.**, Influence of salts and electrostatic interactions between poliovirus and membrane filters, *Appl. Environ. Microbiol.*, 45, 526, 1983.
106. **Hatefi, Y. and Hanstein, W. G.**, Solubilization of particulate proteins and nonelectrolytes by chaotropic agents, *Proc. Natl. Acad. Sci. U.S.A.*, 62, 1129, 1969.
107. **Hatefi, Y. and Hanstein, W. G.**, Destabilization of membranes with chaotropic ions, *Methods Enzymol.*, 31, 770, 1974.
108. **Cookson, J. T. and North, W. J.**, Adsorption of viruses on activated carbon. Equilibrium and kinetics of the attachment of *Escherichia coli* bacteriophage T4 on activated carbon, *Environ. Sci. Technol.*, 1, 46, 1967.
109. **Greenland, D. J.**, Interactions between clays and organic compounds in soils. I. Mechanisms of interaction between clays and defined organic compounds, *Soils Fert.*, 28, 415, 1965.
110. **Bishai, F. R. and Labzoffsky, N. A.**, Stability of different viruses in a newly developed transport medium, *Can. J. Microbiol.*, 20, 75, 1974.
111. **Lipson, S. M. and Stotzky, G.**, Effect of proteins on reovirus adsorption to clay minerals, *Appl. Environ. Microbiol.*, 48, 525, 1984.
112. **Albert, J. T. and Harter, R. D.**, Adsorption of lysozyme and ovalbumin by clay: effect of clay suspension pH and clay mineral type, *Soil Sci.*, 115, 130, 1973.
113. **Kirby, E. P. and McDevitt, P. J.**, The binding of bovine factor XII to kaolin, *Blood*, 61, 652, 1983.
114. **Dales, S.**, Penetration of animal viruses into cells, *Prog. Med. Virol.*, 7, 1, 1965.
115. **Joklik, W. K.**, The multiplication of poxvirus DNA, *Cold Spring Harbor Symp. Quant. Biol.*, 27, 199, 1962.
116. **Miller, C. A. and Raine, C. S.**, Heterogeneity of virus particles in measles virus, *J. Gen. Virol.*, 45, 441, 1979.
117. **Valensin, P. E., DiCairano, M. L., and Bandinelli, M. L.**, Capacita di moltiplicazione di cloni di virus vaccinico a differenti temperature e valori di pH, *Boll. 1st Sieroter. Malanese*, 57, 1, 1978.
118. **Wiehe, M. E. and Scherer, W. F.**, Heterogeneity of envelope polypeptides among strains of Venezuelan encephalitis virus, *Virology*, 94, 474, 1979.
119. **Adams, D. J., Ridinger, D. N., Spendlove, R. S., and Barnett, B. B.**, Protamine precipitation of two reovirus particle types from polluted waters, *Appl. Environ. Microbiol.*, 44, 589, 1982.
120. **Ahmed, R. and Fields, B. N.**, Reassortment of genome segments between reovirus defective interfering particles and infectious virus: construction of temperature-sensitive and attenuated viruses by rescue of mutations from DI particles, *Virology*, 111, 351, 1981.

121. **Minoyama, M., Watanabe, Y., and Graham, A. F.,** Defective virions of reovirus, *J. Virol.*, 6, 226, 1970.

122. **Lance, J. C. and Gerba, C. P.,** Poliovirus movement during high rate land filtration of sewage water, *J. Environ. Qual.*, 9, 31, 1980.

123. **Burge, W. D. and Enkiri, N. K.,** Adsorption kinetics of bacteriophage φX-174 on soil, *J. Environ. Qual.*, 7, 536, 1978.

124. **Smith, R. E., Zweenink, H. J., and Joklik, W. K.,** Polypeptide components of virions, top component and cores of reovirus type 3, *Virology*, 39, 791, 1969.

125. **Sykes, I. K. and Williams, S. T.,** Interaction of actinophages and clays, *J. Gen. Microbiol.*, 108, 97, 1978.

126. **Bystricky, V., Stotzky, G., and Schiffenbauer, M.,** Electron microscopy of T1-bacteriophage adsorbed to clay minerals: application of the critical point drying method, *Can. J. Microbiol.*, 21, 1278, 1975.

127. **Roper, M. M. and Marshall, K. C.,** Modification of the interaction between *Escherichia coli* and bacteriophage in saline sediments, *Microb. Ecol.*, 1, 1, 1974.

128. **Serwer, P., Waterson, R. H., Hayes, S. J., and Allen, J. L.,** Comparison of the physical properties and assembly pathways of the related bacteriophages T7, T3, OII, *J. Mol. Biol.*, 170, 447, 1983.

129. **Stagg, C. H., Wallis, C., and Ward, C. H.,** Interaction of clay-associated bacteriophage MS-2 by chlorine, *Appl. Environ. Microbiol.*, 33, 385, 1977.

130. **Boardman, G. D. and Sproul, O. J.,** Protection of viruses during disinfection by adsorption to particulate matter, *J. Water Pollut. Control Fed.*, 49, 1857, 1977.

131. **Lipson, S. M. and Stotzky, G.,** Adsorption of viruses by clay minerals, *Abstr. Annu. Meet. Am. Soc. Microbiol.*, 188, 1979.

132. **McLaren, A. D. and Estermann, E. F.,** The adsorption and reaction of enzymes and proteins on kaolinite. III. The isolation of enzyme-substrate complexes, *Arch. Biochem. Biophys.*, 61, 158, 1956.

133. **Vilker, V. L., Meronek, G. C., and Butler, P. C.,** Interactions of poliovirus with montmorillonite clay in phosphate-buffered saline, *Environ. Sci. Technol.*, 17, 631, 1983.

134. **Dunn, D. B. and Hitchborn, J. H.,** The use of bentonite in the purification of plant viruses, *Virology*, 25, 171, 1965.

135. **Van der Want, J. P. H.,** Some remarks on a soil-borne potato virus, *Proc. Conf. Potato Dis.*, 71, 1952.

136. **Fraenkel-Conrat, H., Singer, B., and Tsugita, A.,** Purification of viral RNA by means of bentonite, *Virology*, 14, 54, 1961.

137. **Sarkar, S.,** Relative infectivity of tobacco mosaic virus and its nucleic acid, *Virology*, 20, 185, 1963.

138. **Singer, B. and Fraenkel-Conrat, H.,** Effect of bentonite on infectivity and stability of TMV-RNA, *Virology*, 14, 53, 1961.

139. **Schaub, S. A., Kenyon, K. F., Bledsoe, B., and Thomas, R.E.,** Evaluation of the over-land run-off mode of land wastewater treatment for viral removal, *Appl. Environ. Microbiol.*, 39, 127, 1980.

140. **Shirobokov, V. P.,** Use of bentonite overlay for titration of enteroviruses by the plaque method, *Vopr. Virusol.*, 18, 611, 1973.

141. **Yarwood, C. E.,** Bentonite aids virus transmission, *Virology*, 28, 459, 1966.

142. **Dubes, G. R.,** The mechanism of transfection enhancement by bentonite, *Arch. Ges. Virus Forsch.*, 39, 13, 1972.

143. **Harrison, B. D.,** Soil transmission of scottish rasberry leaf-curl disease, *Nature (London)*, 178, 553, 1965.

144. **Miyamoto, Y.,** The nature of soil transmission of soil-borne viruses, *Virology*, 7, 250, 1959.

145. **McKinney, H. H.,** Soil factors in relation to incidence and symptom expression of virus diseases, *Soil Sci..*, 61, 93, 1946.

146. **McKinney, H. H.,** Virus diseases of cereal crops, in Plant Diseases: The Yearbook of Agriculture, U.S. Government Printing Office, Washington, D.C., 1953, 350.

147. **Gerba, C. P.,** Applied and theoretical aspects of virus adsorption to surfaces, *Adv. Appl. Microbiol.*, 30, 133, 1984.

148. **Stotzky, G. and Krasovsky, V. N.,** Ecological factors that affect the survival, establishment growth, and genetic recombination of microbes in natural habitats, in *Molecular Biology, Pathology, and Ecology of Bacterial Plasmids*, Levy, S. B., Clowes, R. C., and Koenig, E. L., Eds., Plenum Press, New York, 1981, 31.

149. **Stotzky, G. and Babich, H.,** Fate of genetically-engineered microbes in natural environments, *Recomb. DNA Tech. Bull.*, 7, 163, 1984.

150. **Stotzky, G., Krasovsky, V. N., and Richter, M. W.,** Conjugation in *Escherichia coli* in sterile and nonsterile natural habitats, *Aust. Microbiol.*, 5, 183, 1984.

151. **Weinberg, S. R. and Stotzky, G.,** Conjugation and genetic recombination of *Escherichia coli* in soil, *Soil Biol. Biochem.*, 4, 171, 1972.

152. **Callow, B. R.,** Further studies on *Staphylococcus* bacteriophage, *J. Infect. Dis.*, 41, 124, 1927.

153. **Lewis, M. R. and Andervont, H. B.,** The adsorption of certain viruses by means of particulate substances, *Am. J. Hyg.*, 7, 505, 1927.

154. **Atherton, J. G. and Bell, S. S.**, Adsorption of viruses on magnetite particles. I. Adsorption of MS2 bacteriophage and the effect of cations, clay and poly-electrolyte, *Water Res.*, 17, 943, 1983.

155. **Barkley, M. B. and Desjardins, P. R.**, Simple, effective method for purifying the AS-1 cyanophage, *Appl. Environ. Microbiol.*, 33, 971, 1977.

156. **Bartell, P., Pierzchalla, A., and Tint, H.**, The adsorption of enteroviruses by activated attapulgite, *J. Am. Pharm. Assoc. Sci. Ed.*, 49, 1, 1960.

157. **Bohn, E. M. and van Alstyne, B.**, The generation of defective interfering rubella virus particles, *Virology*, 111, 549, 1981.

158. **Boyce, D. S., Sproul, O. J., and Buck, C. E.**, The effect of bentonite clay on ozone disinfection of bacteria and viruses in water, *Water Res.*, 15, 759, 1981.

159. **Brakke, M. K.**, Degradation of brome mosaic and tobacco mosaic viruses in bentonite, *Virology*, 46, 575, 1971.

160. **Brakke, M. K. and Van Pelt, N.**, Influence of bentonite, magnesium, and polyamines on degradation and aggregation of tobacco mosaic virus, *Virology*, 39, 516, 1969.

161. **Cubbage, C. P., Gannan, J. J., Cochran, K. W., and Williams, G. W.**, Loss of infectivity of poliovirus 1 in river water under simulated field conditions, *Water Res.*, 13, 1091, 1980.

162. **Evans, H. F., Biship, J. M., and Pape, E. A.**, Methods for the quantitative assessment of nuclear-polyhedrosis virus in soil, *J. Invert. Pathol.*, 35, 1, 1980.

163. **Filder, P. and Kay, D.**, The conditions which govern the adsorption of tryptophan-dependent bacteriophage to kaolinite and bacteria, *J. Gen. Microbiol.*, 30, 183, 1963.

164. **Globa, L. I., Lastovets, L. M., and Rotmistrov, M.**, Adsorption of viruses from wastewater by some clay minerals, *Mikrobiol. Zh.*, 34, 64, 1972.

165. **Globa, L. I., Nikovskaya, G. N., and Rotmistrov, M.**, Effect of dispersion medium salt content on *Escherichia coli* B phage T2 sorption by a natural Cherkassian montmorillonite, *Dopov. Akad. Nauk RSR Ser. B Heol. Khim. Biol. Nauk*, 4, 337, 1978.

166. **Jakowbowski, W.**, Adsorption of poliovirus in seawater by clay minerals and marine sediments, *Bacteriol. Proc.*, 179, 1969.

167. **Kazantseva, V. A., Aizen, M. S., and Drosdov, S. G.**, A method for determination of the amount of enteroviruses present in natural water by concentration of virus particles, *Vopr. Virusol.*, 4, 475, 1978.

168. **Klein, J. D., Strangert, K., and Collier, A. M.**, Stability of respiratory syncytial virus in a new bentonite holding medium, *J. Clin. Microbiol.*, 1, 534, 1975.

169. **Kligler, I. J. and Olitzki, L.**, Studies on protein-free suspensions of viruses. I. The adsorption and elution of bacteriophage and fowl-pox viruses, *Br. J. Exp. Pathol.*, 12, 172, 1931.

170. **Koya, K. V. A. and Chaudhuri, M.**, Virus retention by soil, *Prog. Water Technol.*, 9, 43, 1977.

171. **Lipson, S. M. and Stotzky, G.**, Effect of bacteria on the inactivation and adsorption on clay minerals of reovirus, *Can. J. Microbiol.*, 31, 730, 1985.

172. **Moore, R. S., Taylor, D. H., Sturman, L. S., Reddy, M. M., and Fuhs, G. W.**, Poliovirus adsorption by 34 minerals and soils, *Appl. Environ. Microbiol.*, 42, 963, 1981.

173. **Morton, S. D. and Sawyer, E. W.**, Clay minerals remove organics, viruses and heavy metals from water, *Water Sew. Works*, R116, 1976.

174. **Murphy, W. H., Eylar, O. R., Schmidt, E. L., and Syverton, J. T.**, Adsorption and translocation of mammalian viruses by plants. I. Survival of mouse encephalomyocarditis and poliomyelitis viruses in soil and plant root environment, *Virology*, 6, 612, 1958.

175. **Rotmistrov, M. N., Globa, L. I., Nikovskaya, G. N., and Tarasevich, Y. I.**, Adsorption of *Escherichia coli* B phage by natural Cherkassy montmorillonite, *Vopr. Virusol.*, 1, 108, 1978.

176. **Taylor, J. W., Gary, G. W., and Greenbery, H. B.**, Norwalk-related viral gastroenteritis due to contaminated drinking water, *Amer. J. Epidemiol.*, 114, 584, 1982.

176a. **Sattar, S. A. and Westbood, J. C. N.**, Comparison of talc-celite and polyelectrolyte 60 in virus recovery from sewage: development of technique and experiments with poliovirus (type: Sabin)-contaminated multilitre samples, *Can. J. Microbiol.*, 22, 1620, 1976.

177. **Schaub, S. A., Sorber, C. A., and Taylor, G. W.**, The association of enteric viruses with natural turbidity in the aquatic environment, in *Virus Survival in Water and Wastewater Systems*, Malina, J. F., Jr. and Sagik, B. P., Eds., University of Austin, 1974, 71.

178. **Shirobokov, V. P.**, Concentration and purification of enteroviruses with bentonite, *Vopr. Virusol.*, 2, 228, 1974.

179. **Sobsey, M. D., Dean, C. H., Knuckles, M. E., and Wagner, R. A.**, Interactions and survival of enteric viruses in soil materials, *Appl. Environ. Microbiol.*, 40, 92, 1980.

180. **Tang, F. F.**, Adsorption experiments with the virus of vaccinia, *J. Bacteriol.*, 24, 133, 1932.

181. **Taylor, D. H., Moore, R. S., and Sturman, L. S.**, Influence of pH and electrolyte composition on adsorption of poliovirus by soils and minerals, *Appl. Environ. Microbiol.*, 42, 976, 1981.

182. **Thorup, R. T., Nixon, F. P., Wentworth, D. F., and Sproul, O. J.**, Virus removal by coagulation with polyelectrolytes, *J. Am. Water Works Assoc.*, 62, 97, 1970.

182a. **Lycke, E., Blomberg, J., Berg, G., Erikson, A., and Madsen, L.,** Epidemic acute diarrhea in adults associated with infantile gastroenteritis, *Lancet,* 2, 1056, 1978.

183. **Murphy, A. M., Grohmann, G. S., Christopher, P. J., Lopez, W. A., Davey, G. R., and Millsom, R. H.,** An Australia-wide outbreak of gastroenteritis from oysters by Norwalk virus, *Med. J. Aust.,* 2, 329, 1979.

184. **Grohman, G. S., Greenberg, H. B., Welch, B. M., and Murphy, A. M.,** Oyster-associated gastroenteritis in Australia: the detection of Norwalk virus and its antibody by immune electron microscopy and radioimmunoassay, *J. Med. Virol.,* 6, 11, 1980.

185. **Kaplan, J. E., and Goodman, R. E.,** An outbreak of Norwalk virus associated with a municipal water system, *Am. J. Epidemiol.,* 114, 433, 1981.

186. **Kappus, K. D., Marks, J. S., Holman, R. C., Bryant, J. K., Baker, C., Gary, G. W., and Greenberg, H. B.,** An outbreak of Norwalk gastroenteritis, associated with swimming in a pool and secondary person-to-person spread, *Am. J. Epidemiol.,* 116, 834, 1982.

187. **Hejkel, T. W., Keswick, B., LaBelle, R. L., Gerba, C. P., Sanchez, Y., Dreesman, G., Hafkin, B., and Melnick, J. L.,** Viruses in a community water supply associated with an outbreak of gastroenteritis and infectious hepatitis, *J. Am. Water Works Assoc.,* 74, 318, 1982.

188. **Goodman, R. A., Buehler, J. W., Greenberg, H. B., McKinley, T. W., and Smith, J. D.,** Norwalk gastroenteritis associated with a water system in a rural Georgia community, *Arch. Environ. Health,* 37, 358, 1982.

189. **Anon.,** Community outbreak of Norwalk gastroenteritis - Georgia, *Morbid. Mortal. Wkly Rep.,* 31, 405, 1982.

190. **Anon.,** Enteric illness associated with raw clam consumption - New York, *Morbid. Mortal. Wkly Rep.* 31, 449, 1986.

191. **Tao, H., Changan, W., Zhaoying, F., Zinyi, C., Xuejian, C., Xiaoquang, L., Guangu, C., Henli, Y., Tungxin, C., Weiwe, Y., Shuasen, D., and Weicheng, C.,** Waterborne outbreak of rotavirus diarrhea in adults in China caused by a novel rotavirus, *Lancet,* 1, 1139, 1984.

192. **Clark, C. S., Linnemann, C. C., Gartside, P. S., Phair, J. P., Blacklow, N., and Zeiss, C. R.,** Serological survey of rotavirus, Norwalk agent, and *Prototheca wickerhamii* in wastewater workers, *Am. J. Public Health,* 75, 83, 1985.

193. **O'Mahoney, M. C., Gooch, C. D., Smyth, D. A., Thrussell, A. J., Bartlett, C. L. R., and Noah, N. D.,** Epidemic hepatitis A from cockles, *Lancet,* 1, 518, 1983.

194. **Bergeisen, G. H., Hinds, M. W., and Skaggs, J. W.,** A waterborne outbreak of hepatitis A in Meade County, Kentucky, *Am. J. Public Health,* 75, 161, 1985.

195. **Fass, R., Straussman, Y., Shahar, A., and Mizrahi, A.,** Silicates as nonspecific adsorbents of bacteriophage: a model for purification of water from viruses, *Appl. Environ. Microbiol.,* 39, 227, 1980.

196. **Cliver, D. O.,** Development of a Method to Quantitate Food-borne Viral Infectivity, Rep. No. SAM-TR-68-17, USAF Sch. Aero. Med., Aero. Med. Div., Brooks Air Force Base, Tex., 1968, 1.

197. **Vilker, V. L., Fong, J., and Seyyed-Hoseyni, M.** Poliovirus adsorption to narrow particle size fractions of sand and montmorillonite clay, *J. Coll. Int. Sci.,* 92, 422, 1983.

198. **Aardema, B. W., Lorenz, M. G., and Krumbein, W. E.,** Protection of sediments adsorbed transforming DNA against enzymatic inactivation, *Appl. Environ. Microbiol.,* 46, 417, 1983.

199. **Taylor, D.,** personal communication.

200. **Schiffenbauer, M. and Stotzky, G.,** in preparation.

201. **Yu, B., and Stotzky, G.,** in preparation.

202. **Lipson, S. M. and Stotzky, G.,** Effect of kaolinite on the specific infectivity of reovirus, *Microbiol. Lett.,* 37, 83, 1986.

203. **Lipson, S. M. and Szabo, K.,** Virus update, May to December 1984, Virology Laboratory, Division of Microbiology, Department of Pathology and Laboratories, Nassau County Medical Center, East Meadow, N.Y.

INDEX

A

B

Milton Keynes UK
Ingram Content Group UK Ltd.
UKHW051951071024
449327UK00026B/2258